A Concise Introduction
to Numerical Analysis

A Concise Introduction to Numerical Analysis

A. C. Faul

University of Cambridge, UK

CRC Press
Taylor & Francis Group
Boca Raton London New York

CRC Press is an imprint of the
Taylor & Francis Group, an **informa** business

A CHAPMAN & HALL BOOK

CRC Press
Taylor & Francis Group
6000 Broken Sound Parkway NW, Suite 300
Boca Raton, FL 33487-2742

First issued in paperback 2020

© 2016 by Taylor & Francis Group, LLC
CRC Press is an imprint of Taylor & Francis Group, an Informa business

No claim to original U.S. Government works

ISBN-13: 978-1-4987-1218-7 (hbk)
ISBN-13: 978-0-367-65856-4 (pbk)

Library of Congress Cataloging-in-Publication Data

Names: Faul, A. C. (Anita C.)
Title: A concise introduction to numerical analysis / A.C. Faul.
Description: Boca Raton : Taylor & Francis, 2016. | "A CRC title." | Includes bibliographical references and index.
Identifiers: LCCN 2015041395 | ISBN 9781498712187 (alk. paper)
Subjects: LCSH: Numerical analysis--Textbooks.
Classification: LCC QA297 .F377 2016 | DDC 518--dc23
LC record available at http://lccn.loc.gov/2015041395

**Visit the Taylor & Francis Web site at
http://www.taylorandfrancis.com**

**and the CRC Press Web site at
http://www.crcpress.com**

To Philip, Rosemary,
and Sheila.

Contents

List of Figures

Preface

This book has been developed from the notes of lectures on Numerical Analysis held within the MPhil in Scientific Computing at the University of Cambridge, UK. This course has been taught successfully since 2011. Terms at Cambridge University are very short, only eight weeks in length. Therefore lectures are concise, concentrating on the essential content, but also conveying the underlying connecting principles. On the other hand, the Cambridge Mathematical Tripos was established in around 1790. Thus, knowledge has been distilled literally over centuries.

I have tried to carry over this spirit into the book. Students will get a concise, but thorough introduction to numerical analysis. In addition the algorithmic principles are emphasized to encourage a deeper understanding of why an algorithms is suitable (and sometimes unsuitable) for a particular problem.

The book is also intended for graduate students coming from other subjects, but who will use numerical techniques extensively in their graduate studies. The intake of the MPhil in Scientific Computing is very varied. Mathematicians are actually in the minority and share the classroom with physicists, engineers, chemists, computer scientists, and others. Also some of the MPhil students return to university after a professional career wanting to further and deepen their knowledge of techniques they have found they are lacking in their professional life. Because of this the book makes the difficult balance between being mathematically comprehensive, but also not overwhelming with mathematical detail. In some places where further detail was felt to be out of scope of this book, the reader is referred to further reading.

Techniques are illustrated by MATLAB®[1]implementations. The main purpose is to show the workings of the method and thus MATLAB®'s own implementations are avoided (unless they are used as building blocks of an algorithm). In some cases the listings are printed in the book, but all are available online at https://www.crcpress.com/A-Concise-Introduction-to-Numerical-Analysis/Faul/9781498712187 as part of the package K25104_Downloads.zip. Most implementations are in the form of functions returning the outcome of the algorithm. Also examples for the use of the functions are given.

Exercises are put inline with the text where appropriate. Each chapter ends with a selection of revision exercises which are suitable exam questions. A PDF entitled "Solutions to Odd-Numbered Exercises for a Concise Introduction to Numerical Analysis" can be downloaded at https://www.crcpress.com/A-

Concise-Introduction-to-Numerical-Analysis/Faul/9781498712187 as part of the package `K25104_Downloads.zip`.

Students will find the index comprehensive, making it easy to find the information they are after. Hopefully this will prove the book useful as a reference and make it an essential on the bookshelves of anybody using numerical algorithms.

[1]MATLAB and Simulink are registered trademarks of The Mathworks, Inc. For product information please contact:
The Mathworks, Inc.
3 Apple Hill Drive
Natick, MA 01760-2098 USA
Tel: 508-647-7000
Fax: 508-647-7001
E–mail: info@mathworks.com
Web: www.mathworks.com

Acknowledgments

First and foremost I have to thank Dr Nikos Nikiforakis without whom this book would not have come into being. I first contacted him in 2011 when I was looking for work which I could balance with being a mum of three. My intention was to supervise small groups of students in Numerical Analysis at Cambridge University. He asked whether I could also lecture the subject for the students undertaking the MPhil in Scientific Computing. My involvement in the MPhil and his research group grew from there. Aside from his unwavering belief in what I can accomplish, he has become one of my best friends supporting me through difficult times.

Next thanks are also due to my PhD supervisor Professor Mike Powell who sadly passed away in April 2014. From him I learnt that one should strive for understanding and simplicity. Some of his advice was, "Never cite a paper you haven't understood." My citation lists have been short ever since. I more often saw him working through an algorithm with pen and paper than sitting at a computer. He wanted to know why a particular algorithm was successful. I now ask my students, "How can you improve on something, if you do not understand it?"

Of course I also need to thank all the MPhil students whose questions and quest to find typos have improved this book over several years, especially Peter Wirnsberger, Will Drazin, and Chongli Qin. In addition, Geraint Harcombe contributed significantly to the MATLAB® examples of this book.

I would also like to express my gratitude to Cambridge University and the staff and fellows at Selwyn College for creating such a wonderful atmosphere in which to learn and teach.

This book would not have been written without the support of many people in my private life, foremost my parents Helmut and Marieluise Faul, who instilled a love for knowledge in me. Next, my many friends of whom I would like to mention, especially the ones helping out with childcare, and by name, Annemarie Moore (née Clemence) who was always there when I needed help to clear my head, Sybille Hutter for emotional support, and Karen Higgins who is brilliant at sorting out anything practical.

A.C. Faul

CHAPTER 1

Fundamentals

1.1 Floating Point Arithmetic

We live in a continuous world with infinitely many real numbers. However, a computer has only a finite number of bits. This requires an approximate representation. In the past, several different representations of real numbers have been suggested, but now the most widely used by far is the *floating point representation*. Each floating point representations has a *base* β (which is always assumed to be even) which is typically 2 (binary), 8 (octal), 10 (decimal), or 16 (hexadecimal), and a *precision* p which is the number of digits (of base β) held in a floating point number. For example, if $\beta = 10$ and $p = 5$, the number 0.1 is represented as 1.0000×10^{-1}. On the other hand, if $\beta = 2$ and $p = 20$, the decimal number 0.1 cannot be represented exactly but is approximately $1.1001100110011001100 \times 2^{-4}$. We can write the representation as $\pm d_0.d_1 \cdots d_{p-1} \times \beta^e$, where $d_0.d_1 \cdots d_{p-1}$ is called the *significand* (or *mantissa*) and has p digits and e is the *exponent*. If the leading digit d_0 is non-zero, the number is said to be *normalized*. More precisely $\pm d_0.d_1 \cdots d_{p-1} \times \beta^e$ is the number

$$\pm(d_0 + d_1\beta^{-1} + d_2\beta^{-2} + \cdots + d_{p-1}\beta^{-(p-1)})\beta^e, 0 \le d_i < \beta.$$

If the exponents of two floating point numbers are the same, they are said to be of the same *magnitude*. Let's look at two floating point numbers of the same magnitude which also have the same digits apart from the digit in position p, which has index $p - 1$. We assume that they only differ by one in that digit. These floating point numbers are neighbours in the representation and differ by

$$1 \times \beta^{-(p-1)} \times \beta^e = \beta^{e-p+1}.$$

Thus, if the exponent is large the difference between neighbouring floating point numbers is large, while if the exponent is small the difference between neighbouring floating point numbers is small. This means floating point numbers are more *dense* around zero.

As an example consider the decimal number 3.141592, the first seven digits of π. Converting this into binary, we could approximate it by

$$1.10010010000111111010111111100100011 \times 2^1 =$$
$$(1 + 1 * \tfrac{1}{2} + 0 * \tfrac{1}{4} + 0 * \tfrac{1}{8} + 1 * \tfrac{1}{16} + 0 * \tfrac{1}{32} + 0 * \tfrac{1}{64} + ...) * 2^1 \approx 3.14159200003.$$

If we omit the last digit we arrive at

$$1.10010010000111111010101111110010001 \times 2^1 \approx 3.14159199991.$$

Looking at the conversion to a hexadecimal representation, we get the approximation

$$3.243F5F92 =$$
$$3 + 2 * \tfrac{1}{16} + 4 * \tfrac{1}{16^2} + 3 * \tfrac{1}{16^3} + 15 * \tfrac{1}{16^4} + 5 * \tfrac{1}{16^5} + ... \approx 3.14159200001.$$

Omitting the last digit again gives

$$3.243F5F9 \approx 3.14159199968.$$

Thus the amount of accuracy lost varies with the underlying representation.

The largest and smallest allowable exponents are denoted $e_{\max}$ and $e_{\min}$, respectively. Note that $e_{\max}$ is positive, while $e_{\min}$ is negative. Thus there are $e_{\max} - e_{\min} + 1$ possible exponents, the $+1$ standing for the zero exponent. Since there are β^p possible significands, a floating-point number can be encoded in $[\log_2(e_{\max} - e_{\min} + 1)] + [\log_2(\beta^p)] + 1$ bits where the final $+1$ is for the sign bit.

The storage of floating point numbers varies between machine architectures. A particular computer may store the number using 24 bits for the significand, 1 bit for the sign (of the significand), and 7 bits for the exponent in order to store each floating-point number in 4 bytes (1 byte = 8 bits). This format may be used by two different machines, but they may store the exponent and significand in the opposite order. Calculations will produce the same answer, but the internal bit pattern in each word will be different. Therefore operations on bytes should be avoided and such code optimization left to the compiler.

There are two reasons why a real number cannot be exactly represented as a floating-point number. The first one is illustrated by the decimal number 0.1. Although it has a finite decimal representation, in binary it has an infinite repeating representation. Thus in this representation it lies strictly between two floating point numbers and is neither of them.

Another situation is that a real number is too large or too small. This is also known as being *out of range*. That is, its absolute value is larger than or equal to $\beta \times \beta^{e_{\max}}$ or smaller than $1.0 \times \beta^{e_{\min}}$. We use the terms *overflow* and *underflow* for these numbers. These will be discussed in the next chapter.

1.2   Overflow and Underflow

Both overflow and underflow present difficulties but in rather different ways. The representation of the exponent is chosen in the IEEE binary standard with this in mind. It uses a *biased* representation (as opposed to sign/magnitude and two's complement, for which see [12] I. Koren *Computer Arithmetic Algorithms*). In the case of single precision, where the exponent is stored in 8 bits, the bias is 127 (for double precision, which uses 11 bits, it is 1023). If the exponent bits are interpreted as an unsigned integer k, then the exponent of the floating point number is $k - 127$. This is often called the *unbiased exponent* to distinguish it from the biased exponent k.

In single precision the maximum and minimum allowable values for the unbiased exponent are $e_{\max} = 127$ and $e_{\min} = -126$. The reason for having $|e_{\min}| < e_{\max}$ is so that the reciprocal of the smallest number (i.e., $1/2^{e_{\min}}$) will not overflow. However, the reciprocal of the largest number will underflow, but this is considered less serious than overflow.

The exponents $e_{\max} + 1$ and $e_{\min} - 1$ are used to encode special quantities as we will see below. This means that the unbiased exponents range between $e_{\min} - 1 = -127$ and $e_{\max} + 1 = 128$, whereas the biased exponents range between 0 and 255, which are the non-negative numbers that can be represented using 8 bits. Since floating point numbers are always normalized, the most significant bit of the significand is always 1 when using base $\beta = 2$, and thus this bit does not need to be stored. It is known as the *hidden bit*. Using this trick the significand of the number 1 is entirely zero. However, the significand of the number 0 is also entirely zero. This requires a special convention to distinguish 0 from 1. The method is that an exponent of $e_{\min} - 1$ and a significand of all zeros represents 0. The following table shows which other special quantities are encoded using $e_{\max} + 1$ and $e_{\min} - 1$.

Exponent	Significand	Represents
$e = e_{\min} - 1$	$s = 0$	± 0
$e = e_{\min} - 1$	$s \neq 0$	$0.s \times 2^{e_{\min}}$
$e_{\min} \leq e \leq e_{\max}$	any s	$1.s \times 2^{e}$
$e = e_{\max} + 1$	$s = 0$	∞
$e = e_{\max} + 1$	$s \neq 0$	NaN

Here NAN stands for *not a number* and shows that the result of an operation is undefined.

Overflow is caused by any operation whose result is too large in absolute value to be represented. This can be the result of exponentiation or multiplication or division or, just possibly, addition or subtraction. It is better to highlight the occurrence of overflow with the quantity ∞ than returning the largest representable number. As an example, consider computing $\sqrt{x^2 + y^2}$, when $\beta = 10$, $p = 3$, and $e_{\max} = 100$. If $x = 3.00 \times 10^{60}$ and $y = 4.00 \times 10^{60}$, then x^2, y^2, and $x^2 + y^2$ will each overflow in turn, and be replaced by 9.99×10^{100}. So the final result will be $\sqrt{9.99 \times 10^{100}} = 3.16 \times 10^{50}$, which is drastically wrong: the correct answer is 5×10^{60}. In IEEE arithmetic, the result of x^2 is ∞ and so is y^2, $x^2 + y^2$ and $\sqrt{x^2 + y^2}$. Thus the final result is ∞, indicating that the problems should be dealt with programmatically. A well-written routine will remove possibilities for overflow occurring in the first place.

Exercise 1.1. *Write a C-routine which implements the calculation of* $\sqrt{x^2 + y^2}$ *in a way which avoids overflow. Consider the cases where x and y differ largely in magnitude.*

Underflow is caused by any calculation where result is too small to be distinguished from zero. As with overflow this can be caused by different operations, although addition and subtraction are less likely to cause it. However, in many circumstances continuing the calculation with zero is sufficient. Often it is safe to treat an underflowing value as zero. However, there are several exceptions. For example, suppose a loop terminates after some fixed time has elapsed and the calculation uses a variable time step δ_t which is used to update the elapsed time t by assignments of the form

$$\begin{aligned} \delta_t &:= \delta_t \times update \\ t &:= t + \delta_t \end{aligned} \tag{1.1}$$

Updates such as in Equation (1.1) occur in stiff differential equations. If the variable δ_t ever underflows, then the calculation may go into an infinite loop. A well-thought-through algorithm would anticipate and avoid such problems.

1.3   Absolute, Relative Error, Machine Epsilon

Suppose that x, y are real numbers well away from overflow or underflow. Let x^* denote the floating-point representation of x. We define the *absolute error* ϵ by

$$x^* = x + \epsilon$$

and the *relative error* δ by

$$x^* = x(1 + \delta) = x + x\delta.$$

Thus

$$\epsilon = x\delta \quad \text{or, if} \quad x \neq 0, \quad \delta = \frac{\epsilon}{x}.$$

The absolute and relative error are zero if and only if x can be represented exactly in the chosen floating point representation.

In floating-point arithmetic, relative error seems appropriate because each number is represented to a similar relative accuracy. For example consider $\beta = 10$ and $p = 3$ and the numbers $x = 1.001 \times 10^3$ and $y = 1.001 \times 10^0$ with representations $x^* = 1.00 \times 10^3$ and $y^* = 1.00 \times 10^0$. For x we have an absolute error of $\epsilon_x = 0.001 \times 10^3 = 1$ and for y $\epsilon_y = 0.001 \times 10^0 = 0.001$.

However, the relative errors are

$$\delta_x = \frac{\epsilon_x}{x} = \frac{1}{1.001 \times 10^3} \approx 0.999 \times 10^{-3}$$

$$\delta_y = \frac{\epsilon_y}{y} = \frac{0.001}{1.001 \times 10^0} = \frac{1 \times 10^{-3}}{1.001} \approx 0.999 \times 10^{-3}$$

When $x = 0$ or x is very close to 0, we will need to consider absolute error as well. In the latter case, when dividing by x to obtain the relative error, the resultant relative error will be very large, since the reciprocal of a very small number is very large.

Writing $x^* = x(1 + \delta)$ we see that δ depends on x. Consider two neighbouring numbers in the floating point representation of magnitude β^e. We have seen earlier that they differ by β^{e-p+1}. Any real number x lying between these two floating point numbers is represented by the floating point number it is closer to. Thus it is represented with an absolute error ϵ of less than $1/2 \times \beta^{e-p+1}$ in modulus. To obtain the relative error we need to divide x. However it is sufficient to divide by $1 \times \beta^e$ to obtain a bound for the relative error, since $|x| > 1 \times \beta^e$. Thus

$$|\delta| = \left|\frac{\epsilon}{x}\right| < \left|\frac{1/2 \times \beta^{e-p+1}}{1 \times \beta^e}\right| = 1/2 \times \beta^{-p+1}.$$

Generally, the smallest number u such that $|\delta| \le u$, for all x (excluding x values very close to overflow or underflow) is called the *unit round off*.

On most computers $1^* = 1$. The smallest positive ϵ_m such that

$$(1 + \epsilon_m)^* > 1$$

is called *machine epsilon* or *macheps*. It is often assumed that $u \approx \epsilon_m$ and the terms machine epsilon and unit round off are used interchangeably. Indeed, often people refer to machine epsilon, but mean unit round off.

Suppose $\beta = 10$ and $p = 3$. Consider, for example the number π (whose value is the non-terminating, non-repeating decimal 3.1415926535...), whose representation is 3.14×10^0. There is said to be an error of 0.15926535... *units in the last place*. This is abbreviated *ulp* (plural *ulps*). In general, if the floating-point number $d_0.d_1 \cdots d_{p-1} \times \beta^e$ is used to represent x, then it is in error by

$$\left| d_0.d_1 \cdots d_{p-1} - \frac{x}{\beta^e} \right| \times \beta^{p-1} \quad \text{units in the last place.}$$

It can be viewed as a fraction of the least significant digit.

In particular, when a real number is approximated by the closest floating-point number $d_0.d_1 \cdots d_{p-1} \times \beta^e$, we have already seen that the absolute error might be as large as $1/2 \times \beta^{e-p+1}$. Numbers of the form $d_0.d_1 \cdots d_{p-1} \times \beta^e$ represent real numbers in the interval $[\beta^e, \beta \times \beta^e)$, where the round bracket indicates that this number is not included. Thus the relative error δ corresponding to 0.5 ulps ranges between

$$\frac{1/2 \times \beta^{e-p+1}}{\beta^{e+1}} < \delta \leq \frac{1/2 \times \beta^{e-p+1}}{\beta^{e}} \Rightarrow \frac{1}{2}\beta^{-p} < \delta \leq \frac{1}{2}\beta^{-p+1} \tag{1.2}$$

In particular, the relative error can vary by a factor of β for a fixed absolute error. This factor is called the *wobble*. We can set the unit round off u to $1/2 \times \beta^{-p+1}$. With the choice $\beta = 10$ and $p = 3$ we have $u = 1/2 \times 10^{-2} = 0.005$. The quantities u and ulp can be viewed as measuring units. The absolute error is measured in ulps and the relative error in u.

Continuing with $\beta = 10$ and $p = 3$, consider the real number $x = 12.35$. It is approximated by $x^* = 1.24 \times 10^1$. The absolute error is 0.5 ulps, the relative error is

$$\delta = \frac{0.005 \times 10^1}{1.235 \times 10^1} = \frac{0.05}{12.35} \approx 0.004 = \frac{0.004}{0.005}u = 0.8u.$$

Next we multiply by 8. The exact value is $8x = 98.8$, while the computed value is $8x^* = 8 \times 1.24 \times 10^1 = 9.92 \times 10^1$. The error is now 4.0 ulps. On the other hand, the relative error is

$$\frac{9.92 \times 10^1 - 98.8}{98.8} = \frac{0.4}{98.8} \approx 0.004 = \frac{0.004}{0.005}u = 0.8u.$$

The error measured in ulps has grown 8 times larger. The relative error, however, is the same, because the scaling factor to obtain the relative error has also been multiplied by 8.

1.4   Forward and Backward Error Analysis

Forward error analysis examines how perturbations of the input propagate. For example, consider the function $f(x) = x^2$. Let $x^* = x(1 + \delta)$ be the representation of x, then squaring both sides gives

$$\begin{aligned}
(x^*)^2 &= x^2(1 + \delta)^2 \\
&= x^2(1 + 2\delta + \delta^2) \\
&\approx x^2(1 + 2\delta),
\end{aligned}$$

because δ^2 is small. This means the relative error is approximately doubled. Forward error analysis often leads to pessimistic overestimates of the error, especially when a sequence of calculations is considered and in each calculation the error of the worst case is assumed. When error analyses were first performed, it was feared that the final error could be unacceptable, because of the accumulation of intermediate errors. In practice, however, errors average out. An error in one calculation gets reduced by an error of opposite sign in the next calculation.

Backward error analysis examines the question: How much error in input would be required to explain all output error? It assumes that an approximate solution to a problem is good if it is the exact solution to a nearby problem. Returning to our example, the output error can be written as

$$[f(x)]^* = (x^2)^* = x^2(1+\rho),$$

where ρ denotes the relative error in the output such that $\rho \leq u$. As ρ is small, $1+\rho > 0$. Thus there exists $\tilde{\rho}$ such that $(1+\tilde{\rho})^2 = 1+\rho$ with $|\tilde{\rho}| < |\rho| \leq u$, since $(1+\tilde{\rho})^2 = 1 + 2\tilde{\rho} + \tilde{\rho}^2 = 1 + \tilde{\rho}(2+\tilde{\rho})$. We can now write

$$\begin{aligned} [f(x)]^* &= x^2(1+\tilde{\rho})^2 \\ &= f[x(1+\tilde{\rho})]. \end{aligned}$$

If the backward error is small, we accept the solution, since it is the correct solution to a nearby problem.

Another reason for the preference which is given to backward error analysis is that often the inverse of a calculation can be performed much easier than the calculation itself. Take for example

$$f(x) = \sqrt{x}$$

with the inverse

$$f^{-1}(y) = y^2.$$

The square root of x can be calculated iteratively by the Babylonian method, letting

$$x_{n+1} = \frac{1}{2}\left(x_n + \frac{x}{x_n}\right).$$

We can test whether a good-enough approximation has been reached by checking the difference

$$x_n^2 - x,$$

which is the backward error of the approximation to $\sqrt{x}$. Note that in the above analysis we did not use the representation x^* of x. The assumption is that x is represented correctly. We will continue to do so and concentrate on the errors introduced by performing approximate calculations.

Another example is approximating e^x by

$$f(x) = 1 + x + \frac{x^2}{2!} + \frac{x^3}{3!}.$$

The forward error is simply $f(x) - e^x$. For the backward error we need to find x^* such that $e^{x^*} = f(x)$. In particular,

$$x^* = ln(f(x)).$$

At the point $x = 1$ we have (to seven decimal places) $e^1 = 2.718282$ whereas $f(1) = 1 + 1 + \frac{1}{2} + \frac{1}{6} = 2.666667$. Thus the forward error is $2.666667 - 2.718282 = -0.051615$ while the backward error is $ln(f(1)) - 1 = 0.980829 - 1 = 0.019171$. They are different to each other and cannot be compared.

Next we consider how errors build up using the basic floating point operations: multiplication $\times$, division $/$, and exponentiation $\uparrow$. As a practical example, consider double precision in IBM System/370 arithmetic. Here the value of u is approximately 0.22×10^{-15}. We simplify this by letting all numbers be represented with the same relative error 10^{-15}.

Starting with multiplication, we write

$$
\begin{aligned}
x_1^* &= x_1(1 + \delta_1) \\
x_2^* &= x_2(1 + \delta_2).
\end{aligned}
$$

Then

$$
\begin{aligned}
x_1^* \times x_2^* &= x_1 x_2 (1 + \delta_1)(1 + \delta_2) \\
&= x_1 x_2 (1 + \delta_1 + \delta_2 + \delta_1 \delta_2).
\end{aligned}
$$

The term $\delta_1 \delta_2$ can be neglected, since it is of magnitude 10^{-30} in our example. The worst case is, when δ_1 and δ_2 have the same sign, i.e., the relative error in $x_1^* \times x_2^*$ is no worse than $|\delta_1| + |\delta_2|$. If we perform one million floating-point multiplications, then at worst the relative error will have built up to $10^6 \times 10^{-15} = 10^{-9}$.

Division can be easily analyzed in the same way by using the binomial expansion to write

$$
\frac{1}{x_2^*} = \frac{1}{x_2}(1 + \delta_2)^{-1} = \frac{1}{x_2}(1 - \delta_2 + ...).
$$

The omitted terms are of magnitude 10^{-30} or smaller and can be neglected. Again, the relative error in x_1^*/x_2^* is no worse than $|\delta_1| + |\delta_2|$.

We can compute $x_1^* \uparrow n$, for any integer n by repeated multiplication or division. Consequently we can argue that the relative error in $x_1^* \uparrow n$ is no worse than $n|\delta_1|$.

This leaves addition $+$ and subtraction $-$ with which we will deal in the next section. Here the error build-up depends on absolute accuracy, rather than relative accuracy.

1.5   Loss of Significance

Consider

$$
\begin{aligned}
x_1^* + x_2^* &= x_1(1 + \delta_1) + x_2(1 + \delta_2) \\
&= x_1 + x_2 + (x_1 \delta_1 + x_2 \delta_2) \\
&= x_1 + x_2 + (\epsilon_1 + \epsilon_2).
\end{aligned}
$$

Note how the error build-up in addition and subtraction depends on the absolute errors ϵ_1 and ϵ_2 in representing x_1, x_2, respectively. In the worst case scenario ϵ_1 and ϵ_2 have the same sign, i.e., the absolute error in $x_1^* + x_2^*$ is no worse than $|\epsilon_1| + |\epsilon_2|$.

Using the fact that $(-x_2)* = -x_2 - \epsilon_2$ we get that the absolute error in $x_1^* - x_2^*$ is also no worse than $|\epsilon_1| + |\epsilon_2|$. However, the relative error is

$$
\frac{|\epsilon_1| + |\epsilon_2|}{x_1 - x_2}.
$$

Suppose we calculate $\sqrt{10} - \pi$ using a computer with precision $p = 6$. Then $\sqrt{10} \approx 3.16228$ with absolute error of about 2×10^{-6} and $\pi \approx 3.14159$ with absolute error of about 3×10^{-6}. The absolute error in the result $\sqrt{10} - \pi \approx 0.02069$ is about 5×10^{-6}. However, calculating the relative error, we get approximately

$$5 \times 10^{-6}/0.02068\ldots \approx 2 \times 10^{-4}.$$

This means that the relative error in the subtraction is about 100 times as big as the relative error in $\sqrt{10}$ or π.

This problem is known as *loss of significance*. It can occur whenever two similar numbers of equal sign are subtracted (or two similar numbers of opposite sign are added). If possible it should be avoided programmatically. We will see an example of how to do so later when discussing robustness.

As another example with the same precision $p = 6$, consider the numbers $x_1 = 1.00000$ and $x_2 = 9.99999 \times 10^{-1}$. The true solution of $x_1 - x_2 = 0.000001$. However, when calculating the difference, the computer first adjusts the magnitude such that both x_1 and x_2 have the same magnitude. This way x_2 becomes 0.99999. Note that we have lost one digit in x_2. The computed result is $1.00000 - 0.99999 = 0.00001$ and the absolute error is $|0.000001 - 0.00001| = 0.000009$. The relative error is $0.000009/0.000001 = 9$. We see that the relative error has become as large as the base minus one. The following theorem generalizes this for any base.

Theorem 1.1. *Using a floating-point format with parameters β and p, and computing differences using p digits, the relative error of the result can be as large as $\beta - 1$.*

Proof. Consider the expression $x - y$ when $x = 1.00\ldots0$ and $y = \rho.\rho\rho\ldots\rho \times \beta^{-1}$, where $\rho = \beta - 1$. Here y has p digits (all equal to ρ). The exact difference is $x - y = \beta^{-p}$. However, when computing the answer using only p digits, the rightmost digit of y gets shifted off, and so the computed difference is β^{-p+1}. Thus the absolute error is $\beta^{-p+1} - \beta^{-p} = \beta^{-p}(\beta - 1)$, and the relative error is

$$\frac{\beta^{-p}(\beta - 1)}{\beta^{-p}} = \beta - 1.$$

$\square$

The problem is solved by the introduction of a *guard digit*. That is, after the smaller number is shifted to have the same exponent as the larger number, it is truncated to $p+1$ digits. Then the subtraction is performed and the result rounded to p digits.

Theorem 1.2. *If x and y are positive floating-point numbers in a format with parameters β and p, and if subtraction is done with $p+1$ digits (i.e., one guard digit), then the relative error in the result is less than $(\frac{\beta}{2} + 1)\beta^{-p}$.*

Proof. Without loss of generality, $x > y$, since otherwise we can exchange their roles. We can also assume that x is represented by $x_0.x_1 \ldots x_{p-1} \times \beta^0$, since both numbers can be scaled by the same factor. If y is represented as $y_0.y_1 \ldots y_{p-1}$, then the difference is exact, since both numbers have the same magnitude. So in this case there is no error.

Let y be represented by $y_0.y_1 \ldots y_{p-1} \times \beta^{-k-1}$, where $k \geq 0$. That is, y is at least one magnitude smaller than x. To perform the subtraction the digits of y are shifted to the right in the following way.

$$
\begin{array}{ccccccccc}
\overbrace{}^{k} & & & & & & & & \\
0. & 0 & \ldots & 0 & y_0 & \ldots & y_{p-k-1} & y_{p-k} & \cdots & y_{p-1} \\
\uparrow & \uparrow & & \uparrow & \uparrow & & \uparrow & \uparrow & & \uparrow \\
\beta^0 & \beta^{-1} & \ldots & \beta^{-k} & \beta^{-k-1} & \ldots & \beta^{-p} & \beta^{-p-1} & \ldots & \beta^{-p-k}
\end{array}
$$

$$\underbrace{\phantom{0. \quad 0 \quad \ldots \quad 0 \quad y_0 \quad \ldots \quad y_{p-k-1} \quad y_{p-k} \quad \cdots \quad y_{p-1}}}_{p+1 \text{ digits}}$$

The lower row gives the power of β associated with the position of the digit. Let $\hat{y}$ be y truncated to $p+1$ digits. Then

$$
\begin{aligned}
y - \hat{y} &= y_{p-k}\beta^{-p-1} + y_{p-k+1}\beta^{-p-2} + \cdots + y_{p-1}\beta^{-p-k} \\
&\leq (\beta - 1)(\beta^{-p-1} + \beta^{-p-2} + \ldots + \beta^{-p-k}).
\end{aligned}
$$

From the definition of guard digit, the computed value of $x - y$ is $x - \hat{y}$ rounded to be a floating-point number, that is, $(x - \hat{y}) + \alpha$, where the rounding error α satisfies

$$|\alpha| \leq \frac{\beta}{2}\beta^{-p}.$$

The exact difference is $x - y$, so the absolute error is $|(x - y) - (x - \hat{y} + \alpha)| = |\hat{y} - y - \alpha|$. The relative error is the absolute error divided by the true solution

$$\frac{|\hat{y} - y - \alpha|}{x - y}.$$

We now find a bound for the relative error. If $x - y \geq 1$, then the relative error is bounded by

$$
\begin{aligned}
\frac{|\hat{y} - y - \alpha|}{x - y} &\leq \frac{|y - \hat{y}| + |\alpha|}{1} \\
&\leq \beta^{-p}\left[(\beta - 1)(\beta^{-1} + \ldots + \beta^{-k}) + \frac{\beta}{2}\right] \\
&\leq \beta^{-p}\left[1 + \beta^{-1} + \ldots + \beta^{-k+1} - \beta^{-1} - \ldots - \beta^{-k} + \frac{\beta}{2}\right] \\
&\leq \beta^{-p}(1 + \tfrac{\beta}{2}),
\end{aligned}
$$

which is the bound given in the theorem.

If $x - y < 1$, we need to consider two cases. Firstly, if $x - \hat{y} < 1$, then no rounding was necessary, because the first digit in $x - \hat{y}$ is zero and can be shifted off to the left, making room to keep the $p + 1^{\text{th}}$ digit. In this case $\alpha = 0$. Defining $\rho = \beta - 1$, we find the smallest $x - y$ can be by letting x be as small as possible, which is $x = 1$, and y as large as possible, which is $y = \rho.\rho \dots \rho \times \beta^{-k-1}$. The difference is then

$$1 - 0.\overbrace{0 \dots 0}^{k}\overbrace{\rho \dots \rho}^{p} = 0.\overbrace{\rho \dots \rho}^{k}\overbrace{0 \dots 0}^{p-1}1 > (\beta - 1)(\beta^{-1} + \dots + \beta^{-k}).$$

In this case the relative error is bounded by

$$\frac{|y - \hat{y}| + |\alpha|}{(\beta - 1)(\beta^{-1} + \dots + \beta^{-k})} < \frac{(\beta - 1)\beta^{-p}(\beta^{-1} + \dots + \beta^{-k})}{(\beta - 1)(\beta^{-1} + \dots + \beta^{-k})} = \beta^{-p}.$$

The other case is when $x - y < 1$, but $x - \hat{y} \geq 1$. The only way this could happen is if $x - \hat{y} = 1$, in which case $\alpha = 0$, then the above equation applies and the relative error is bounded by $\beta^{-p} < \beta^{-p}(1 + \frac{\beta}{2})$. □

Thus we have seen that the introduction of guard digits alleviates the loss of significance. However, careful algorithm design can be much more effective, as the following exercise illustrates.

Exercise 1.2. *Consider the function* $\sin x$, *which has the series expansion*

$$\sin x = x - \frac{x^3}{3!} + \frac{x^5}{5!} - \dots,$$

which converges, for any x, *to a value in the range* $-1 \leq \sin x \leq 1$. *Write a MATLAB-routine which examines the behaviour when summing this series of terms with alternating signs as it stands for different starting values of* x. *Decide on a convergence condition for the sum. For different* x, *how many terms are needed to achieve convergence, and how large is the relative error then? (It gets interesting when* x *is relatively large, e.g., 20). How can this sum be safely implemented?*

1.6 Robustness

An algorithm is described as *robust* if for any valid data input which is reasonably representable, it completes successfully. This is often achieved at the expense of time. Robustness is best illustrated by example. We consider the quadratic equation $ax^2 + bx + c = 0$. Solving it seems to be a very elementary problem. Since it is often part of a larger calculation, it is important that it is implemented in a robust way, meaning that it will not fail and give reasonably accurate answers for any coefficients a, b and c which are not too close to overflow or underflow. The standard formula for the two roots is

$$x = \frac{-b \pm \sqrt{b^2 - 4ac}}{2a}.$$

A problem arises if b^2 is much larger than $|4ac|$. In the worst case the difference in the magnitudes of b^2 and $|4ac|$ is so large that $b^2 - 4ac$ evaluates to b^2 and the square root evaluates to b, and one of the calculated roots lies at zero. Even if the difference in magnitude is not that large, one root is still small. Without loss of generality we assume $b > 0$ and the small root is given by

$$x = \frac{-b + \sqrt{b^2 - 4ac}}{2a}. \tag{1.3}$$

We note that there is loss of significance in the numerator. As we have seen before, this can lead to a large relative error in the result compared to the relative error in the input. The problem can be averted by manipulating the formula in the following way:

$$\begin{aligned}
x &= \frac{-b + \sqrt{b^2 - 4ac}}{2a} \times \frac{-b - \sqrt{b^2 - 4ac}}{-b - \sqrt{b^2 - 4ac}} \\
&= \frac{b^2 - (b^2 - 4ac)}{2a(-b - \sqrt{b^2 - 4ac})} \\
&= \frac{-2c}{b + \sqrt{b^2 - 4ac}}.
\end{aligned} \tag{1.4}$$

Now quantities of similar size are added instead of subtracted.

Taking $a = 1$, $b = 100000$ and $c = 1$ and as accuracy 2×10^{-10}, Equation (1.3) gives $x = -1.525878906 \times 10^{-5}$ for the smaller root while (1.4) gives $x = -1.000000000 \times 10^{-5}$, which is the best this accuracy allows.

In general, adequate analysis has to be conducted to find cases where numerical difficulties will be encountered, and a robust algorithm must use an appropriate method in each case.

1.7 Error Testing and Order of Convergence

Often an algorithm first generates an approximation to the solution and then improves this approximation again and again. This is called an *iterative* numerical process. Often the calculations in each iteration are the same. However, sometimes the calculations are adjusted to reach the solution faster. If the process is successful, the approximate solutions will converge to a solution. Note that it is *a* solution, not *the* solution. We will see this beautifully illustrated when considering the fractals generated by Newton's and Halley's methods.

More precisely, *convergence* of a sequence is defined as follows. Let x_0, x_1, x_2, ... be a sequence (of approximations) and let x be the true solution. We define the absolute error in the nth iteration as

$$\epsilon_n = x_n - x.$$

The sequence converges to the *limit* x of the sequence if

$$\lim_{n \to \infty} \epsilon_n = 0$$

Note that convergence of a sequence is defined in terms of *absolute error*.

There are two forms of error testing, one using a target absolute accuracy ϵ_t, the other using a target relative error δ_t. In the first case the calculation is terminated when

$$|\epsilon_n| \le \epsilon_t. \tag{1.5}$$

In the second case the calculation is terminated when

$$|\epsilon_n| \le \delta_t |x_n|. \tag{1.6}$$

Both methods are flawed under certain circumstances. If x is large, say 10^{20}, and $u = 10^{-16}$, then ϵ_n is never likely to be much less than 10^4, so condition (1.5) is unlikely to be satisfied if ϵ_t is chosen too small even when the process converges. On the other hand, if $|x_n|$ is very small, then $\delta_t |x_n|$ may underflow and test (1.6) may never be satisfied (unless the error becomes exactly zero).

As (1.5) is useful when (1.6) is not, and vice versa, they are combined into a *mixed error test*. A target error η_t is prescribed and the calculation is terminated when

$$|\epsilon_n| \le \eta_t (1 + |x_n|). \tag{1.7}$$

If $|x_n|$ is small, η_t is regarded as target absolute error, or if $|x_n|$ is large η_t is regarded as target relative error.

Tests such as (1.7) are used in modern numerical software, but we have not addressed the problem of estimating ϵ_n, since the true value x is unknown. The simplest formula is

$$\epsilon_n \approx x_n - x_{n-1} \tag{1.8}$$

However, theoretical research has shown that in a wide class of numerical methods, cases arise where adjacent values in an approximation sequence have the same value, but are both the incorrect answer. Test (1.8) will cause the algorithm to terminate too early with an incorrect solution.

A safer estimate is

$$\epsilon_n \approx |x_n - x_{n-1}| + |x_{n-1} - x_{n-2}|, \tag{1.9}$$

but again research has shown that even the approximations of three consecutive iterations can all be the same for certain methods, so (1.9) might not work either. However, in many problems convergence can be tested independently, for example when the inverse of a function can be easily calculated (calculating the kth power as compared to taking the kth root). Error and convergence testing should always be fitted to the underlying problem.

We now turn our attention to the speed of convergence. It is defined by comparing the errors of two subsequent iterations. If there exist constants p and C such that

$$\lim_{n \to \infty} \left| \frac{\epsilon_{n+1}}{\epsilon_n^p} \right| = C,$$

then the process has *order of convergence* p, where $p \ge 1$. This is often expressed as

$$|\epsilon_{n+1}| = O(|\epsilon_n|^p),$$

known as the *O-notation*. Or in other words, order p convergence implies

$$|\epsilon_{n+1}| \approx C|\epsilon_n|^p$$

for sufficiently large n. Thus the error in the next iteration step is approximately the pth power of the current iteration error times a constant C. For $p = 1$ we need $C < 1$ to have a reduction in error. This is not necessary for $p > 1$, because as long as the current error is less than one, taking any power greater or equal to one leads to a reduction. For $p = 0$, the O-notation becomes $O(1)$, which signifies that something remains constant. Of course in this case there is no convergence.

The following categorization for various rates of convergence is in use:

1. $p = 1$: linear convergence. Each iteration produces the same reduction in absolute error. This is generally regarded as being too slow for practical methods.

2. $p = 2$: quadratic convergence. Each iteration squares the absolute error.

3. $1 < p < 2$: superlinear convergence. This is not as good as quadratic but the minimum acceptable rate in practice.

4. Exponential rate of convergence.

So far we have only considered the concept of convergence for numbers $x_1, x_2, x_3, \ldots$ and x. However, in numerical calculations the true solution might be a complex structure. For example, it might be approximations to the solution of a partial differential equation on an irregular grid. In this case a measurement of how close the approximation is to the true solution has to be defined. This can look very different depending on the underlying problem which is being solved.

1.8   Computational Complexity

A well-designed algorithm should not only be robust, and have a fast rate of convergence, but should also have a reasonable *computational complexity*. That is, the computation time shall not increase prohibitingly with the size of the problem, because the algorithm is then too slow to be used for large problems.

Suppose that some operation, call it $\odot$, is the most expensive in a particular algorithm. Let n be the size of the problem. If the number of operations of the algorithm can be expressed as $O[f(n)]$ operations of type $\odot$, then we say that the *computational complexity* is $f(n)$. In other words, we neglect the less expensive operations. However, less expensive operations cannot be neglected, if a large number of them need to be performed for each expensive operation.

For example, in matrix calculations the most expensive operations are multiplications of array elements and array references. Thus in this case the operation ⊙ may be defined to be a combination of one multiplication and one or more array references. Let's consider the multiplication of $n \times n$ matrices $A = (A_{ij})$ and $B = (B_{ij})$ to form a product $C = (C_{ij})$. For each element in C, we have to calculate

$$C_{ij} = \sum_{k=1}^{n} A_{ik} B_{kj},$$

which requires n multiplications (plus two array references per multiplication). Since there are n^2 elements in C, the computational complexity is $n^2 \times n = n^3$.

Note that processes of lower complexity are absorbed into higher complexity ones and do not change the overall computational complexity of an algorithm. This is the case, unless the processes of lower complexity are performed a large number of times.

For example, if an n^2 process is performed each time an n^3 process is performed then, because of

$$O(n^2) + O(n^3) = O(n^3)$$

the overall computational complexity is still n^3. If, however, the n^2 process was performed n^2 times each time the n^3 process was performed, then the computational complexity would be $n^2 \times n^2 = n^4$.

Exercise 1.3. *Let A be an $n \times n$ nonsingular band matrix that satisfies the condition $A_{ij} = 0$ for $|i-j| > r$, where r is small, and let Gaussian elimination (introduced in Linear Systems 2.2) be used to solve $A\mathbf{x} = \mathbf{b}$. Deduce that the total number of additions and multiplications of the complete calculation can be bounded by a constant multiple of nr^2.*

1.9   Condition

The *condition* of a problem is inherent to the problem whichever algorithm is used to solve it. The *condition number* of a numerical problem measures the asymptotically worst case of how much the outcome can change in proportion to small perturbations in the input data. A problem with a low condition number is said to be well-conditioned, while a problem with a high condition number is said to be ill-conditioned. The condition number is a property of the problem and not of the different algorithms that can be used to solve the problem.

As an example consider the problem where a graph crosses the line $x = 0$. Naively one could draw the graph and measure the coordinates of the crossover points. Figure 1.1 illustrates two cases. In the left-hand problem it would be easier to measure the crossover points, while in the right-hand problem the crossover points lie in a region of candidates. A better (or worse) algorithm would be to use a higher (or lower) resolution. In the chapter on non-linear systems we will encounter many methods to find the roots of a function that is the points where the graph of a function crosses the line $x = 0$.

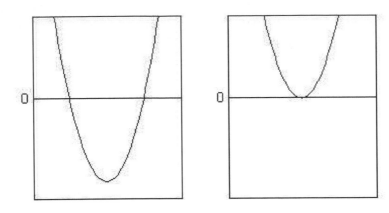

Figure 1.1 Example of a well-conditioned and an ill-conditioned problem

As a further example we consider the problem of evaluating a differentiable function f at a given point x, i.e., calculating $f(x)$. Let $\hat{x}$ be a point close to x. The *relative change* in x is

$$\frac{x - \hat{x}}{x}$$

and the relative change in $f(x)$ is

$$\frac{f(x) - f(\hat{x})}{f(x)}.$$

The condition number K is defined as the limit of the relative change in $f(x)$ divided by the relative change in x as $\hat{x}$ tends to x.

$$
\begin{aligned}
K(x) &= \lim_{\hat{x} \to x} \left| \frac{f(x) - f(\hat{x})}{f(x)} \right| \Big/ \left| \frac{x - \hat{x}}{x} \right| \\
&= \lim_{\hat{x} \to x} \left| \frac{f(x) - f(\hat{x})}{f(x)} \right| \left| \frac{x}{x - \hat{x}} \right| \\
&= \left| \frac{x}{f(x)} \right| \lim_{\hat{x} \to x} \left| \frac{f(x) - f(\hat{x})}{x - \hat{x}} \right| \\
&= \left| \frac{x f'(x)}{f(x)} \right|,
\end{aligned}
$$

since the limit is exactly the definition of the derivative of f at x.

To illustrate we consider the functions $f(x) = \sqrt{x}$ and $f(x) = \frac{1}{1-x}$. In the first case we get

$$K(x) = \left| \frac{x f'(x)}{f(x)} \right| = \frac{x(\frac{1}{2} x^{-1/2})}{\sqrt{x}} = \frac{1}{2}.$$

So K is constant for all non-negative x and thus taking square roots has the same condition for all inputs x. Whatever the value of x is, a perturbation in the input leads to a perturbation of the output which is half of the input perturbation.

Exercise 1.4. *Perform a forward and backward error analysis for* $f(x) = \sqrt{x}$. *You should find that the relative error is reduced by half in the process.*

However, when $f(x) = \frac{1}{1-x}$ we get

$$K(x) = \left| \frac{xf'(x)}{f(x)} \right| = \left| \frac{x[1/(1-x)^2]}{1/(1-x)} \right| = \left| \frac{x}{1-x} \right|.$$

Thus for values of x close to 1, $K(x)$ can get large. For example if $x = 1.000001$, then $K(1.000001) = 1.000001 \times 10^6$ and thus the relative error will increase by a factor of about 10^6.

Exercise 1.5. *Examine the condition of the evaluating* $\cos x$.

As another example we examine the condition number of an $n \times n$ matrix $A = (A_{ij})$ associated with the linear equation $A\mathbf{x} = \mathbf{b}$. Here the right-hand side $\mathbf{b}$ can be any non-zero vector. Many numerical algorithms reduce to solving a system of equations for many different right-hand sides $\mathbf{b}$. Therefore, knowing how accurately this can be done is essential. Remember that the condition number is a property of the matrix, not the algorithm or the machine accuracy of the computer used to solve the corresponding system. The condition number is the rate at which the solution $\mathbf{x}$ will change with changes in the right-hand side $\mathbf{b}$. A large condition number implies that even a small error in $\mathbf{b}$ may cause a large error in $\mathbf{x}$.

The condition number is defined to be the maximum ratio of the relative change in $\mathbf{x}$ divided by the relative change in $\mathbf{b}$ for any $\mathbf{b}$. Let $\mathbf{e}$ be a perturbation in $\mathbf{b}$. Thus the relative change in $\mathbf{b}$ is

$$\|\mathbf{e}\| / \|\mathbf{b}\| = \|\mathbf{e}\| / \|A\mathbf{x}\|,$$

where $\| \cdot \|$ denotes any *vector norm*. We will see below how different norms lead to different condition numbers.

Assuming that A is invertible, then $\mathbf{x}$ is given by $A^{-1}\mathbf{b}$ and the perturbation in $\mathbf{x}$ is $A^{-1}\mathbf{e}$. Thus the relative change in x is $\|A^{-1}\mathbf{e}\| / \|\mathbf{x}\|$. Hence the condition number is

$$K(A) = \max_{\mathbf{x}, \mathbf{e} \neq 0} \frac{\|A^{-1}\mathbf{e}\|}{\|\mathbf{x}\|} \Big/ \frac{\|\mathbf{e}\|}{\|A\mathbf{x}\|} = \max_{\mathbf{e} \neq 0} \frac{\|A^{-1}\mathbf{e}\|}{\|\mathbf{e}\|} \times \max_{\mathbf{x} \neq 0} \frac{\|A\mathbf{x}\|}{\|\mathbf{x}\|}.$$

Now the definition of a *matrix norm* derived from a vector norm is

$$\|A\| = \max_{\mathbf{v} \neq 0} \frac{\|A\mathbf{v}\|}{\|\mathbf{v}\|}.$$

We see that the calculation of the condition number involves the definition of the matrix norm. Thus the condition number of an invertible matrix is

$$K(A) = \|A^{-1}\| \times \|A\|.$$

Of course, this definition depends on the choice of norm. We just give a brief outline of different vector norms and the condition numbers induced by these. For more information, see [7] J. W. Demmel *Applied Numerical Linear Algebra*.

1. If $\| \cdot \|$ is the standard 2-*norm* also known as *Euclidean norm* defined as

$$\|\mathbf{x}\|_2 := \left(\sum_{i=1}^{n} x_i^2 \right)^{1/2},$$

then the induced matrix norm is

$$\|A\|_2 = \sigma_{\max}(A),$$

and

$$K_2(A) = \frac{\sigma_{\max}(A)}{\sigma_{\min}(A)},$$

where $\sigma_{\max}(A)$ and $\sigma_{\min}(A)$ are the maximal and minimal singular values of A, respectively. (The singular values of a matrix A are the square roots of the eigenvalues of the matrix A^*A where A^* denotes the complex conjugate transpose of A.) In particular A is called a *normal matrix* if $A^*A = AA^*$. In this case

$$K_2(A) = \left| \frac{\lambda_{\max}(A)}{\lambda_{\min}(A)} \right|$$

where $\lambda_{\max}(A)$ and $\lambda_{\min}(A)$ are the maximum and minimum modulus eigenvalues of A. If A is *unitary*, i.e., multiplying the matrix with its conjugate transpose results in the identity matrix, then $K_2(A) = 1$.

2. If $\| \cdot \|$ is the ∞-*norm* defined as

$$\|\mathbf{x}\|_\infty := \max_{i=1,\dots,n} |x_i|,$$

then the induced matrix norm is

$$\|A\|_\infty = \max_{i=1,\dots,n} \sum_{j=1}^{n} |A_{ij}|,$$

which is the maximum absolute row sum of the matrix.

Exercise 1.6. *If A is lower triangular and non-singular, and using the ∞-norm, show that*

$$K_\infty(A) \geq \frac{\max_{i=1,\dots n}(|A_{ii}|)}{\min_{i=1,\dots n}(|A_{ii}|)}$$

As an example we consider the matrix equation

$$\left(\begin{array}{cc} 404 & 60 \\ 60 & 4 \end{array}\right)\left(\begin{array}{c} x_1 \\ x_2 \end{array}\right) = \left(\begin{array}{c} b_1 \\ b_2 \end{array}\right).$$

The matrix has eigenvalues $204 \pm 20\sqrt{109}$ that is $\lambda_1 \approx 412.8$ and $\lambda_2 \approx 4.8$. Recall that the condition number is defined as the maximum ratio of the relative change in $\mathbf{x}$ over the relative change $\mathbf{e}$ in $\mathbf{b}$:

$$K(A) = \max_{\mathbf{x},\mathbf{e}\neq 0} \frac{\|A^{-1}\mathbf{e}\|}{\|\mathbf{x}\|} / \frac{\|\mathbf{e}\|}{\|A\mathbf{x}\|} = \max_{\mathbf{e}\neq 0} \frac{\|A^{-1}\mathbf{e}\|}{\|\mathbf{e}\|} \times \max_{\mathbf{x}\neq 0} \frac{\|A\mathbf{x}\|}{\|\mathbf{x}\|}.$$

Now assuming the error $\mathbf{e}$ on the right-hand side is aligned with the eigenvector which belongs to the smaller eigenvalue, then this is multiplied by a factor of $|1/\lambda_2| = 1/4.8$. On the other hand a small error in the solution which is aligned with the eigenvector belonging to the larger eigenvalue takes us away from the right-hand side $\mathbf{b}$ by a factor of 412.8. This is the worst-case scenario and gives

$$K(A) = \frac{\lambda_1}{\lambda_2} \approx \frac{412.8}{4.8}.$$

Exercise 1.7. *Let*

$$A = \left(\begin{array}{cc} 1000 & 999 \\ 999 & 998 \end{array}\right).$$

Calculate A^{-1}, the eigenvalues and eigenvectors of A, $K_2(A)$, and $K_\infty(A)$. What is special about the vectors $\left(\begin{array}{c} 1 \\ 1 \end{array}\right)$ and $\left(\begin{array}{c} -1 \\ 1 \end{array}\right)$?

An example of notoriously ill-conditioned matrices are the Hilbert matrices. These are square matrices with entries

$$H_{i,j} = \frac{1}{i+j-1}.$$

If n is the size of the matrix, then the condition number is of order $O((1 + \sqrt{2})^{4n}/\sqrt{n})$.

1.10 Revision Exercises

Exercise 1.8. *(a) Define overflow and underflow.*

(b) Describe loss of significance and give an example.

(c) Show that using a floating-point format with base β and p digits in the significand and computing differences using p digits, the relative error of the result can be as large as $\beta - 1$.

(d) Describe (without proof) how computing differences can be improved.

(e) *Consider the quadratic expression $Q(x) = ax^2 + bx + c$ in which a, b, c and x are all represented with the same relative error δ. In computing bx and ax^2, estimate the relative error, and hence the absolute error of both expressions. Hence deduce an estimate for the absolute error in computing $Q(x)$.*

(f) *Comment on the suitability of the formula*

$$x = \frac{-b \pm \sqrt{b^2 - 4ac}}{2a}$$

for computing the roots of $Q(x)$ in floating point arithmetic. Derive an alternative formula and describe how it can be used in practice.

Exercise 1.9. (a) *Define absolute error, relative error, and state their relationship.*

(b) *Show how the relative error builds up in multiplication and division.*

(c) *Explain forward and backward error analysis using the example of approximating*

$$\cos x \approx f(x) = 1 - x^2/2.$$

(d) *Considering the binary floating point representation of numbers, explain the concept of the hidden bit.*

(e) *Explain the biased representation of the exponent in binary floating point representation.*

(f) *How are 0, ∞, and NaN represented?*

(g) *How are the numbers 2^k for positive and negative k represented?*

Exercise 1.10. (a) *Define absolute error, relative error, machine epsilon, and unit round-off.*

(b) *Although machine epsilon is defined in terms of absolute error, which assumption makes it useful as a measurement of relative error?*

(c) *What does it mean if a floating point number is said to be normalized? What is the hidden bit and how is it used?*

(d) *What does NaN stand for? Give an example of an operation which yields an NaN value.*

(e) *Given $e_{\max} = 127$ and $e_{\min} = -126$, one bit for the sign and 23 bits for the significand, show the bit pattern representing each of the following* numbers. *State the sign, the exponent, and the significand. You may use* $0 \ldots 0$ *to represent a long string of zeros.*

0

$-\infty$

-1.0

$1.0 + machine\ epsilon$

4.0

$4.0 + machine\ epsilon$

NaN

x^*, *the smallest representable number greater than* 2^{16}

In the last case, what is the value of the least significant bit in the significand of x^* *and what is the relative error if rounding errors cause* $x = 2^{16}$ *to become* x^*?

Exercise 1.11. *(a) Define absolute error, relative error, and state their relationship.*

(b) Explain absolute error test and relative error test and give examples of circumstances when they are unsuitable. What is a mixed error test?

(c) Explain loss of significance.

(d) Let $x_1 = 3.0001$ *be the true value approximated by* $x_1^* = 3.0001 + 10^{-5}$ *and* $x_2 = -3.0000$ *be the true value approximated by* $x_2^* = -3.0000 + 10^{-5}$. *State the absolute and relative errors in* x_1^* *and* x_2^*. *Calculate the absolute error and relative error in approximating* $x_1 + x_2$ *by* $x_1^* + x_2^*$. *How many times bigger is the relative error in the sum compared to the relative error in* x_1^* *and* x_2^*?

(e) Let
$$f(x) = x - \sqrt{x^2 + 1}, \qquad x \geq 0. \qquad (1.10)$$
Explain when and why loss of significance occurs in the evaluation of f.

(f) Derive an alternative formula for evaluating f *which avoids loss of significance.*

(g) Test your alternative by considering a decimal precision $p = 16$ *and* $x = 10^8$. *What answer does your alternative formula give compared to the original formula?*

(h) Explain condition and condition number in general terms.

(i) Derive the condition number for evaluating a differentiable function f *at a point* x, *i.e., calculating* $f(x)$.

(j) Considering $f(x)$ *as defined in (1.10), find the smallest interval in which the condition number lies. Is the problem well-conditioned or ill-conditioned?*

CHAPTER 2

Linear Systems

2.1 Simultaneous Linear Equations

Here we consider the solution of simultaneous linear equations of the form

$$A\mathbf{x} = \mathbf{b}, \tag{2.1}$$

where A is a matrix of coefficients, $\mathbf{b}$ is a given vector, and $\mathbf{x}$ is the vector of unknowns to be determined. In the first instance, we assume A is square with n rows and n columns, and $\mathbf{x}, \mathbf{b} \in \mathbb{R}^n$. At least one element of $\mathbf{b}$ is non-zero. The equations have a unique solution if and only if A is a *non-singular* matrix, i.e., the inverse A^{-1} exists. The solution is then $\mathbf{x} = A^{-1}\mathbf{b}$. There is no need to calculate A^{-1} explicitly, since the vector $A^{-1}\mathbf{b}$ needs to be calculated and the calculation of A^{-1} would be an intermediate step. The calculation of a matrix inverse is usually avoided unless the elements of the inverse itself are required for other purposes, since this can lead to unnecessary loss of accuracy.

If A is singular, there exist non-zero vectors $\mathbf{v}$ such that

$$A\mathbf{v} = \mathbf{0}.$$

These vectors lie in the *null space* of A. That is the space of all vectors mapped to zero when multiplied by A. If $\mathbf{x}$ is a solution of (2.1) then so is $\mathbf{x} + \mathbf{v}$, since

$$A(\mathbf{x} + \mathbf{v}) = A\mathbf{x} + A\mathbf{v} = \mathbf{b} + \mathbf{0} = \mathbf{b}.$$

In this case there are infinitely many solutions.

The result of A applied to all vectors in $\mathbb{R}^n$ is called the *image* of A. If $\mathbf{b}$ does not lie in the image of A, then no vector $\mathbf{x}$ satisfies (2.1) and there is no solution. In this case the equations are *inconsistent*. This situation can also occur when A is singular.

The solution of Equation (2.1) is trivial if the matrix A is either lower triangular or upper triangular, i.e.,

$$
\begin{pmatrix}
a_{1,1} & 0 & \cdots & & 0 \\
\vdots & \ddots & \ddots & & \vdots \\
a_{n-1,1} & \cdots & a_{n-1,n-1} & & 0 \\
a_{n,1} & \cdots & & \cdots & a_{n,n}
\end{pmatrix}
\quad \text{or} \quad
\begin{pmatrix}
a_{1,1} & \cdots & \cdots & a_{1,n} \\
0 & a_{2,2} & \cdots & a_{2,n} \\
\vdots & \ddots & \ddots & \vdots \\
0 & \cdots & 0 & a_{n,n}
\end{pmatrix} .
$$

If any of the diagonal elements $a_{i,i}$ is zero, then A is singular and there is no unique solution. There might be no solution or infinitely many.

Considering the upper triangular form, the solution is obtained by the *back substitution algorithm*. The last equation contains only one unknown x_n, which can be calculated by

$$
x_n = \frac{b_n}{a_{n,n}} .
$$

Having determined x_n, the second to last equation contains only the unknown x_{n-1}, which can then be calculated and so on. The algorithm can be summarized as

$$
x_i = \frac{b_i - \sum_{j=i+1}^{n} a_{i,j} x_j}{a_{i,i}}, \qquad i = n, n-1, \ldots, 1.
$$

For the lower triangular form the *forward substitution algorithm* of similar form is used. Here is an implementation in MATLAB.

```
function [x] = Forward(A,b)
% Solves the lower triangular system of equations Ax = b
% A input argument, square lower triangular matrix
% b input argument
% x solution

[n,m]=size(A); % finding the size of A
if n≠ m
    error('input is not a square matrix');
end
if size(b,1) ≠ n
    error('input dimensions do not match');
end
x = zeros(n,1); % initialise x to the same dimension
if abs(A(1,1)) > 1e−12 % not comparing to zero because of possible
                        % rounding errors
    x(1) = b(1)/A(1,1); % solve for the first element of x
else
    disp('input singular'); % A is singular if any of the diagonal
                             % elements are zero
    return;
end
for k=2:n % the loop considers one row after the other
    if abs(A(k,k))>1e−12 % not comparing to zero because of possible
                          % rounding errors
        temp = 0;
        for j=1:k−1
            temp = temp + A(k,j) * x(j); % Multiply the elements of
```

```
                                    % the k-th row of A before the
                                    % diagonal by the   elements of x
                                    % already calculated
            end
        x(k) = (b(k)-temp)/A(k,k); % solve for the k-th element of x
    else
            error('input singular'); % A is singular if any of the
                                    % diagonal elements are zero
        end
end
```

Exercise 2.1. *Implement backward substitution.*

2.2   Gaussian Elimination and Pivoting

Given a set of simultaneous equations $A\mathbf{x} = \mathbf{b}$, the solution $\mathbf{x}$ is unchanged if any of the following operations is performed:

1. Multiplication of an equation by a non-zero constant.

2. Addition of the multiple of one equation to another.

3. Interchange of two equations.

The same operations have to be performed on both sides of the equal sign. These operations are used to convert the system of equations to the trivial case, i.e., upper or lower triangular form. This is called *Gaussian elimination*. By its nature there are a many different ways to go about this. The usual strategy is called *pivotal strategy* and we see below that this in general avoids the accumulation of errors and in some situations is crucially important.

A *pivot* entry is usually required to be at least distinct from zero and often well away from it. Finding this element is called *pivoting*. Once the pivot element is found, interchange of rows (and possibly columns) may follow to bring the pivot element into a certain position. Pivoting can be viewed as sorting rows (and possibly columns) in a matrix. The swapping of rows is equivalent to multiplying A by a permutation matrix. In practice the matrix elements are, however, rarely moved, since this would cost too much time. Instead the algorithms keep track of the permutations. Pivoting increases the overall computational cost of an algorithm. However, sometimes pivoting is necessary for the algorithm to work at all, at other times it increases the numerical *stability*. We illustrate this with two examples.

Consider the three simultaneous equations where the diagonal of the matrix consists entirely of zeros,

$$
\begin{pmatrix} 0 & 1 & 1 \\ 1 & 0 & 1 \\ 1 & 1 & 0 \end{pmatrix} \begin{pmatrix} x_1 \\ x_2 \\ x_3 \end{pmatrix} = \begin{pmatrix} 1 \\ 2 \\ 4 \end{pmatrix}.
$$

which we convert into upper triangular form to use back substitution. The first equation cannot form the first row of the upper triangle because its first coefficient is zero. Both the second and third row are suitable since their first element is 1. However, we also need a non-zero element in the $(2, 2)$ position and therefore the first step is to exchange the first two equations, hence

$$\begin{pmatrix} 1 & 0 & 1 \\ 0 & 1 & 1 \\ 1 & 1 & 0 \end{pmatrix} \begin{pmatrix} x_1 \\ x_2 \\ x_3 \end{pmatrix} = \begin{pmatrix} 2 \\ 1 \\ 4 \end{pmatrix}.$$

After subtracting the new first and the second equation from the third, we arrive at

$$\begin{pmatrix} 1 & 0 & 1 \\ 0 & 1 & 1 \\ 0 & 0 & -2 \end{pmatrix} \begin{pmatrix} x_1 \\ x_2 \\ x_3 \end{pmatrix} = \begin{pmatrix} 2 \\ 1 \\ 1 \end{pmatrix}.$$

Back substitution then gives $x_1 = \frac{5}{2}$, $x_2 = \frac{3}{2}$, and $x_3 = -\frac{1}{2}$.

This trivial example shows that a subroutine to deal with the general case is much less straightforward, since any number of coefficients may be zero at any step.

The second example shows that not only zero coefficients cause problems. Consider the pair of equations

$$\begin{pmatrix} 0.0002 & 1.044 \\ 0.2302 & -1.624 \end{pmatrix} \begin{pmatrix} x_1 \\ x_2 \end{pmatrix} = \begin{pmatrix} 1.046 \\ 0.678 \end{pmatrix}.$$

The exact solution is $x_1 = 10$, $x_2 = 1$. However, if we assume the accuracy is restricted to four digits in every operation and multiply the first equation by $0.2302/0.0002 = 1151$ and subtract from the second, the last equation becomes

$$-1204x_2 = -1203,$$

which gives the solution $x_2 = 0.9992$. Using this number to solve for x_1 in the first equation gives the answer $x_1 = 14.18$, which is quite removed from the true solution. A small non-zero number as pivotal value was not suitable. In fact, the problem is that 0.002 is small compared to 1.044.

A successful pivotal strategy requires the comparison of the relative size of a coefficient to the other coefficients in the same equation. We can calculate the relative size by dividing each coefficient by the largest absolute value in the row. This is known as *scaled pivoting*.

To summarize, suppose an $n \times n$ set of linear equations is solved by Gaussian elimination. At the first step there are n possible equations and one is chosen as pivot and moved to the first row. Then zeros are introduced in the first column below the first row. This leaves at the second step $n - 1$ equations to be transformed. At the start of the kth step there are $n - k + 1$ equations remaining. From these a pivot is selected. There are different ways to go about this. In *partial pivoting* these $n - k + 1$ equations are scaled by dividing by the modulus of the largest coefficient of each. Then the pivotal equation is chosen as the one with the largest (scaled) coefficient of x_k and moved into the kth row. In *total* (or *complete* or *maximal*) *pivoting* the pivotal equation and pivotal variable are selected by choosing the largest (unscaled) coefficient of any of the remaining variables. This is moved into the (k, k) position. This can involve the exchange of columns as well as rows. If columns are exchanged the order of the unknowns needs to be adjusted accordingly. Total pivoting is more expensive, but for certain systems it may be required for acceptable accuracy.

2.3   LU Factorization

Another possibility to solve a linear system is to factorize A into a lower triangular matrix L (i.e., $L_{i,j} = 0$ for $i < j$) and a upper triangular matrix U (i.e., $U_{i,j} = 0$ for $i > j$), that is, $A = LU$. This is called *LU factorization*. The linear system then becomes $L(U\mathbf{x}) = \mathbf{b}$, which we decompose into $L\mathbf{y} = \mathbf{b}$ and $U\mathbf{x} = \mathbf{y}$. Both these systems can be solved easily by back substitution.

Other applications of the LU factorization are

1. Calculation of determinant:

$$\det A = (\det L)(\det U) = (\prod_{k=1}^{n} L_{k,k})(\prod_{k=1}^{n} U_{k,k}).$$

2. Non-singularity testing: $A = LU$ is non-singular if and only if all the diagonal elements of L and U are nonzero.

3. Calculating the inverse: The inverse of triangular matrices can be easily calculated directly. Subsequently $A^{-1} = U^{-1}L^{-1}$.

In the following we derive how to obtain the LU factorization. We denote the columns of L by $\mathbf{l}_1, \ldots, \mathbf{l}_n$ and the rows of U by $\mathbf{u}_1^T, \ldots, \mathbf{u}_n^T$. Thus

$$A = LU = (\mathbf{l}_1 \ldots \mathbf{l}_n) \begin{pmatrix} \mathbf{u}_1^T \\ \vdots \\ \mathbf{u}_n^T \end{pmatrix} = \sum_{k=1}^{n} \mathbf{l}_k \mathbf{u}_k^T.$$

Assume that A is nonsingular and that the factorization exists. Hence the diagonal elements of L are non-zero. Since $\mathbf{l}_k \mathbf{u}_k^T$ stays the same if $\mathbf{l}_k$ is replaced by $\alpha \mathbf{l}_k$ and $\mathbf{u}_k$ by $\alpha^{-1} \mathbf{u}_k$, where $\alpha \neq 0$, we can assume that all diagonal elements of L are equal to 1.

Since the first $k - 1$ components of $\mathbf{l}_k$ and $\mathbf{u}_k$ are zero, each matrix $\mathbf{l}_k \mathbf{u}_k^T$ has zeros in the first $k - 1$ rows and columns. It follows that $\mathbf{u}_1^T$ is the first row of A and $\mathbf{l}_1$ is the first column of A divided by $A_{1,1}$ so that $L_{1,1} = 1$.

Having found $\mathbf{l}_1$ and $\mathbf{u}_1$, we form the matrix $A_1 = A - \mathbf{l}_1 \mathbf{u}_1^T = \sum_{k=2}^{n} \mathbf{l}_k \mathbf{u}_k^T$.

The first row and column of A_1 are zero and it follows that $\mathbf{u}_2^T$ is the second row of A_1, while $\mathbf{l}_2$ is the second column of A_1 scaled so that $L_{2,2} = 1$.

To formalize the LU algorithm, set $A_0 := A$. For all $k = 1, \ldots, n$ set $\mathbf{u}_k^T$ to the kth row of A_{k-1} and $\mathbf{l}_k$ to the kth column of A_{k-1}, scaled so that $L_{k,k} = 1$. Further calculate $A_k := A_{k-1} - \mathbf{l}_k \mathbf{u}_k^T$ before incrementing k by 1.

Exercise 2.2. *Calculate the LU factorization of the matrix*

$$A = \begin{pmatrix} 8 & 6 & -2 & 1 \\ 8 & 8 & -3 & 0 \\ -2 & 2 & -2 & 1 \\ 4 & 3 & -2 & 5 \end{pmatrix},$$

where all the diagonal elements of L are 1. Choose one of these factorizations to find the solution to $A\mathbf{x} = \mathbf{b}$ for $\mathbf{b}^T = (-2 \ 0 \ 2 \ -1)$.

All elements of the first k rows and columns of A_k are zero. Therefore we can use the storage of the original A to accumulate L and U. The full LU accumulation requires $O(n^3)$ operations.

Looking closer at the equation $A_k = A_{k-1} - \mathbf{l}_k \mathbf{u}_k^T$, we see that the jth row of A_k is the jth row of A_{k-1} minus $L_{j,k}$ times $\mathbf{u}_k^T$ which is the kth row of A_{k-1}. This is an elementary row operation. Thus calculating $A_k = A_{k-1} - \mathbf{l}_k \mathbf{u}_k^T$ is equivalent to performing $n - k$ row operations on the last $n - k$ rows. Moreover, the elements of $\mathbf{l}_k$ which are the multipliers $L_{k,k}, L_{k+1,k}, \ldots, L_{n,k}$ are chosen so that the kth column of A_k is zero. Hence we see that the LU factorization is analogous to Gaussian elimination for solving $A\mathbf{x} = \mathbf{b}$. The main difference however is that the LU factorization does not consider $\mathbf{b}$ until the end. This is particularly useful when there are many different vectors $\mathbf{b}$, some of which might not be known at the outset. For each different $\mathbf{b}$, Gaussian elimination would require $O(n^3)$ operations, whereas with LU factorization $O(n^3)$ operations are necessary for the initial factorization, but then the solution for each $\mathbf{b}$ requires only $O(n^2)$ operations.

The algorithm can be rewritten in terms of the elements of A, L, and U. Since

$$A_{i,j} = \sum_{a=1}^{n} \left(\mathbf{l}_a \mathbf{u}_a^T \right)_{i,j} = \sum_{a=1}^{n} L_{i,a} U_{a,j},$$

and since $U_{a,j} = 0$ for $a > j$ and $L_{i,a} = 0$ for $a > i$, we have

$$A_{i,j} = \sum_{a=1}^{\min(i,j)} L_{i,a} U_{a,j}.$$

At the kth step the elements $L_{i,a}$ are known for $a < i$ and the elements $U{a,j}$ are known for $a < j$. For $i = j$,

$$A_{i,i} = \sum_{a=1}^{i} L_{i,a} U_{a,i},$$

and since $L_{i,i} = 1$, we can solve for

$$U_{i,i} = A_{i,i} - \sum_{a=1}^{i-1} L_{i,a} U_{a,i}.$$

If $U_{i,i} = 0$, then U has a zero on the diagonal and thus is singular, since U is upper triangular. The matrix U inherits the rank of A, while L is always non-singular, since its diagonal consists entirely of 1.

For $j > i$, $A_{i,j}$ lies above the diagonal and

$$A_{i,j} = \sum_{a=1}^{i} L_{i,a} U_{a,j} = \sum_{a=1}^{i-1} L_{i,a} U_{a,j} + U_{i,j}$$

and we can solve for

$$U_{i,j} - A_{i,j} - \sum_{a=1}^{i-1} L_{i,a} U_{a,j}.$$

Similarly, for $j < i$, $A_{i,j}$ lies to the left of the diagonal and

$$A_{i,j} = \sum_{a=1}^{j} L_{i,a} U_{a,j} = \sum_{a=1}^{j-1} L_{i,a} U_{a,j} + L_{i,j} U_{j,j}$$

which gives

$$L_{i,j} = \frac{A_{i,j} - \sum_{a=1}^{j-1} L_{i,a} U_{a,j}}{U_{j,j}}.$$

Note that the last formula is only valid when $U_{j,j}$ is non-zero. That means when A is non-singular. If A is singular, other strategies are necessary, such as pivoting, which is described below.

We have already seen the equivalence of LU factorization and Gaussian elimination. Therefore the concept of pivoting also exists for LU factorization. It is necessary for such cases as when, for example, $A_{1,1} = 0$. In this case we need to exchange rows of A to be able to proceed with the factorization. Specifically, pivoting means that having obtained A_{k-1}, we exchange two rows of A_{k-1} so that the element of largest magnitude in the kth column is in the pivotal position (k, k), i.e.,

$$|(A_{k-1})_{k,k}| \geq |(A_{k-1})_{j,k}|, \qquad j = 1, \ldots, n.$$

Since the exchange of rows can be regarded as the pre-multiplication of the relevant matrix by a permutation matrix, we need to do the same exchange in the portion of L that has been formed already (i.e., the first $k - 1$ columns):

$$A_{k-1}^{\text{new}} = PA_{k-1} = PA - P\sum_{j=1}^{k-1} \mathbf{l}_j \mathbf{u}_j^T = PA - \sum_{j=1}^{k-1} P\mathbf{l}_j \mathbf{u}_j^T.$$

We also need to record the permutations of rows to solve for **b**.

Pivoting copes with zeros in the pivot position unless the entire k^{th} column of A_{k-1} is zero. In this particular case we let $\mathbf{l}_k$ be the kth unit vector while $\mathbf{u}_k^T$ is the kth row of A_{k-1} as before. With this choice we retain that the matrix $\mathbf{l}_k \mathbf{u}_k^T$ has the same k^{th} row and column as A_{k-1}.

An important advantage of pivoting is that $|L_{i,j}| \leq 1$ for all $i, j = 1, \ldots, n$. This avoids the chance of large numbers occurring that might lead to ill conditioning and to accumulation of round-off error.

Exercise 2.3. *By using pivoting if necessary an LU factorization is calculated of an $n \times n$ matrix A, where L has ones on the diagonal and the moduli of all off-diagonal elements do not exceed 1. Let α be the largest moduli of the elements of A. Prove by induction that elements of U satisfy $|U_{i,j}| \leq 2^{i-1}\alpha$. Construct 2×2 and 3×3 nonzero matrices A that give $|U_{2,2}| = 2\alpha$ and $|U_{3,3}| = 4\alpha$.*

Depending on the architecture different formulations of the algorithm are easier to implement. For example MATLAB and its support for matrix calculations and manipulations lends itself to the first formulation, as the following example shows.

```
function [L,U]=LU(A)
% Computes the LU factorisation
% A input argument, square matrix
% L square matrix of the same dimensions as A containing the lower
% triangular portion of the LU factorisation
% U square matrix of the same dimensions as A containing the upper
% triangular portion of the LU factorisation

[n,m]=size(A); % finding the size of A
if n≠ m
    error('input is not a square matrix');
end
L=eye(n); % initialising L to the identity matrix
U=A; % initialising U to be A
for k=1:n % loop calculates one column of L and one row of U at a time
    % Note that U holds in its lower portion a modified portion of A
    for j=k+1:n
        % if U(k,k) = 0, do nothing, because L is already initialised
        % to the identity matrix and thus the k-th column is the k-th
        %   standard basis vector
        if abs(U(k,k)) > 1e-12 % not comparing to zero because of
                                % possible rounding errors
            L(j,k)=U(j,k)/U(k,k); % let the k-th column of L be the k-th
                                   % column of the current U scaled by
                                   % the diagonal element
        end
        U(j,:)=U(j,:)-L(j,k)*U(k,:); % adjust U by subtracting the
                                      % outer product of of the k-th
                                      % column of L and the k-th row
                                      % of U
    end
```

```
end
```

Exercise 2.4. *Implement the LU factorization with pivoting.*

It is often required to solve *very, very* large systems of equations $A\mathbf{x} = \mathbf{b}$ where nearly all of the elements of A are zero (for example, arising in the solution of partial differential equations). A system of the size $n = 10^5$ would be a small system in this context. Such a matrix is called sparse and this sparsity should be exploited for an efficient solution. In particular, we wish the matrices L and U to inherit as much as possible of the sparsity of A and for the cost of computation to be determined by the number of nonzero entries, rather than by n.

Exercise 2.5. *Let A be a real nonsingular $n \times n$ matrix that has the factorization $A = LU$, where L is lower triangular with ones on its diagonal and U is upper triangular. Show that for $k = 1, \ldots, n$ the first k rows of U span the same subspace as the first k rows of A. Show also that the first k columns of A are in the k-dimensional subspace spanned by the first k columns of L.*

From the above exercise, this useful theorem follows.

Theorem 2.1. *Let $A = LU$ be an LU factorization (without pivoting) of a sparse matrix. Then all leading zeros in the rows of A to the left of the diagonal are inherited by L and all the leading zeros in the columns of A above the diagonal are inherited by U.*

Therefore we should use the freedom to exchange rows and columns in a preliminary calculation so that many of the zero elements are leading zero elements in rows and columns. This will minimize fill-in. We illustrate this with the following example where we first calculate the LU factorization without exchange of rows and columns.

$$
\begin{pmatrix}
-3 & 1 & 1 & 2 & 0 \\
1 & -3 & 0 & 0 & 1 \\
1 & 0 & 2 & 0 & 0 \\
2 & 0 & 0 & 3 & 0 \\
0 & 1 & 0 & 0 & 3
\end{pmatrix}
=
\begin{pmatrix}
1 & 0 & 0 & 0 & 0 \\
-\frac{1}{3} & 1 & 0 & 0 & 0 \\
-\frac{1}{3} & -\frac{1}{8} & 1 & 0 & 0 \\
-\frac{2}{3} & -\frac{1}{4} & \frac{6}{19} & 1 & 0 \\
0 & -\frac{3}{8} & \frac{1}{19} & \frac{4}{81} & 1
\end{pmatrix}
\begin{pmatrix}
-3 & 1 & 1 & 2 & 0 \\
0 & -\frac{8}{3} & \frac{1}{3} & \frac{2}{3} & 1 \\
0 & 0 & \frac{19}{8} & \frac{3}{4} & \frac{1}{8} \\
0 & 0 & 0 & \frac{81}{19} & \frac{4}{19} \\
0 & 0 & 0 & 0 & \frac{272}{81}
\end{pmatrix}.
$$

We see that the fill-in is significant. If, however, we symmetrically exchange rows and columns, swapping the first and third, second and fourth, and fourth and fifth, we get

$$
\begin{pmatrix}
2 & 0 & 1 & 0 & 0 \\
0 & 3 & 2 & 0 & 0 \\
1 & 2 & -3 & 0 & 1 \\
0 & 0 & 0 & 3 & 1 \\
0 & 0 & 1 & 1 & -3
\end{pmatrix}
=
\begin{pmatrix}
1 & 0 & 0 & 0 & 0 \\
0 & 1 & 0 & 0 & 0 \\
\frac{1}{2} & \frac{2}{3} & 1 & 0 & 0 \\
0 & 0 & 0 & 1 & 0 \\
0 & 0 & -\frac{6}{29} & \frac{1}{3} & 1
\end{pmatrix}
\begin{pmatrix}
2 & 0 & 1 & 0 & 0 \\
0 & 3 & 2 & 0 & 0 \\
0 & 0 & -\frac{29}{6} & 0 & 1 \\
0 & 0 & 0 & 3 & 1 \\
0 & 0 & 0 & 0 & -\frac{272}{87}
\end{pmatrix}.
$$

There has been much research on how best to exploit sparsity with the help of graph theory. However, this is beyond this course. The above theorem can also be applied to *banded matrices*.

Definition 2.1. *The matrix A is a banded matrix if there exists an integer $r < n$ such that $A_{i,j} = 0$ for $|i - j| > r$, $i, j = 1, \ldots, n$. In other words, all the nonzero elements of A reside in a band of width $2r + 1$ along the main diagonal.*

For banded matrices, the factorization $A = LU$ implies that $L_{i,j} = U_{i,j} = 0$ for $|i - j| > r$ and the banded structure is inherited by both L and U.

In general, the expense of calculating an LU factorization of an $n \times n$ dense matrix A is $O(n^3)$ operations and the expense of solving $Ax = b$ using that factorization is $O(n^2)$. However, in the case of a banded matrix, we need just $O(r^2 n)$ operations to factorize and $O(rn)$ operations to solve the linear system. If r is a lot smaller than n, then this is a substantial saving.

2.4 Cholesky Factorization

Let A be an $n \times n$ symmetric matrix, i.e., $A_{i,j} = A_{j,i}$. We can take advantage of the symmetry by expressing A in the form of $A = LDL^T$ where L is lower triangular with ones on its diagonal and D is a diagonal matrix. More explicitly, we can write the factorization which is known as *Cholesky factorization* as

$$
A = (\mathbf{l}_1 \ \ldots \ \mathbf{l}_n)
\begin{pmatrix}
D_{1,1} & 0 & \cdots & 0 \\
0 & D_{2,2} & \ddots & \vdots \\
\vdots & \ddots & \ddots & 0 \\
0 & \cdots & 0 & D_{n,n}
\end{pmatrix}
\begin{pmatrix}
\mathbf{1}_1^T \\
\mathbf{1}_2^T \\
\vdots \\
\mathbf{1}_n^T
\end{pmatrix}
= \sum_{k=1}^{n} D_{k,k} \mathbf{l}_k \mathbf{l}_k^T.
$$

Again $\mathbf{l}_k$ denotes the kth column of L. The analogy to the LU algorithm is obvious when letting $U = DL^T$. However, this algorithm exploits the symmetry and requires roughly half the storage. To be more specific, we let $A_0 = A$ at the beginning and for $k = 1, \ldots, n$ we let $\mathbf{l}_k$ be the kth column of A_{k-1} scaled such that $L_{k,k} = 1$. Set $D_{k,k} = (A_{k-1})_{k,k}$ and calculate $A_k = A_{k-1} - D_{k,k} \mathbf{l}_k \mathbf{l}_k^T$.

An example of such a factorization is

$$
\begin{pmatrix} 4 & 1 \\ 1 & 4 \end{pmatrix} = \begin{pmatrix} 1 & 0 \\ \frac{1}{4} & 1 \end{pmatrix} \begin{pmatrix} 4 & 0 \\ 0 & \frac{15}{4} \end{pmatrix} \begin{pmatrix} 1 & \frac{1}{4} \\ 0 & 1 \end{pmatrix}.
$$

Recall that A is positive definite if $\mathbf{x}^T A \mathbf{x} > 0$ for all $\mathbf{x} \neq 0$.

Theorem 2.2. *Let A be a real $n \times n$ symmetric matrix. It is positive definite if and only if an LDL^T factorization exists where the diagonal elements of D are all positive.*

Proof. One direction is straightforward. Suppose $A = LDL^T$ with $D_{i,i} > 0$ for $i = 1,\ldots,n$ and let $\mathbf{x} \neq 0 \in \mathbb{R}^n$. Since L is nonsingular (ones on the diagonal), $\mathbf{y} := L^T\mathbf{x}$ is nonzero. Hence $\mathbf{x}^T A\mathbf{x} = \mathbf{y}^T D\mathbf{y} = \sum_{i=1}^{n} D_{i,i} y_i^2 > 0$ and A is positive definite.

For the other direction, suppose A is positive definite. Our aim is to show that an LDL^T factorization exists. Let $\mathbf{e}_k \in \mathbb{R}^n$ denote the kth unit vector. Firstly, $\mathbf{e}_1^T A\mathbf{e}_1 = A_{1,1} > 0$ and $\mathbf{l}_1$ and $D_{1,1}$ are well-defined. In the following we show that $(A_{k-1})_{k,k} > 0$ for $k = 1,\ldots,n$ and that hence $\mathbf{l}_k$ and $D_{k,k}$ are well-defined and the factorization exists. We have already seen that this is true for $k = 1$. We continue by induction, assuming that $A_{k-1} = A - \sum_{i=1}^{k-1} D_{i,i}\mathbf{l}_i\mathbf{l}_i^T$ has been computed successfully. Let $\mathbf{x} \in \mathbb{R}^n$ be such that $x_{k+1} = x_{k+2} = \cdots = x_n = 0$, $x_k = 1$ and $x_j = -\sum_{i=j+1}^{k} L_{i,j}x_i$ for $j = k-1, k-2, \ldots, 1$. With this choice of $\mathbf{x}$ we have

$$\mathbf{l}_j^T\mathbf{x} = \sum_{i=1}^{n} L_{i,j}x_i = \sum_{i=j}^{k} L_{i,j}x_i = x_j + \sum_{i=j+1}^{k} L_{i,j}x_i = 0 \qquad j = 1,\ldots,k-1.$$

Since the first $k-1$ rows and columns of A_{k-1} and the last $n-k$ components of $\mathbf{x}$ vanish and $x_k = 1$, we have

$$\begin{aligned}(A_{k-1})_{k,k} &= \mathbf{x}^T A_{k-1}\mathbf{x} = \mathbf{x}^T\left(A - \sum_{j=1}^{k-1} D_{j,j}\mathbf{l}_j\mathbf{l}_j^T\right)\mathbf{x} \\ &= \mathbf{x}^T A\mathbf{x} - \sum_{j=1}^{k-1} D_{j,j}(\mathbf{l}_j^T\mathbf{x})^2 = \mathbf{x}^T A\mathbf{x} > 0.\end{aligned}$$

$\square$

The conclusion of this theorem is that we can check whether a symmetric matrix is positive definite by attempting to calculate its LDL^T factorization.

Definition 2.2 (Cholesky factorization). *Define $D^{1/2}$ as the diagonal matrix where $(D^{1/2})_{k,k} = \sqrt{D_{k,k}}$. Hence $D^{1/2}D^{1/2} = D$. Then for positive definite A, we can write*

$$A = (LD^{1/2})(D^{1/2}L^T) = (LD^{1/2})(LD^{1/2})^T.$$

Letting $\tilde{L} := LD^{1/2}$, $A = \tilde{L}\tilde{L}^T$ is known as the Cholesky factorization.

Exercise 2.6. *Calculate the Cholesky factorization of the matrix*

$$\begin{pmatrix} 1 & 1 & 0 & \cdots & \cdots & 0 \\ 1 & 2 & 1 & \ddots & & \vdots \\ 0 & 1 & 2 & 1 & \ddots & \vdots \\ \vdots & \ddots & 1 & 3 & 1 & 0 \\ \vdots & & \ddots & 1 & 3 & 1 \\ 0 & \cdots & & 0 & 1 & \lambda \end{pmatrix}.$$

Deduce from the factorization the value of λ which makes the matrix singular.

2.5 QR Factorization

In the following we examine another way to factorize a matrix. However, first we need to recall a few concepts.

For all $\mathbf{x}, \mathbf{y} \in \mathbb{R}^n$, the *scalar product* is defined by

$$\langle \mathbf{x}, \mathbf{y} \rangle = \langle \mathbf{y}, \mathbf{x} \rangle = \sum_{i=1}^{n} x_i y_i = \mathbf{x}^T \mathbf{y} = \mathbf{y}^T \mathbf{x}.$$

The scalar product is a *linear operation*, i.e., for $\mathbf{x}, \mathbf{y}, \mathbf{z} \in \mathbb{R}^n$ and $\alpha, \beta \in \mathbb{R}$

$$\langle \alpha \mathbf{x} + \beta \mathbf{y}, \mathbf{z} \rangle = \alpha \langle \mathbf{x}, \mathbf{z} \rangle + \beta \langle \mathbf{y}, \mathbf{z} \rangle.$$

The *norm* or *Euclidean length* of $\mathbf{x} \in \mathbb{R}^n$ is defined as

$$\|\mathbf{x}\| = \left(\sum_{i=1}^{n} x_i^2 \right)^{1/2} = \langle \mathbf{x}, \mathbf{x} \rangle^{1/2} \geq 0.$$

The norm of $\mathbf{x}$ is zero if and only if $\mathbf{x}$ is the zero vector.

Two vectors $\mathbf{x}, \mathbf{y} \in \mathbb{R}^n$ are called *orthogonal to each other* if

$$\langle \mathbf{x}, \mathbf{y} \rangle = 0.$$

Of course the zero vector is orthogonal to every vector including itself.

A set of vectors $\mathbf{q}_1, \ldots, \mathbf{q}_m \in \mathbb{R}^n$ is called *orthonormal* if

$$\langle \mathbf{q}_k, \mathbf{q}_l \rangle = \begin{cases} 1, & k = l, \\ 0, & k \neq l, \end{cases} \qquad k, l = 1, \ldots, m.$$

Let $Q = (\mathbf{q}_1 \ \cdots \ \mathbf{q}_n)$ be an $n \times n$ real matrix. It is called *orthogonal* if its columns are orthonormal. It follows from $(Q^T Q)_{k,l} = \langle \mathbf{q}_k, \mathbf{q}_l \rangle$ that $Q^T Q = I$ where I is the *unit* or *identity matrix*. Thus Q is nonsingular and the inverse exists, $Q^{-1} = Q^T$. Furthermore, $QQ^T = QQ^{-1} = I$. Therefore the rows of an orthogonal matrix are also orthonormal and Q^T is also an orthogonal matrix. Further, $1 = \det I = \det(QQ^T) = \det Q \det Q^T = (\det Q)^2$ and we deduce $\det Q = \pm 1$.

Lemma 2.1. *If P, Q are orthogonal, then so is PQ.*

Proof. Since $P^T P = Q^T Q = I$, we have

$$(PQ)^T (PQ) = (Q^T P^T)(PQ) = Q^T (P^T P) Q = Q^T Q = I$$

and hence PQ is orthogonal. ◻

We will require the following lemma to construct orthogonal matrices.

Lemma 2.2. *Let $\mathbf{q}_1, \ldots, \mathbf{q}_m \in \mathbb{R}^n$ be orthonormal and $m < n$. Then there exists $\mathbf{q}_{m+1} \in \mathbb{R}^n$ such that $\mathbf{q}_1, \ldots, \mathbf{q}_{m+1}$ is orthonormal.*

Proof. The proof is constructive. Let Q be the $n \times m$ matrix whose columns are $\mathbf{q}_1, \ldots, \mathbf{q}_m$. Since

$$\sum_{i=1}^{n} \sum_{j=1}^{m} Q_{i,j}^2 = \sum_{j=1}^{m} \|\mathbf{q}_j\|^2 = m < n,$$

we can find a row of Q where the sum of the squares is less than 1, in other words there exists $l \in \{1, \ldots, n\}$ such that $\sum_{j=1}^{m} Q_{l,j}^2 < 1$. Let $\mathbf{e}_l$ denote the lth unit vector. Then $Q_{l,j} = \langle \mathbf{q}_j, \mathbf{e}_l \rangle$. Further, set $\mathbf{w} = \mathbf{e}_l - \sum_{j=1}^{m} \langle \mathbf{q}_j, \mathbf{e}_l \rangle \mathbf{q}_j$. The scalar product of $\mathbf{w}$ with each $\mathbf{q}_i$ for $i = 1, \ldots, m$ is then

$$\langle \mathbf{q}_i, \mathbf{w} \rangle = \langle \mathbf{q}_i, \mathbf{e}_l \rangle - \sum_{j=1}^{m} \langle \mathbf{q}_j, \mathbf{e}_l \rangle \langle \mathbf{q}_i, \mathbf{q}_j \rangle = 0.$$

Thus $\mathbf{w}$ is orthogonal to $\mathbf{q}_1, \ldots, \mathbf{q}_m$. Calculating the norm of $\mathbf{w}$ we get

$$
\begin{aligned}
\|\mathbf{w}\|^2 &= \langle \mathbf{e}_l, \mathbf{e}_l \rangle - 2 \sum_{j=1}^{m} \langle \mathbf{q}_j, \mathbf{e}_l \rangle^2 + \sum_{j=1}^{m} \langle \mathbf{q}_j, \mathbf{e}_l \rangle \sum_{k=1}^{m} \langle \mathbf{q}_k, \mathbf{e}_l \rangle \langle \mathbf{q}_j, \mathbf{q}_k \rangle \\
&= \langle \mathbf{e}_l, \mathbf{e}_l \rangle - \sum_{j=1}^{m} \langle \mathbf{q}_j, \mathbf{e}_l \rangle^2 = 1 - \sum_{j=1}^{m} Q_{l,j}^2 > 0.
\end{aligned}
$$

Thus $\mathbf{w}$ is nonzero and we define $\mathbf{q}_{m+1} = \mathbf{w}/\|\mathbf{w}\|$. $\qquad\square$

Definition 2.3 (QR factorization). *The QR factorization of an $n \times m$ matrix A ($n > m$) has the form $A = QR$, where Q is an $n \times n$ orthogonal matrix and R is an $n \times m$ upper triangular matrix (i.e., $R_{i,j} = 0$ for $i > j$). The matrix R is said to be in the* standard form *if the number of leading zeros in each row increases strictly monotonically until all the rows of R are zero.*

Denoting the columns of A by $\mathbf{a}_1, \ldots, \mathbf{a}_m \in \mathbb{R}^n$ and the columns of Q by $\mathbf{q}_1, \ldots, \mathbf{q}_n \in \mathbb{R}^n$, we can write the factorization as

$$
(\mathbf{a}_1 \ \cdots \ \mathbf{a}_m) = (\mathbf{q}_1 \ \cdots \ \mathbf{q}_n)
\begin{pmatrix}
R_{1,1} & R_{1,2} & \cdots & R_{1,m} \\
0 & R_{2,2} & & \vdots \\
\vdots & \ddots & \ddots & \vdots \\
0 & \cdots & 0 & R_{m,m} \\
0 & \cdots & \cdots & 0 \\
\vdots & & & \vdots \\
0 & \cdots & \cdots & 0
\end{pmatrix}
$$

We see that

$$\mathbf{a}_k = \sum_{i=1}^{k} R_{i,k} \mathbf{q}_i \qquad k = 1, \ldots, n. \tag{2.2}$$

In other words, the kth column of A is a linear combination of the first k columns of Q.

As with the factorizations we have encountered before, the QR factorization can be used to solve $A\mathbf{x} = \mathbf{b}$ by factorizing first $A = QR$ (note $m = n$ here). Then we solve $Q\mathbf{y} = \mathbf{b}$ by $\mathbf{y} = Q^T\mathbf{b}$ and $R\mathbf{x} = \mathbf{y}$ by back substitution.

In the following sections we will examine three algorithms to generate a QR factorization.

2.6   The Gram–Schmidt Algorithm

The first algorithm follows the construction process in the proof of Lemma 2.2. Assuming $\mathbf{a}_1$ is not the zero vector, we obtain $\mathbf{q}_1$ and $R_{1,1}$ from Equation (2.2) for $k = 1$. Since $\mathbf{q}_1$ is required to have unit length we let $\mathbf{q}_1 = \mathbf{a}_1/\|\mathbf{a}_1\|$ and $R_{1,1} = \|\mathbf{a}_1\|$.

Next we form the vector $\mathbf{w} = \mathbf{a}_2 - \langle\mathbf{q}_1, \mathbf{a}_2\rangle\mathbf{q}_1$. It is by construction orthogonal to $\mathbf{q}_1$, since we subtract the component in the direction of $\mathbf{q}_1$. If $\mathbf{w} \neq 0$, we set $\mathbf{q}_2 = \mathbf{w}/\|\mathbf{w}\|$. With this choice $\mathbf{q}_1$ and $\mathbf{q}_2$ are orthonormal. Furthermore,

$$\langle\mathbf{q}_1, \mathbf{a}_2\rangle\mathbf{q}_1 + \|\mathbf{w}\|\mathbf{q}_2 = \langle\mathbf{q}_1, \mathbf{a}_2\rangle\mathbf{q}_1 + \mathbf{w} = \mathbf{a}_2.$$

Hence we let $R_{1,2} = \langle\mathbf{q}_1, \mathbf{a}_2\rangle$ and $R_{2,2} = \|\mathbf{w}\|$.

This idea can be extended to all columns of A. More specifically let A be an $n \times m$ matrix which is not entirely zero. We have two counters, k is the number of columns which have already be generated, j is the number of columns of A which have already been considered. The individual steps are as follows:

1. Set $k := 0, j := 0$.

2. Increase j by 1. If $k = 0$, set $\mathbf{w} := \mathbf{a}_j$, otherwise (i.e., when $k \geq 1$) set $\mathbf{w} := \mathbf{a}_j - \sum_{i=1}^{k}\langle\mathbf{q}_i, \mathbf{a}_j\rangle\mathbf{q}_i$. By this construction $\mathbf{w}$ is orthogonal to $\mathbf{q}_1, \ldots, \mathbf{q}_k$.

3. If $\mathbf{w} = 0$, then $\mathbf{a}_j$ lies in the space spanned by $\mathbf{q}_1, \ldots, \mathbf{q}_k$ or is zero. If $\mathbf{a}_j$ is zero, set the jth column of R to zero. Otherwise, set $R_{i,j} := \langle\mathbf{q}_i, \mathbf{a}_j\rangle$ for $i = 1, \ldots, k$ and $R_{i,j} = 0$ for $i = k+1, \ldots, n$. Note that in this case no new column of Q is constructed. If $\mathbf{w} \neq 0$, increase k by one and set $\mathbf{q}_k := \mathbf{w}/\|\mathbf{w}\|$, $R_{i,j} := \langle\mathbf{q}_i, \mathbf{a}_j\rangle$ for $i = 1, \ldots, k-1$, $R_{k,j} := \|\mathbf{w}\|$ and $R_{i,j} := 0$ for $i = k+1, \ldots n$. By this construction, each column of Q has unit length and $\mathbf{a}_j = \sum_{i=1}^{k} R_{i,j}\mathbf{q}_i$ as required and R is upper triangular, since $k \leq j$.

4. Terminate if $j = m$, otherwise go to 2.

Since the columns of Q are orthonormal, there are at most n of them. In other words k cannot exceed n. If it is less than n, then additional columns can be chosen such that Q becomes an $n \times n$ orthogonal matrix.

The following example illustrates the workings of the Gram–Schmidt algorithm, when A is singular. Let

$$A = \begin{pmatrix} 1 & 2 & 1 \\ 0 & 0 & 1 \\ 0 & 0 & 1 \end{pmatrix}.$$

The first column of A has already length 1, and thus no normalization is necessary. It becomes the first column $\mathbf{q}_1$ of Q, and we set $R_{1,1} = 1$ and $R_{2,1} = R_{3,1} = 0$. Next we calculate

$$\mathbf{w} = \mathbf{a}_2 - \langle \mathbf{a}_2, \mathbf{q}_1 \rangle \mathbf{q}_1 = \begin{pmatrix} 2 \\ 0 \\ 0 \end{pmatrix} - 2 \begin{pmatrix} 1 \\ 0 \\ 0 \end{pmatrix} = 0.$$

In this case, we set $R_{1,2} = 2$ and $R_{2,2} = R_{3,2} = 0$. No column of Q has been generated. Next

$$\mathbf{w} = \mathbf{a}_3 - \langle \mathbf{a}_3, \mathbf{q}_1 \rangle \mathbf{q}_1 = \begin{pmatrix} 1 \\ 1 \\ 1 \end{pmatrix} - \begin{pmatrix} 1 \\ 0 \\ 0 \end{pmatrix} = \begin{pmatrix} 0 \\ 1 \\ 1 \end{pmatrix}.$$

Now the second column of Q can be generated as $\mathbf{q}_2 = \mathbf{w}/\sqrt{2}$. We set $R_{1,3} = 1$, $R_{2,3} = \sqrt{2}$ and $R_{3,3} = 0$. Since we have considered all columns of A, but Q is not square yet, we need to pad it out. The vector $(0, 1/\sqrt{2}, -1/\sqrt{2})^T$ has length 1 and is orthogonal to $\mathbf{q}_1$ and $\mathbf{q}_2$. We can check

$$\begin{pmatrix} 1 & 0 & 0 \\ 0 & 1/\sqrt{2} & 1/\sqrt{2} \\ 0 & 1/\sqrt{2} & -1/\sqrt{2} \end{pmatrix} \begin{pmatrix} 1 & 2 & 1 \\ 0 & 0 & \sqrt{2} \\ 0 & 0 & 0 \end{pmatrix} = \begin{pmatrix} 1 & 2 & 1 \\ 0 & 0 & 1 \\ 0 & 0 & 1 \end{pmatrix}.$$

Exercise 2.7. *Let $\mathbf{a}_1, \mathbf{a}_2$ and $\mathbf{a}_3$ denote the columns of the matrix*

$$A = \begin{pmatrix} 3 & 6 & -1 \\ -6 & -6 & 1 \\ 2 & 1 & -1 \end{pmatrix}.$$

Using the Gram–Schmidt procedure, generate orthonormal vectors $\mathbf{q}_1, \mathbf{q}_2$ and $\mathbf{q}_3$ and real numbers $R_{i,j}$ such that $\mathbf{a}_i = \sum_{j=1}^{i} R_{i,j} \mathbf{q}_j$, $i = 1, 2, 3$. Thus express A as the product $A = QR$, where Q is orthogonal and R is upper triangular.

The disadvantage of this algorithm is that it is ill-conditioned. Small errors in the calculation of inner products spread rapidly, which then can lead to loss of orthogonality. Errors accumulate fast and the off-diagonal elements of $Q^T Q$ (which should be the identity matrix) may become large.

However, orthogonality conditions are retained well, if two given orthogonal matrices are multiplied. Therefore algorithms which compute Q as the product of simple orthogonal matrices are very effective. In the following we encounter one of these algorithms.

2.7 Givens Rotations

Given a real $n \times m$ matrix A, we let $A_0 = A$ and seek a sequence $\Omega_1, \ldots, \Omega_k$ of $n \times n$ orthogonal matrices such that the matrix $A_i := \Omega_i A_{i-1}$ has more zeros below the diagonal than A_{i-1} for $i = 1, \ldots, k$. The insertion of zeros shall be in such a way that A_k is upper triangular. We then set $R = A_k$. Hence $\Omega_k \cdots \Omega_1 A = R$ and $Q = (\Omega_k \cdots \Omega_1)^{-1} = (\Omega_k \cdots \Omega_1)^T = \Omega_1^T \cdots \Omega_k^T$. Therefore $A = QR$ and Q is orthogonal and R is upper triangular.

Definition 2.4 (Givens rotation). *An $n \times n$ orthogonal matrix Ω is called a Givens rotation, if it is the same as the identity matrix except for four elements and we have $\det \Omega = 1$. Specifically we write $\Omega^{[p,q]}$, where $1 \leq p < q \leq n$, for a matrix such that*

$$\Omega^{[p,q]}_{p,p} = \Omega^{[p,q]}_{q,q} = \cos\theta, \qquad \Omega^{[p,q]}_{p,q} = \sin\theta, \qquad \Omega^{[p,q]}_{q,p} = -\sin\theta$$

for some $\theta \in [-\pi, \pi]$.

Letting $n = 4$ we have for example

$$\Omega^{[1,2]} = \begin{pmatrix} \cos\theta & \sin\theta & 0 & 0 \\ -\sin\theta & \cos\theta & 0 & 0 \\ 0 & 0 & 1 & 0 \\ 0 & 0 & 0 & 1 \end{pmatrix}, \qquad \Omega^{[2,4]} = \begin{pmatrix} 1 & 0 & 0 & 0 \\ 0 & \cos\theta & 0 & \sin\theta \\ 0 & 0 & 1 & 0 \\ 0 & -\sin\theta & 0 & \cos\theta \end{pmatrix}.$$

Geometrically these matrices correspond to the underlying coordinate system being rotated along a two-dimensional plane, which is called a *Euler rotation* in mechanics. Orthogonality is easily verified using the identity $\cos^2\theta + \sin^2\theta = 1$.

Theorem 2.3. *Let A be an $n \times m$ matrix. Then for every $1 \leq p < q \leq n$, we can choose indices i and j, where i is either p or q and j is allowed to range over $1, \ldots, m$, such that there exists $\theta \in [-\pi, \pi]$ such that $(\Omega^{[p,q]} A)_{i,j} = 0$. Moreover, all the rows of $\Omega^{[p,q]} A$, except for the pth and the qth, remain unchanged, whereas the pth and the qth rows of $\Omega^{[p,q]} A$ are linear combinations of the pth and q^{th} rows of A.*

Proof. If $A_{p,j} = A_{q,j} = 0$, there are already zeros in the desired positions and no action is needed. Let $i = q$ and set

$$\cos\theta := A_{p,j} / \sqrt{A_{p,j}^2 + A_{q,j}^2}, \qquad \sin\theta := A_{q,j} / \sqrt{A_{p,j}^2 + A_{q,j}^2}.$$

Note that if $A_{q,j} = 0$, $A_{i,j}$ is already zero, since $i = q$. In this case $\cos\theta = 1$ and $\sin\theta = 0$ and $\Omega^{[p,q]}$ is the identity matrix. On the other hand, if $A_{p,j} = 0$, then $\cos\theta = 0$ and $\sin\theta = 1$ and $\Omega^{[p,q]}$ is the permutation matrix which swaps the pth and q^{th} rows to bring the already existing zero in the desired position. Let $A_{p,j} \neq 0$ and $A_{q,j} \neq 0$. Considering the qth row of $\Omega^{[p,q]} A$, we see

$$(\Omega^{[p,q]} A)_{q,k} = -\sin\theta A_{p,k} + \cos\theta A_{q,k},$$

for $k = 1, \ldots, m$. It follows that for $k = j$

$$(\Omega^{[p,q]} A)_{q,j} = \frac{-A_{q,j} A_{p,j} + A_{p,j} A_{q,j}}{\sqrt{A_{p,j}^2 + A_{q,j}^2}} = 0.$$

On the other hand, when $i = p$, we let

$$\cos\theta := A_{q,j}/\sqrt{A_{p,j}^2 + A_{q,j}^2}, \qquad \sin\theta := -A_{p,j}/\sqrt{A_{p,j}^2 + A_{q,j}^2}.$$

Looking at the pth row of $\Omega^{[p,q]} A$, we have

$$(\Omega^{[p,q]} A)_{p,k} = \cos\theta A_{p,k} + \sin\theta A_{q,k},$$

for $k = 1, \ldots, m$. Therefore for $k = j$ $(\Omega^{[p,q]} A)_{p,j} = 0$.

Since $\Omega^{[p,q]}$ equals the identity matrix apart from the $(p, p), (p, q), (q, p)$ and (q, q) entries it follows that only the pth and q^{th} rows change and are linear combinations of each other. $\qquad\square$

As an example we look at a 3×3 matrix

$$A = \begin{pmatrix} 4 & -2\sqrt{2} & 5 \\ 3 & \sqrt{2} & 5 \\ 0 & 1 & \sqrt{2} \end{pmatrix}.$$

We first pick $\Omega^{[1,2]}$ such that $(\Omega^{[1,2]} A)_{2,1} = 0$. Therefore

$$\Omega^{[1,2]} = \begin{pmatrix} \frac{4}{5} & \frac{3}{5} & 0 \\ -\frac{3}{5} & \frac{4}{5} & 0 \\ 0 & 0 & 1 \end{pmatrix}.$$

The resultant matrix is then

$$\Omega^{[1,2]} A = \begin{pmatrix} 5 & -\sqrt{2} & 7 \\ 0 & 2\sqrt{2} & 1 \\ 0 & 1 & \sqrt{2} \end{pmatrix}.$$

Since $(\Omega^{[1,2]} A)_{3,1}$ is already zero, we do not need $\Omega^{[1,3]}$ to introduce a zero there. Next we pick $\Omega^{[2,3]}$ such that $(\Omega^{[2,3]} \Omega^{[1,2]} A)_{3,2} = 0$. Thus

$$\Omega^{[2,3]} = \begin{pmatrix} 1 & 0 & 0 \\ 0 & \frac{2}{3}\sqrt{2} & \frac{1}{3} \\ 0 & -\frac{1}{3} & \frac{2}{3}\sqrt{2} \end{pmatrix}.$$

The final triangular matrix is then

$$\Omega^{[2,3]} \Omega^{[1,2]} A = \begin{pmatrix} 5 & -\sqrt{2} & 7 \\ 0 & 3 & \sqrt{2} \\ 0 & 0 & 1 \end{pmatrix}.$$

Exercise 2.8. *Calculate the QR factorization of the matrix in Exercise 2.7 by using three Givens rotations.*

For a general $n \times m$ matrix A, let l_i be the number of leading zeros in the ith row of A for $i = 1, \ldots, n$. We increase the number of leading zeros until A is upper triangular. The individual steps of the *Givens algorithm* are

1. Stop if the sequence of leading zeros $l_1, \ldots, l_n$ is strictly monotone for $l_i \leq m$. That is, every row has at least one more leading zero than the row above it until all entries are zero.

2. Pick any row indices $1 \leq p < q \leq n$ such that either $l_p > l_q$ or $l_p = l_q < m$. In the first case the pth row has more leading zeros than the qth row while in the second case both rows have the same number of leading zeros.

3. Replace A by $\Omega^{[p,q]} A$ such that the $(q, l_q + 1)$ entry becomes zero. In the case $l_p > l_q$, we let $\theta = \pm \pi$, such that the pth and q^{th} row are effectively swapped. In the case $l_p = l_q$ this calculates a linear combination of the pth and q^{th} row.

4. Update the values of l_p and l_q and go to step 1.

The final matrix A is the required matrix R. Since the number of leading zeros increases strictly monotonically until all rows of R are zero, R is in standard form.

The number of rotations needed is at most the number of elements below the diagonal, which is

$$(n - 1) + (n - 2) + \cdots + (n - m) = mn - \sum_{j=1}^{m} j = mn - \frac{m(m + 1)}{2} = O(mn).$$

(For $m = n$ this becomes $n(n - 1)/2 = O(n^2)$). Each rotation replaces two rows by their linear combinations, which requires $O(m)$ operations. Hence the total cost is $O(m^2 n)$.

When solving a linear system, we multiply the right-hand side by the same rotations. The cost for this is $O(mn)$, since for each rotation two elements of the vector are combined.

However, if the orthogonal matrix Q is required explicitly, we begin by letting Ω be the $m \times m$ unit matrix. Each time A is pre-multiplied by $\Omega^{[p,q]}$, Ω is also pre-multiplied by the same rotation. Thus the final Ω is the product of all rotations and we have $Q = \Omega^T$. The cost for obtaining Q explicitly is $O(m^2 n)$, since in this case the rows have length m.

In the next section we encounter another class of orthogonal matrices.

2.8 Householder Reflections

Definition 2.5 (Householder reflections). *Let $\mathbf{u} \in \mathbb{R}^n$ be a non-zero vector. The $n \times n$ matrix of the form*

$$I - 2\frac{\mathbf{u}\mathbf{u}^T}{\|\mathbf{u}\|^2}$$

is called a Householder reflection.

A *Householder reflection* describes a reflection about a hyperplane which is orthogonal to the vector $\mathbf{u}/\|\mathbf{u}\|$ and which contains the origin. Each such matrix is symmetric and orthogonal, since

$$
\left(I - 2\frac{\mathbf{u}\mathbf{u}^T}{\|\mathbf{u}\|^2}\right)^T \left(I - 2\frac{\mathbf{u}\mathbf{u}^T}{\|\mathbf{u}\|^2}\right) = \left(I - 2\frac{\mathbf{u}\mathbf{u}^T}{\|\mathbf{u}\|^2}\right)^2
$$

$$
= I - 4\frac{\mathbf{u}\mathbf{u}^T}{\|\mathbf{u}\|^2} + 4\frac{\mathbf{u}(\mathbf{u}^T\mathbf{u})\mathbf{u}^T}{\|\mathbf{u}\|^4} = I.
$$

We can use Householder reflections instead of Givens rotations to calculate a QR factorization.

With each multiplication of an $n \times m$ matrix A by a Householder reflection we want to introduce zeros under the diagonal in an entire column. To start with we construct a reflection which transforms the first nonzero column $\mathbf{a} \in \mathbb{R}^n$ of A into a multiple of the first unit vector $\mathbf{e}_1$. In other words we want to choose $\mathbf{u} \in \mathbb{R}^n$ such that the last $n-1$ entries of

$$
\left(I - 2\frac{\mathbf{u}\mathbf{u}^T}{\|\mathbf{u}\|^2}\right)\mathbf{a} = \mathbf{a} - 2\frac{\mathbf{u}^T\mathbf{a}}{\|\mathbf{u}\|^2}\mathbf{u} \tag{2.3}
$$

vanish. Since we are free to choose the length of $\mathbf{u}$, we normalize it such that $\|u\|^2 = 2\mathbf{u}^T\mathbf{a}$, which is possible since $\mathbf{a} \neq 0$. The right side of Equation (2.3) then simplifies to $\mathbf{a} - \mathbf{u}$ and we have $u_i = a_i$ for $i = 2, \ldots, n$. Using this we can rewrite the normalization as

$$
2u_1a_1 + 2\sum_{i=2}^{n} a_i^2 = u_1^2 + \sum_{i=2}^{n} a_i^2.
$$

Gathering the terms and extending the sum, we have

$$
u_1^2 - 2u_1a_1 + a_1^2 - \sum_{i=1}^{n} a_i^2 = 0 \Leftrightarrow (u_1 - a_1)^2 = \sum_{i=1}^{n} a_i^2.
$$

Thus $u_1 = a_1 \pm \|\mathbf{a}\|$. In numerical applications it is usual to let the sign be the same sign as a_1 to avoid $\|\mathbf{u}\|$ becoming too small, since a division by a very small number can lead to numerical difficulties.

Assume the first $k-1$ columns have been transformed such that they have zeros below the diagonal. We need to choose the next Householder reflection to transform the kth column such that the first $k-1$ columns remain in this form. Therefore we let the first $k-1$ entries of $\mathbf{u}$ be zero. With this choice the first $k-1$ rows and columns of the outer product $\mathbf{u}\mathbf{u}^T$ are zero and the top left $(k-1)\times(k-1)$ submatrix of the Householder reflection is the identity matrix. Let $u_k = a_k \pm \sqrt{\sum_{i=k}^n a_i^2}$ and $u_i = a_i$ for $i = k+1,\ldots,n$. This introduces zeros below the diagonal in the kth column.

The end result after processing all columns of A in sequence is an upper triangular matrix R in standard form.

For large n no explicit matrix multiplications are executed. Instead we use

$$\left(I - 2\frac{\mathbf{u}\mathbf{u}^T}{\|\mathbf{u}\|^2}\right)A = A - 2\frac{\mathbf{u}(\mathbf{u}^T A)}{\|\mathbf{u}\|^2}.$$

So we first calculate $\mathbf{w}^T := \mathbf{u}^T A$ and then $A - \frac{2}{\|\mathbf{u}\|^2}\mathbf{u}\mathbf{w}^T$.

Exercise 2.9. *Calculate the QR factorization of the matrix in Exercise 2.7 by using two Householder reflections. Show that for a general $n \times m$ matrix A the computational cost is $O(m^2 n)$.*

How do we choose between Givens rotations and Householder reflections? If A is dense, Householder reflections are generally more effective. However, Givens rotations should be chosen if A has many leading zeros. For example, if an $n \times n$ matrix A has zeros under the first subdiagonal, these can be removed by just $n-1$ Givens rotations, which costs only $O(n^2)$ operations.

QR factorizations are used to solve over-determined systems of equations as we will see in the next section.

2.9 Linear Least Squares

Consider a system of linear equations $A\mathbf{x} = \mathbf{b}$ where A is an $n \times m$ matrix and $\mathbf{b} \in \mathbb{R}^n$.

In the case $n < m$ there are not enough equations to define a unique solution. The system is called *under-determined*. All possible solutions form a vector space of dimension r, where $r \leq m - n$. This problem seldom arises in practice, since generally we choose a solution space in accordance with the available data. An example, however, are cubic splines, which we will encounter later.

In the case $n > m$ there are more equations than unknowns. The system is called *over-determined*. This situation may arise where a simple data model is fitted to a large number of data points. Problems of this form occur frequently when we collect n observations which often carry measurement errors, and we want to build an m-dimensional linear model where generally m is much smaller than n. In statistics this is known as *linear regression*. Many machine learning algorithms have been developed to address this problem (see for example [2] C. M. Bishop *Pattern Recognition and Machine Learning*).

We consider the simplest approach, that is, we seek $\mathbf{x} \in \mathbb{R}^m$ that minimizes the Euclidean norm $\|A\mathbf{x} - \mathbf{b}\|$. This is known as the *least-squares problem*.

Theorem 2.4. *Let A be any $n \times m$ matrix $(n > m)$ and let $\mathbf{b} \in \mathbb{R}^n$. The vector $\mathbf{x} \in \mathbb{R}^m$ minimizes $\|A\mathbf{x} - \mathbf{b}\|$ if and only if it minimizes $\|\Omega A\mathbf{x} - \Omega\mathbf{b}\|$ for an arbitrary $n \times n$ orthogonal matrix Ω.*

Proof. Given an arbitrary vector $\mathbf{v} \in \mathbb{R}^m$, we have

$$\|\Omega\mathbf{v}\|^2 = \mathbf{v}^T\Omega^T\Omega\mathbf{v} = \mathbf{v}^T\mathbf{v} = \|\mathbf{v}\|^2.$$

Hence multiplication by orthogonal matrices preserves the length. In particular, $\|\Omega A\mathbf{x} - \Omega\mathbf{b}\| = \|A\mathbf{x} - \mathbf{b}\|$. □

Suppose for simplicity that the rank of A is m, the largest it can be. That is, there are no linear dependencies in the model space. Suppose that we have a QR factorization of A with R in standard form. Because of the theorem above, letting $\Omega := Q^T$, we have $\|A\mathbf{x} - \mathbf{b}\|^2 = \|Q^T(A\mathbf{x} - \mathbf{b})\|^2 = \|R\mathbf{x} - Q^T\mathbf{b}\|^2$. We can write R as

$$R = \begin{pmatrix} R_1 \\ 0 \end{pmatrix},$$

where R_1 is $m \times m$ upper triangular. Since R and thus R_1 has the same rank as A, it follows that R_1 is nonsingular and therefore has no zeros on the diagonal. Further, let us partition $Q^T\mathbf{b}$ as

$$Q^T\mathbf{b} = \begin{pmatrix} \mathbf{b}_1 \\ \mathbf{b}_2, \end{pmatrix},$$

where $\mathbf{b}_1$ is a vector of length m and $\mathbf{b}_2$ is a vector of length $n - m$. The minimization problem then becomes

$$\|R\mathbf{x} - Q^T\mathbf{b}\|^2 = \|R_1\mathbf{x} - \mathbf{b}_1\|^2 + \|\mathbf{b}_2\|^2.$$

In order to obtain a minimum we can force the first term to be zero. The second term remains. Therefore the solution to the least squares problem can be calculated from $R_1\mathbf{x} = \mathbf{b}_1$ and the minimum is

$$\min_{\mathbf{x}\in\mathbb{R}^n} \|A\mathbf{x} - \mathbf{b}\| = \|\mathbf{b}_2\|.$$

We cannot do any better since multiplication with any orthogonal matrix preserves the Euclidean norm. Thus, QR factorization is a convenient way to solve the linear least squares problem.

2.10   Singular Value Decomposition

In this section we examine another possible factorization known as the *singular value decomposition (SVD)*, but we see that it is not a good choice in many circumstances.

For any non-singular square $n \times n$ matrix A, the matrix $A^T A$ is symmetric and positive definite which can been easily seen from

$$\mathbf{x}^T A^T A\mathbf{x} = \|A\mathbf{x}\|^2 > 0$$

for any nonzero $\mathbf{x} \in \mathbb{R}^n$.

This result generalizes as follows: for any $n \times m$ real matrix A, the $m \times m$ matrix $A^T A$ is symmetric and positive semi-definite. The latter means that there might be vectors such that $\mathbf{x}^T A^T A \mathbf{x} = 0$. Hence it has real, non-negative eigenvalues $\lambda_1, \ldots, \lambda_m$, which we arrange in decreasing order. Note that some of the eigenvalues might be zero. Let D be the diagonal matrix with entries $D_{i,i} = \lambda_i$, $i = 1, \ldots, m$. From the *spectral theorem* (see for example [4] C. W. Curtis *Linear Algebra: an introductory approach*) we know that a orthonormal basis of eigenvectors $\mathbf{v}_1, \ldots, \mathbf{v}_m$ can be constructed for $A^T A$. These are also known as *right singular vectors*. We write these as an orthogonal matrix $V = (\mathbf{v}_1, \ldots, \mathbf{v}_m)$. Therefore we have

$$V^T A^T A V = D = \left(\begin{array}{cc} D_1 & 0 \\ 0 & 0 \end{array} \right),$$

where D_1 is a $k \times k$ matrix ($k \leq \min\{m, n\}$) and contains the nonzero portion of D. Since D_1 is a diagonal matrix with positive elements, $D_1^{1/2}$ and its inverse $D_1^{-1/2}$ can be formed easily.

We partition V accordingly into a $m \times k$ matrix V_1 and $m \times (m-k)$ matrix V_2. That is, V_1 contains the eigenvectors corresponding to nonzero eigenvalues, while V_2 contains the eigenvectors corresponding to zero eigenvalues. We arrive at

$$\left(\begin{array}{c} V_1^T \\ V_2^T \end{array} \right) A^T A \left(V_1 \ V_2 \right) = \left(\begin{array}{cc} V_1^T A^T A V_1 & V_1^T A^T A V_2 \\ V_2^T A^T A V_1 & V_2^T A^T A V_2 \end{array} \right),$$

where $V_1^T A^T A V_1 = D_1$ and $V_2^T A^T A V_2 = 0$ (which means $A V_2 = 0$).

Now let $U_1 = A V_1 D_1^{-1/2}$, which is an $n \times k$ matrix. We then have

$$U_1 D^{1/2} V_1^T = A V_1 D_1^{-1/2} D^{1/2} V_1^T = A.$$

Additionally, the columns of U_1 are orthonormal, since

$$U_1^T U_1 = (D_1^{-1/2})^T V_1^T A^T A V_1 D_1^{-1/2} = D_1^{-1/2} D_1 D_1^{-1/2} = I.$$

Moreover, $A A^T$ is an $n \times n$ matrix with

$$\begin{aligned} U_1^T A A^T U_1 &= (D_1^{-1/2})^T (V_1^T A^T A)(A^T A V_1) D_1^{-1/2} \\ &= D_1^{-1/2} (V_1 D_1)^T (V_1 D_1) D_1^{-1/2} \\ &= D_1^{-1/2} D_1 V_1^T V_1 D_1 D_1^{-1/2} = D_1, \end{aligned}$$

since the columns of V_1 are orthonormal eigenvectors of $A^T A$. Thus the columns of U_1 are eigenvectors of $A A^T$. They are also called *left singular vectors*.

Thus we have arrived at a factorization of A of the form

$$A = U_1 D^{1/2} V_1^T,$$

where U_1 and V_1 are rectangular matrices whose columns are orthonormal to each other. However, it is more desirable to have square orthogonal matrices. V_1 is already part of such a matrix and we extend U_1 in a similar fashion. More specifically, we choose a $n \times (n-k)$ matrix U_2 such that

$$U = (U_1 \ U_2)$$

is an $n \times n$ matrix for which $U^T U = I$, i.e., U is orthogonal.

Since the eigenvalues are non-negative, we can define $\sigma_i = \sqrt{\lambda_i}$ for $i = 1, \ldots, k$. Let S be the $n \times m$ matrix with diagonal $S_{i,i} = \sigma_i$, $i = 1, \ldots, k$. Note that the last diagonal entries of S may be zero.

We have now constructed the *singular value decomposition* of A

$$A = U S V^T.$$

The singular value decomposition can be used to solve $A\mathbf{x} = \mathbf{b}$ for a general $n \times m$ matrix A and $\mathbf{b} \in \mathbb{R}^n$, since it is equivalent to

$$S V^T \mathbf{x} = U^T \mathbf{b}.$$

For overdetermined systems $n > m$, consider the linear least squares problem, which we can rewrite using the orthogonality of U and V.

$$\begin{aligned}
\|A\mathbf{x} - \mathbf{b}\|^2 &= \|U^T(AVV^T\mathbf{x} - \mathbf{b})\|^2 = \|SV^T\mathbf{x} - U^T\mathbf{b}\|^2 \\
&= \sum_{i=1}^k (\sigma_i(V^T\mathbf{x})_i - \mathbf{u}_i^T\mathbf{b})^2 + \sum_{i=k+1}^n (\mathbf{u}_i^T\mathbf{b})^2,
\end{aligned}$$

where $\mathbf{u}_i$ denotes the ith column of U.

We achieve a minimum if we force the first sum to become zero:

$$(V^T\mathbf{x})_i = \frac{\mathbf{u}_i^T\mathbf{b}}{\sigma_i}.$$

Using the orthogonality of V again, we deduce

$$\mathbf{x} = \sum_{i=1}^k \frac{\mathbf{u}_i^T\mathbf{b}}{\sigma_i} \mathbf{v}_i,$$

where $\mathbf{v}_i$ denotes the ith column of V. Thus $\mathbf{x}$ is a linear combination of the first k eigenvectors of $A^T A$ with the given coefficients. This method is however in general not suitable to solve a system of linear equations, since it involves finding the eigenvalues and eigenvectors of $A^T A$, which in itself is a difficult problem.

The following theorem shows another link between the linear least squares problem with the matrix $A^T A$.

Theorem 2.5. $\mathbf{x} \in \mathbb{R}^m$ *is a solution to the linear least squares problem if and only if we have* $A^T(A\mathbf{x} - \mathbf{b}) = 0$.

Proof. If $\mathbf{x}$ is a solution then it minimizes the function

$$f(\mathbf{x}) = \|A\mathbf{x} - \mathbf{b}\|^2 = \mathbf{x}^T A^T A \mathbf{x} - 2\mathbf{x}^T A^T \mathbf{b} + \mathbf{b}^T \mathbf{b}.$$

Hence the gradient $\nabla f(\mathbf{x}) = 2A^T A \mathbf{x} - 2A^T \mathbf{b}$ vanishes. Therefore $A^T (A\mathbf{x} - \mathbf{b}) = 0$.

Conversely, suppose that $A^T (A\mathbf{x} - \mathbf{b}) = 0$ and let $\mathbf{z} \in \mathbb{R}^m$. We show that the norm of $A\mathbf{z} - \mathbf{b}$ is greater or equal to the norm of $A\mathbf{x} - \mathbf{b}$. Letting $\mathbf{y} = \mathbf{z} - \mathbf{x}$, we have

$$\begin{aligned}
\|A\mathbf{z} - \mathbf{b}\|^2 &= \|A(\mathbf{x} + \mathbf{y}) - \mathbf{b}\|^2 \\
&= \|A\mathbf{x} - \mathbf{b}\|^2 + 2\mathbf{y}^T A^T (A\mathbf{x} - \mathbf{b}) + \|A\mathbf{y}\|^2 \\
&= \|A\mathbf{x} - \mathbf{b}\|^2 + \|A\mathbf{y}\|^2 \geq \|A\mathbf{x} - \mathbf{b}\|^2
\end{aligned}$$

and $\mathbf{x}$ is indeed optimal. □

Corollary 2.1. $\mathbf{x} \in \mathbb{R}^m$ *is a solution to the linear least squares problem if and only if the vector* $A\mathbf{x} - \mathbf{b}$ *is orthogonal to all columns of* A.

Hence another way to solve the linear least squares problem is to solve the $m \times m$ system of linear equations $A^T A \mathbf{x} = A^T \mathbf{b}$. This is called the *method of normal equations*.

There are several disadvantages with this approach and also the singular value decomposition. Firstly, $A^T A$ might be singular, secondly a sparse A might be replaced by a dense $A^T A$, and thirdly calculating $A^T A$ might lead to loss of accuracy; for example, due to overflow when large elements are multiplied. These are problems with forming $A^T A$ and this is before we attempt to calculate the eigenvalues and eigenvectors of $A^T A$ for the singular value decomposition.

In the next sections we will encounter iterative methods for linear systems.

2.11   Iterative Schemes and Splitting

Given a linear system of the form $A\mathbf{x} = \mathbf{b}$, where A is an $n \times n$ matrix and $\mathbf{x}, \mathbf{b} \in \mathbb{R}^n$, solving it by factorization is frequently very expensive for large n. However, we can rewrite it in the form

$$(A - B)\mathbf{x} = -B\mathbf{x} + \mathbf{b},$$

where the matrix B is chosen in such a way that $A - B$ is non-singular and the system $(A - B)\mathbf{x} = \mathbf{y}$ is easily solved for any right-hand side $\mathbf{y}$. A simple iterative scheme starts with an estimate $\mathbf{x}^{(0)} \in \mathbb{R}^n$ of the solution (this could be arbitrary) and generates the sequence $\mathbf{x}^{(k)}$, $k = 1, 2, \ldots$, by solving

$$(A - B)\mathbf{x}^{(k+1)} = -B\mathbf{x}^{(k)} + \mathbf{b}. \tag{2.4}$$

This technique is called *splitting*. If the sequence converges to a limit, $\lim_{k \to \infty} \mathbf{x}^{(k)} = \hat{\mathbf{x}}$, then taking the limit on both sides of Equation (2.4) gives $(A - B)\hat{\mathbf{x}} = -B\hat{\mathbf{x}} + \mathbf{b}$. Hence $\hat{\mathbf{x}}$ is a solution of $A\mathbf{x} = \mathbf{b}$.

What are the necessary and sufficient conditions for convergence? Suppose that A is non-singular and therefore has a unique solution $\mathbf{x}^*$. Since $\mathbf{x}^*$ solves $A\mathbf{x} = \mathbf{b}$, it also satisfies $(A - B)\mathbf{x}^* = -B\mathbf{x}^* + \mathbf{b}$. Subtracting this equation from (2.4) gives

$$(A - B)(\mathbf{x}^{(k+1)} - \mathbf{x}^*) = -B(\mathbf{x}^{(k)} - \mathbf{x}^*).$$

We denote $\mathbf{x}^{(k)} - \mathbf{x}^*$ by $\mathbf{e}^{(k)}$. It is the error in the kth iteration. Since $A - B$ is non-singular, we can write

$$\mathbf{e}^{(k+1)} = -(A - B)^{-1}B\mathbf{e}^{(k)}.$$

The matrix $H := -(A-B)^{-1}B$ is known as the *iteration matrix*. In practical applications H is not calculated. We analyze its properties theoretically in order to determine whether or not we have convergence. We will encounter such analyses later on.

Definition 2.6 (Spectral radius). *Let $\lambda_1, \ldots, \lambda_n$ be the eigenvalues of the $n \times n$ matrix H. Then its spectral radius $\rho(H)$ is defined as*

$$\rho(H) = \max_{i=1,\ldots,n} (|\lambda_i|).$$

Note that even if H is real, its eigenvalues might be complex.

Theorem 2.6. *We have $\lim_{k\to\infty} \mathbf{x}^{(k)} = \mathbf{x}^*$ for all $\mathbf{x}^{(0)} \in \mathbb{R}^n$ if and only if $\rho(H) < 1$.*

Proof. For the first direction we assume $\rho(H) \geq 1$. Let λ be an eigenvalue of H such that $|\lambda| = \rho(H)$ and let $\mathbf{v}$ be the corresponding eigenvector, that is $H\mathbf{v} = \lambda\mathbf{v}$. If $\mathbf{v}$ is real , we let $\mathbf{x}^{(0)} = \mathbf{x}^* + \mathbf{v}$, hence $\mathbf{e}^{(0)} = \mathbf{v}$. It follows by induction that $\mathbf{e}^{(k)} = \lambda^k\mathbf{v}$. This cannot tend to zero since $|\lambda| \geq 1$.

If λ is complex, then $\mathbf{v}$ is complex. In this case we have a complex pair of eigenvalues, λ and its complex conjugate $\overline{\lambda}$, which has $\overline{\mathbf{v}}$ as its eigenvector. The vectors $\mathbf{v}$ and $\overline{\mathbf{v}}$ are linearly independent. We let $\mathbf{x}^{(0)} = \mathbf{x}^* + \mathbf{v} + \overline{\mathbf{v}} \in \mathbb{R}^n$, hence $\mathbf{e}^{(0)} = \mathbf{v} + \overline{\mathbf{v}} \in \mathbb{R}^n$. Again by induction we have $\mathbf{e}^{(k)} = \lambda^k\mathbf{v} + \overline{\lambda}^k\overline{\mathbf{v}} \in \mathbb{R}^n$. Now

$$
\begin{aligned}
\|\mathbf{e}^{(k)}\| &= \|\lambda^k\mathbf{v} + \overline{\lambda}^k\overline{\mathbf{v}}\| \\
&= |\lambda^k|\|e^{i\theta_k}\mathbf{v} + e^{-i\theta_k}\overline{\mathbf{v}}\|,
\end{aligned}
$$

where we have changed to polar coordinates. Now θ_k lies in the closed interval $[-\pi, \pi]$ for all $k = 0, 1, \ldots$. The function in $\theta \in [-\pi, \pi]$ given by $\|e^{i\theta}\mathbf{v} + e^{-i\theta}\overline{\mathbf{v}}\|$ is continuous and has a minimum with value, say, μ. This has to be positive since $\mathbf{v}$ and $\overline{\mathbf{v}}$ are linearly independent. Hence $\|\mathbf{e}^{(k)}\| \geq \mu|\lambda^k|$ and $\mathbf{e}^{(k)}$ cannot tend to zero, since $|\lambda| \geq 1$.

The other direction is beyond the scope of this course, but can be found in [20] R. S. Varga *Matrix Iterative Analysis*, which is regarded as a classic in its field. □

Exercise 2.10. *The iteration* $\mathbf{x}^{(k+1)} = H\mathbf{x}^{(k)} + \mathbf{b}$ *is calculated for* $k = 0, 1, \ldots$, *where* H *is given by*

$$H = \begin{pmatrix} \alpha & \gamma \\ 0 & \beta \end{pmatrix},$$

with $\alpha, \beta, \gamma \in \mathbb{R}$ *and* γ *large and* $|\alpha| < 1, |\beta| < 1$. *Calculate* H^k *and show that its elements tend to zero as* $k \to \infty$. *Further deduce the equation* $\mathbf{x}^{(k)} - \mathbf{x}^* = H^k(\mathbf{x}^{(0)} - \mathbf{x}^*)$ *where* $\mathbf{x}^*$ *satisfies* $\mathbf{x}^* = H\mathbf{x}^* + \mathbf{b}$. *Hence deduce that the sequence* $\mathbf{x}^{(k)}$, $k = 0, 1, \ldots$, *tends to* $\mathbf{x}^*$.

Exercise 2.11. *Starting with an arbitrary* $\mathbf{x}^{(0)}$ *the sequence* $\mathbf{x}^{(k)}$, $k = 1, 2, \ldots$, *is calculated by*

$$\begin{pmatrix} 1 & 1 & 1 \\ 0 & 1 & 1 \\ 0 & 0 & 1 \end{pmatrix} \mathbf{x}^{(k+1)} + \begin{pmatrix} 0 & 0 & 0 \\ \alpha & 0 & 0 \\ \gamma & \beta & 0 \end{pmatrix} \mathbf{x}^{(k)} = \mathbf{b}$$

in order to solve the linear system

$$\begin{pmatrix} 1 & 1 & 1 \\ \alpha & 1 & 1 \\ \gamma & \beta & 1 \end{pmatrix} \mathbf{x} = \mathbf{b},$$

where $\alpha, \beta, \gamma \in \mathbb{R}$ *are constant. Find all values for* α, β, γ *such that the sequence converges for every* $\mathbf{x}^{(0)}$ *and* $\mathbf{b}$. *What happens when* $\alpha = \beta = \gamma = -1$ *and* $\alpha = \beta = 0$?

In some cases, however, iteration matrices can arise where we will have convergence, but it will be very, very slow. An example for this situation is

$$H = \begin{pmatrix} 0.99 & 10^6 & 10^{12} \\ 0 & 0 & 10^{20} \\ 0 & 0 & 0.99 \end{pmatrix}.$$

2.12 Jacobi and Gauss–Seidel Iterations

Both the *Jacobi* and *Gauss–Seidel* methods are splitting methods and can be used whenever A has nonzero diagonal elements. We write A in the form $A = L + D + U$, where L is the subdiagonal (or strictly lower triangular), D is the diagonal, and U is the superdiagonal (or strictly upper triangular) portion of A.

Jacobi method

We choose $A - B = D$, the diagonal part of A, or in other words we let $B = L + U$. The iteration step is given by

$$D\mathbf{x}^{(k+1)} = -(L + U)\mathbf{x}^{(k)} + \mathbf{b}.$$

Gauss–Seidel method

We set $A - B = L + D$, the lower triangular portion of A, or in other words $B = U$. The sequence $\mathbf{x}^{(k)}, k = 1, \ldots$, is generated by

$$(L + D)\mathbf{x}^{(k+1)} = -U\mathbf{x}^{(k)} + \mathbf{b}.$$

Note that there is no need to invert $L + D$; we calculate the components of $x^{(k+1)}$ in sequence using the components we have just calculated by forward substitution:

$$A_{i,i}x_i^{(k+1)} = -\sum_{j<i} A_{i,j}x_j^{(k+1)} - \sum_{j>i} A_{i,j}x_j^{(k)} + b_i, \qquad i = 1, \ldots, n.$$

As we have seen in the previous section, the sequence $\mathbf{x}^{(k)}$ converges to the solution if the spectral radius of the iteration matrix, $H_J = -D^{-1}(L+U)$ for Jacobi or $H_{GS} = (L + D)^{-1}U$ for Gauss–Seidel, is less than one. We will show this is true for two important classes of matrices. One is the class of positive definite matrices and the other is given in the following definition.

Definition 2.7 (Strictly diagonally dominant matrices). *A matrix A is called strictly diagonally dominant by rows if*

$$|A_{i,i}| > \sum_{\substack{j=1 \\ j \neq i}}^{n} |A_{i,j}|$$

for i, 1, . . . , n.

For the first class, the following theorem holds:

Theorem 2.7 (Householder–John theorem). *If A and B are real matrices such that both A and $A - B - B^T$ are symmetric and positive definite, then the spectral radius of $H = -(A - B)^{-1}B$ is strictly less than one.*

Proof. Let λ be an eigenvalue of H and $\mathbf{v} \neq 0$ the corresponding eigenvector. Note that λ and thus $\mathbf{v}$ might have nonzero imaginary parts. From $H\mathbf{v} = \lambda\mathbf{v}$ we deduce $-B\mathbf{v} = \lambda(A - B)\mathbf{v}$. λ cannot equal one since otherwise A would map $\mathbf{v}$ to zero and be singular. Thus we deduce

$$\overline{\mathbf{v}}^T B\mathbf{v} = \frac{\lambda}{\lambda - 1}\overline{\mathbf{v}}^T A\mathbf{v}. \tag{2.5}$$

Writing $\mathbf{v} = \mathbf{u} + i\mathbf{w}$, where $\mathbf{u}, \mathbf{w} \in \mathbb{R}^n$, we deduce $\overline{\mathbf{v}}^T A\mathbf{v} = \mathbf{u}^T A\mathbf{u} + \mathbf{w}^T A\mathbf{w}$. Hence, positive definiteness of A and $A - B - B^T$ implies $\overline{\mathbf{v}}^T A\mathbf{v} > 0$ and $\overline{\mathbf{v}}^T (A - B - B^T)\mathbf{v} > 0$. Inserting (2.5) and its conjugate transpose into the latter, and we arrive at

$$0 < \overline{\mathbf{v}}^T A\mathbf{v} - \overline{\mathbf{v}}^T B\mathbf{v} - \overline{\mathbf{v}}^T B^T\mathbf{v} = \left(1 - \frac{\lambda}{\lambda - 1} - \frac{\overline{\lambda}}{\overline{\lambda} - 1}\right)\overline{\mathbf{v}}^T A\mathbf{v} = \frac{1 - |\lambda|^2}{|\lambda - 1|^2}\overline{\mathbf{v}}^T A\mathbf{v}.$$

The denominator does not vanish since $\lambda \neq 1$. Hence $|\lambda - 1|^2 > 0$. Since $\overline{\mathbf{v}}^T A \mathbf{v} > 0$, $1 - |\lambda|^2$ has to be positive. Therefore we have $|\lambda| < 1$ for every eigenvalue of H as required. □

For the second class we use the following simple, but very useful theorem.

Theorem 2.8 (Gerschgorin theorem). *All eigenvalues of an $n \times n$ matrix A are contained in the union of the Gerschgorin discs Γ_i, $i = 1, \ldots, n$ defined in the complex plane by*

$$\Gamma_i := \{z \in \mathbb{C} : |z - A_{i,i}| \leq \sum_{\substack{j=1 \\ j \neq i}}^{n} |A_{i,j}|\}.$$

It follows from the Gerschgorin theorem that strictly diagonally dominant matrices are nonsingular, since then none of the Gerschgorin discs contain zero.

Theorem 2.9. *If A is strictly diagonally dominant, then both the Jacobi and the Gauss–Seidel methods converge.*

Proof. For the Gauss–Seidel method the eigenvalues of the iteration matrix $H_{GS} = -(L + D)^{-1}U$ are solutions to

$$\det[H_{GS} - \lambda I] = \det[-(L + D)^{-1}U - \lambda I] = 0.$$

Since $L + D$ is non-singular, $\det(L + D) \neq 0$ and the equation can be multiplied by this to give

$$\det[U + \lambda D + \lambda L] = 0. \tag{2.6}$$

Now for $|\lambda| \geq 1$ the matrix $\lambda L + \lambda D + U$ is strictly diagonally dominant, since A is strictly diagonally dominant. Hence $\lambda L + \lambda D + U$ is non-singular and thus its determinant does not vanish. Therefore Equation (2.6) does not have a solution with $|\lambda| \geq 1$. Therefore $|\lambda| < 1$ and we have convergence.

The same argument holds for the iteration matrix for the Jacobi method. □

Exercise 2.12. *Prove that the Gauss–Seidel method to solve $A\mathbf{x} = \mathbf{b}$ converges whenever the matrix A is symmetric and positive definite. Show however by a 3×3 counterexample that the Jacobi method for such an A does not necessarily converge.*

Corollary 2.2.   *1. If A is symmetric and positive definite, then the Gauss–Seidel method converges.*

    2. If both A and $2D - A$ are symmetric and positive definite, then the Jacobi method converges.

Proof.   1. This is the subject of the above exercise.

2. For the Jacobi method we have $B = A - D$. If A is symmetric, then $A - B - B^T = 2D - A$. This matrix is the same as A, just the off diagonal elements have the opposite sign. If both A and $2D - A$ are positive definite, the Householder–John theorem applies.

□

We have already seen one example where convergence can be very slow. The next section shows how to improve convergence.

2.13 Relaxation

The efficiency of the splitting method can be improved by *relaxation*. Here, instead of iterating $(A - B)\mathbf{x}^{(k+1)} = -B\mathbf{x}^{(k)} + \mathbf{b}$, we first calculate $(A - B)\tilde{\mathbf{x}}^{(k+1)} = -B\mathbf{x}^{(k)} + \mathbf{b}$ as an intermediate value and then let

$$\mathbf{x}^{(k+1)} = \omega\tilde{\mathbf{x}}^{(k+1)} + (1 - \omega)\mathbf{x}^{(k)}$$

for $k = 0, 1, \ldots$, where $\omega \in \mathbb{R}$ is called the *relaxation parameter*. Of course $\omega = 1$ corresponds to the original method without relaxation. The parameter ω is chosen such that the spectral radius of the relaxed method is smaller. The smaller the spectral radius, the faster the iteration converges. Letting $\mathbf{c} = (A - B)^{-1}\mathbf{b}$, the relaxation iteration matrix H_ω can then be deduced from

$$\mathbf{x}^{(k+1)} = \omega\tilde{\mathbf{x}}^{(k+1)} + (1 - \omega)\mathbf{x}^{(k)} = \omega H\mathbf{x}^{(k)} + (1 - \omega)\mathbf{x}^{(k)} + \omega\mathbf{c}$$

as

$$H_\omega = \omega H + (1 - \omega)I.$$

It follows that the eigenvalues of H_ω and H are related by $\lambda_\omega = \omega\lambda + (1 - \omega)$. The best choice for ω would be to minimize $\max\{|\omega\lambda_i + (1 - \omega)|, i = 1, \ldots, n\}$ where $\lambda_1, \ldots, \lambda_n$ are the eigenvalues of H. However, the eigenvalues of H are often unknown, but sometimes there is information (for example, derived from the Gerschgorin theorem), which makes it possible to choose a good if not optimal value for ω. For example, it might be known that all the eigenvalues are real and lie in the interval $[a, b]$, where $-1 < a < b < 1$. Then the interval containing the eigenvalues of H_ω is given by $[\omega a + (1 - \omega), \omega b + (1 - \omega)]$. An optimal choice for ω is the one which centralizes this interval around the origin:

$$-[\omega a + (1 - \omega)] = \omega b + (1 - \omega).$$

It follows that

$$\omega = \frac{2}{2 - (a + b)}.$$

The eigenvalues of the relaxed iteration matrix lie in the interval $[-\frac{b-a}{2-(a+b)}, \frac{b-a}{2-(a+b)}]$. Note that if the interval $[a, b]$ is already symmetric about zero, i.e., $a = -b$, then $\omega = 1$ and no relaxation is performed. On the other hand consider the case where all eigenvalues lie in a small interval close to 1. More specifically, let $a = 1 - 2\epsilon$ and $b = 1 - \epsilon$; then $\omega = \frac{2}{3\epsilon}$ and the new interval is $[-\frac{1}{3}, \frac{1}{3}]$.

Exercise 2.13. *The Gauss–Seidel method is used to solve $A\mathbf{x} = \mathbf{b}$, where*

$$A = \begin{pmatrix} 100 & -11 \\ 9 & 1 \end{pmatrix}.$$

Find the eigenvalues of the iteration matrix. Then show that with relaxation the spectral radius can be reduced by nearly a factor of 3. In addition show that after one iterations with the relaxed method the error $\|\mathbf{x}^{(k)} - \mathbf{x}^*\|$ is reduced by more than a factor of 3. Estimate the number of iterations the original Gauss–Seidel would need to achieve a similar decrease in the error.

2.14 Steepest Descent Method

In this section we look at an alternative approach to construct iterative methods to solve $A\mathbf{x} = \mathbf{b}$ in the case where A is symmetric and positive definite. We consider the quadratic function

$$F(\mathbf{x}) = \frac{1}{2}\mathbf{x}^T A\mathbf{x} - \mathbf{x}^T \mathbf{b}, \qquad \mathbf{x} \in \mathbb{R}^n.$$

It is a multivariate function which can be written as

$$F(x_1, \ldots, x_n) = \frac{1}{2}\sum_{i,j=1}^{n} A_{ij}x_i x_j - \sum_{i=1}^{n} b_i x_i.$$

Note that the first sum is a double sum. A multivariate function has an extremum at the point where the derivatives in each of the directions x_i, $i = 1, \ldots, n$ vanish. The vector of derivatives is called the *gradient* and is denoted $\nabla F(\mathbf{x})$. So the extremum occurs when the gradient vanishes, or in other words when $\mathbf{x}$ satisfies $\nabla F(\mathbf{x}) = 0$. The derivative of $F(\mathbf{x})$ in the direction of x_k is

$$\frac{d}{dx_k}F(x_1, \ldots, x_n) = \frac{1}{2}\left(\sum_{i=1}^{n} A_{ik}x_i + \sum_{j=1}^{n} A_{kj}x_j\right) - b_k$$

$$= \sum_{j=1}^{n} A_{kj}x_j - b_k,$$

where we used the symmetry of A in the last step. This is one component of the gradient vector and thus

$$\nabla F(\mathbf{x}) = A\mathbf{x} - \mathbf{b}.$$

Thus finding the extremum is equivalent to $\mathbf{x}$ being a solution of $A\mathbf{x} = \mathbf{b}$. In our case the extremum is a minimum, since A is positive definite.

Let $\mathbf{x}^*$ denote the solution. The location of the minimum of F does not change if the constant $\frac{1}{2}\mathbf{x}^{*T}A\mathbf{x}^*$ is added. Using $\mathbf{b} = A\mathbf{x}^*$ we see that it is equivalent to minimize

$$\frac{1}{2}\mathbf{x}^T A\mathbf{x} - \mathbf{x}^T A\mathbf{x}^* + \frac{1}{2}\mathbf{x}^{*T} A\mathbf{x}^* = \frac{1}{2}(\mathbf{x}^* - \mathbf{x})^T A(\mathbf{x}^* - \mathbf{x}).$$

The latter expression can be viewed as the square of a norm defined by $\|\mathbf{x}\|_A := \sqrt{\mathbf{x}^T A\mathbf{x}}$, which is well-defined, since A is positive definite. Thus we are minimizing $\|\mathbf{x}^* - \mathbf{x}\|_A$. Every iteration constructs an approximation $\mathbf{x}^{(k)}$ which is closer to $\mathbf{x}^*$ in the norm defined by A. Since this is equivalent to minimizing F, the constructed sequence should satisfy the condition

$$F(\mathbf{x}^{(k+1)}) < F(\mathbf{x}^{(k)}).$$

That is, the value of F at the new approximation should be less than the value of F at the current approximation, since we are looking for a minimum. Both Jacobi and Gauss–Seidel methods do so.

Generally decent methods have the following form.

1. Pick any starting vector $\mathbf{x}^{(0)} \in \mathbb{R}^n$.

2. For any $k = 0, 1, 2, \ldots$ the calculation stops if the norm of the gradient $\|A\mathbf{x}^{(k)} - \mathbf{b}\| = \|\nabla F(\mathbf{x}^{(k)})\|$ is acceptably small.

3. Otherwise a *search direction* $\mathbf{d}^{(k)}$ is generated that satisfies the *descent condition*

$$\left[\frac{d}{d\omega}F(\mathbf{x}^{(k)} + \omega\mathbf{d}^{(k)})\right]_{\omega=0} < 0.$$

   In other words, if we are walking in the search direction, the values of F become smaller.

4. The value $\omega^{(k)} > 0$ which minimizes $F(\mathbf{x}^{(k)} + \omega\mathbf{d}^{(k)})$ is calculated and the new approximation is

$$\mathbf{x}^{(k+1)} = \mathbf{x}^{(k)} + \omega^{(k)}\mathbf{d}^{(k)}. \tag{2.7}$$

   Return to 2.

We will see specific choices for the search direction $\mathbf{d}^{(k)}$ later. First we look at which value $\omega^{(k)}$ takes. Using the definition of F gives

$$
\begin{aligned}
F(\mathbf{x}^{(k)} + \omega \mathbf{d}^{(k)}) &= \frac{1}{2}(\mathbf{x}^{(k)} + \omega \mathbf{d}^{(k)})^T A(\mathbf{x}^{(k)} + \omega \mathbf{d}^{(k)}) - (\mathbf{x}^{(k)} + \omega \mathbf{d}^{(k)})^T \mathbf{b} \\
&= \frac{1}{2}\left[\mathbf{x}^{(k)^T} A \mathbf{x}^{(k)} + \mathbf{x}^{(k)^T} A \omega \mathbf{d}^{(k)} + \omega \mathbf{d}^{(k)^T} A \mathbf{x}^{(k)} \right. \\
&\quad \left. + \omega^2 \mathbf{d}^{(k)^T} A \mathbf{d}^{(k)} \right] - \mathbf{x}^{(k)^T} \mathbf{b} - \omega \mathbf{d}^{(k)^T} \mathbf{b} \\
&= F(\mathbf{x}^{(k)}) + \omega \mathbf{d}^{(k)^T} \mathbf{g}^{(k)} + \frac{1}{2}\omega^2 \mathbf{d}^{(k)^T} A \mathbf{d}^{(k)},
\end{aligned}
$$
(2.8)

where we used the symmetry of A and where $\mathbf{g}^{(k)} = A\mathbf{x}^{(k)} - \mathbf{b}$ denotes the gradient $\nabla F(\mathbf{x}^{(k)})$. Differentiating with respect to ω and equating to zero, leads to

$$
\omega^{(k)} = -\frac{\mathbf{d}^{(k)^T} \mathbf{g}^{(k)}}{\mathbf{d}^{(k)^T} A \mathbf{d}^{(k)}}.
$$
(2.9)

Looking at (2.8) more closely, the descent direction has to satisfy $\mathbf{d}^{(k)^T} \mathbf{g}^{(k)} < 0$, otherwise no reduction will be achieved. It is possible to satisfy this condition, since the method terminates when $\mathbf{g}^{(k)}$ is zero.

Multiplying both sides of (2.7) by A from the left and subtracting $\mathbf{b}$, successive gradients satisfy

$$
\mathbf{g}^{(k+1)} = \mathbf{g}^{(k)} + \omega^{(k)} A \mathbf{d}^{(k)}.
$$
(2.10)

Multiplying this now by $\mathbf{d}^{(k)^T}$ from the left and using (2.9), we see that this equates to zero. Thus the new gradient is orthogonal to the previous search direction. Thus the descent method follows the search direction until the gradient becomes perpendicular to the search direction.

Making the choice $\mathbf{d}^{(k)} = -\mathbf{g}^{(k)}$ leads to the *steepest descent method*. Locally the gradient of a function shows the direction of the sharpest increase of F at this point. With this choice we have

$$
\mathbf{x}^{(k+1)} = \mathbf{x}^{(k)} - \omega^{(k)} \mathbf{g}^{(k)}
$$

and

$$
\omega^{(k)} = \frac{\mathbf{g}^{(k)^T} \mathbf{g}^{(k)}}{\mathbf{g}^{(k)^T} A \mathbf{g}^{(k)}} = \frac{\|\mathbf{g}^{(k)}\|^2}{\|\mathbf{g}^{(k)}\|_A^2},
$$

which is the square of the Euclidean norm of the gradient divided by the square of the norm defined by A of the gradient.

It can be proven that the infinite sequence $\mathbf{x}^{(k)}$, $k = 0, 1, 2, \ldots$ converges to the solution of $A\mathbf{x} = \mathbf{b}$ (see for example [9] R. Fletcher *Practical Methods of Optimization*). However, convergence can be unacceptably slow.

We look at the *contour lines* of F in two dimensions ($n = 2$). These are lines where F takes the same value. They are ellipses with the minimum lying at the intersection of the axes of the ellipses. The gradients and thus the search directions are perpendicular to the contour lines. That is, they are perpendicular to the tangent to the contour line at that point. When the current search direction becomes tangential to another contour line, the new approximation is reached and the next search direction is perpendicular to the previous one. Figure 2.1 illustrates this. The resultant zigzag search path is typical. The value $F(\mathbf{x}^{(k+1)})$ is decreased locally relative to $F(\mathbf{x}^{(k)})$, but the global decrease with respect to $F(\mathbf{x}^{(0)})$ can be small.

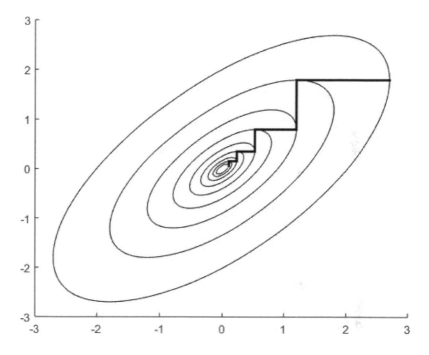

Figure 2.1 Worst-case scenario for the steepest descent method

Let λ_n be the largest eigenvalue of A and λ_1 the smallest. The rate of convergence is then

$$\left(\frac{\lambda_n - \lambda_1}{\lambda_n + \lambda_1} \right)^2$$

(see [16] J. Nocedal, S. Wright *Numerical Optimization*). Returning to the example in two dimensions, if $\lambda_1 = \lambda_2$, then the rate of convergence is zero and the minimum is reached in one step. This is because in this case the contour lines are circles and thus all gradients point directly to the centre where the minimum lies. The bigger the difference in the biggest and smallest eigenvalue, the more elongated the ellipses are and the slower the convergence.

In the following section we see that the use of *conjugate directions* improves the steepest descent method and performs very well.

2.15 Conjugate Gradients

We have already seen how the positive definiteness of A can be used to define a norm. This can be extended to have a concept similar to orthogonality.

Definition 2.8. *The vectors* $\mathbf{u}$ *and* $\mathbf{v}$ *are conjugate with respect to the positive definite matrix* A, *if they are nonzero and satisfy* $\mathbf{u}^T A \mathbf{v} = 0$.

Conjugacy plays such an important role, because through it search directions can be constructed such that the new gradient $\mathbf{g}^{(k+1)}$ is not just orthogonal to the current search direction $\mathbf{d}^{(k)}$ but to all previous search directions. This avoids revisiting search directions as in Figure 2.1.

Specifically, the first search direction is chosen as $\mathbf{d}^{(0)} = -\mathbf{g}^{(0)}$. The following search directions are then constructed by

$$\mathbf{d}^{(k)} = -\mathbf{g}^{(k)} + \beta^{(k)}\mathbf{d}^{(k-1)}, \qquad k = 1, 2, \ldots,$$

where $\beta^{(k)}$ is chosen such that the *conjugacy condition* $\mathbf{d}^{(k)}{}^T A \mathbf{d}^{(k-1)} = 0$ is satisfied. This yields

$$\beta^{(k)} = \frac{\mathbf{g}^{(k)}{}^T A \mathbf{d}^{(k-1)}}{\mathbf{d}^{(k-1)}{}^T A \mathbf{d}^{(k-1)}}, \qquad k = 1, 2, \ldots.$$

This gives the *conjugate gradient method*.

The values of $\mathbf{x}^{(k+1)}$ and $\omega^{(k)}$ are calculated as before in (2.7) and (2.9).

These search directions satisfy the descent condition. From Equation (2.8) we have seen that the descent condition is equivalent to $\mathbf{d}^{(k)}{}^T \mathbf{g}^{(k)} < 0$. Using the formula for $\mathbf{d}^{(k)}$ above and the fact that the new gradient is orthogonal to the previous search direction, i.e., $\mathbf{d}^{(k-1)}{}^T \mathbf{g}^{(k)} = 0$, we see that

$$\mathbf{d}^{(k)}{}^T \mathbf{g}^{(k)} = -\|\mathbf{g}^{(k)}\|^2 < 0.$$

Theorem 2.10. *For every integer* $k \geq 1$ *until* $\|\mathbf{g}^{(k)}\|$ *is small enough, we have the following properties:*

1. *The space spanned by the gradients* $\mathbf{g}^{(j)}$,$j = 0, \ldots, k-1$, *is the same as the space spanned by the search direction* $\mathbf{d}^{(j)}$,$j = 0, \ldots, k-1$,

2. $\mathbf{d}^{(j)}{}^T \mathbf{g}^{(k)} = 0$, *for* $j = 0, 1, \ldots, k-1$,

3. $\mathbf{d}^{(j)}{}^T A \mathbf{d}^{(k)} = 0$ *for* $j = 0, 1, \ldots, k-1$ *and*

4. $\mathbf{g}^{(j)}{}^T \mathbf{g}^{(k)} = 0$ *for* $j = 0, 1, \ldots, k-1$.

Proof. We use induction on $k \geq 1$. The assertions are easy to verify for $k = 1$. Indeed,(1) follows from $\mathbf{d}^{(0)} = -\mathbf{g}^{(0)}$, and (2) and (4) follow, since the new gradient $\mathbf{g}^{(1)}$ is orthogonal to the search direction $\mathbf{d}^{(0)} = -\mathbf{g}^{(0)}$; and (3) follows, since $\mathbf{d}^{(1)}$ is constructed in such a way that it is conjugate to $\mathbf{d}^{(0)}$.

We assume that the assertions are true for some $k \geq 1$ and prove that they remain true when k is increased by one.

For (1), the definition of $\mathbf{d}^{(k)} = -\mathbf{g}^{(k)} + \beta^{(k)}\mathbf{d}^{(k-1)}$ and the inductive hypothesis show that any vector in the span of $\mathbf{d}^{(0)}, \ldots, \mathbf{d}^{(k)}$ also lies in the span of $\mathbf{g}^{(0)}, \ldots, \mathbf{g}^{(k)}$.

We have seen that the search directions satisfy $\mathbf{g}^{(k+1)} = \mathbf{g}^{(k)} + \omega^{(k)}A\mathbf{d}^{(k)}$. Multiplying this by $\mathbf{d}^{(j)T}$ from the left for $j = 0, 1, \ldots, k-1$, the first term of the sum vanishes due to (2) and the second term vanishes due to (3). For $j = k$, the choice of $\omega^{(k)}$ ensures orthogonality. Thus $\mathbf{d}^{(j)T}\mathbf{g}^{(k+1)} = 0$ for $j = 0, 1, \ldots, k$.

This also proves (4), since $\mathbf{d}^{(0)}, \ldots, \mathbf{d}^{(k)}$ span the same space as $\mathbf{g}^{(0)}, \ldots, \mathbf{g}^{(k)}$.

Turning to (3), the next search direction is given by $\mathbf{d}^{(k+1)} = -\mathbf{g}^{(k+1)} + \beta^{(k+1)}\mathbf{d}^{(k)}$. The value of $\beta^{(k+1)}$ gives $\mathbf{d}^{(k)T}A\mathbf{d}^{(k+1)} = 0$. It remains to show that $\mathbf{d}^{(j)T}A\mathbf{d}^{(k+1)} = 0$ for $j = 0, \ldots, k-1$. Inserting the definition of $\mathbf{d}^{(k+1)}$ and using $\mathbf{d}^{(j)T}A\mathbf{d}^{(k)} = 0$, it is sufficient to show $\mathbf{d}^{(j)T}A\mathbf{g}^{(k+1)} = 0$. However,

$$A\mathbf{d}^{(j)} = \frac{1}{\omega^{(j)}}(\mathbf{g}^{(j+1)} - \mathbf{g}^{(j)})$$

and the assertion follows from the mutual orthogonality of the gradients. □

The last assertion of the theorem establishes that if the algorithm is applied in exact arithmetic, then termination occurs after at most n iterations, since there can be at most n mutually orthogonal non-zero vectors in an n-dimensional space. Figure 2.2 illustrates this, showing that the conjugate gradient method converges after two steps in two dimensions.

In the following we reformulate and simplify the conjugate gradient method to show it in standard from. Firstly since $\mathbf{g}^{(k)} - \mathbf{g}^{(k-1)} = \omega^{(k-1)}A\mathbf{d}^{(k-1)}$, the expression for $\beta^{(k)}$ becomes

$$\beta^{(k)} = \frac{\mathbf{g}^{(k)T}(\mathbf{g}^{(k)} - \mathbf{g}^{(k-1)})}{\mathbf{d}^{(k-1)T}(\mathbf{g}^{(k)} - \mathbf{g}^{(k-1)})} = \frac{\|\mathbf{g}^{(k)}\|^2}{\|\mathbf{g}^{(k-1)}\|^2},$$

where $\mathbf{d}^{(k-1)} = -\mathbf{g}^{(k-1)} + \beta^{(k-1)}\mathbf{d}^{(k-2)}$ and the orthogonality properties (2) and (4) of Theorem 2.10 are used.

We write $-\mathbf{r}^{(k)}$ instead of $\mathbf{g}^{(k)}$, where $\mathbf{r}^{(k)}$ is the *residual* $\mathbf{b} - A\mathbf{x}^{(k)}$. The zero vector is chosen as the initial approximation $\mathbf{x}^{(0)}$.

The algorithm is then as follows

1. Set $k = 0, \mathbf{x}^{(0)} = 0, \mathbf{r}^{(0)} = \mathbf{b}$ and $\mathbf{d}^{(0)} = \mathbf{r}^{(0)}$.

2. Stop if $\|\mathbf{r}^{(k)}\|$ is sufficiently small.

3. If $k \geq 1$, set $\mathbf{d}^{(k)} = \mathbf{r}^{(k)} + \beta^{(k)}\mathbf{d}^{(k-1)}$, where $\beta^{(k)} = \|\mathbf{r}^{(k)}\|^2/\|\mathbf{r}^{(k-1)}\|^2$.

4. Calculate $\mathbf{v}^{(k)} = A\mathbf{d}^{(k)}$ and $\omega^{(k)} = \|\mathbf{r}^{(k)}\|^2/\mathbf{d}^{(k)T}\mathbf{v}^{(k)}$.

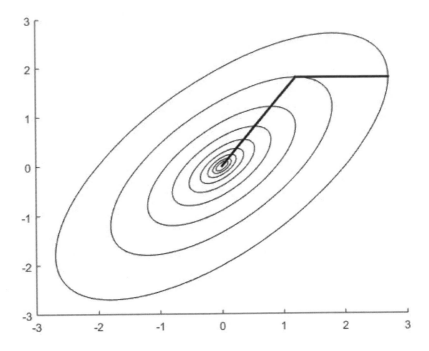

Figure 2.2 Conjugate gradient method applied to the same problem as in Figure 2.1

5. Set $\mathbf{x}^{(k+1)} = \mathbf{x}^{(k)} + \omega^{(k)}\mathbf{d}^{(k)}$ and $\mathbf{r}^{(k+1)} = \mathbf{r}^{(k)} - \omega^{(k)}\mathbf{v}^{(k)}$.

6. Increase k by one and go back to 2.

The number of operations per iteration is $2n$ each to calculate $\mathbf{d}^{(k)^T}\mathbf{v}^{(k)}$, $\|\mathbf{r}^{(k)}\|^2$, $\mathbf{d}^{(k)} = \mathbf{r}^{(k)} + \beta^{(k)}\mathbf{d}^{(k-1)}$, $\mathbf{x}^{(k+1)} = \mathbf{x}^{(k)} + \omega^{(k)}\mathbf{d}^{(k)}$ and $\mathbf{r}^{(k+1)} = \mathbf{r}^{(k)} - \omega^{(k)}\mathbf{v}^{(k)}$. However, the number of operations is dominated by the matrix multiplication $A\mathbf{d}^{(k)}$, which is $O(n^2)$ if A is dense. In the cases where A is sparse this can be reduced and the conjugate gradient method becomes highly suitable.

Only in exact arithmetic is the number of iterations at most n. The conjugate gradient method is sensitive to even small perturbations. In practice, most directions will not be conjugate and the exact solution is not reached. An acceptable approximation is however usually reached within a small (compared to the problem size) number of iterations. The speed of convergence is typically linear, but it depends on the condition number of the matrix A. The larger the condition number the slower the convergence. In the following section we analyze this further and show how to improve the conjugate gradient method.

Exercise 2.14. *Let A be positive definite and let the standard conjugate gradient method be used to solve $A\mathbf{x} = \mathbf{b}$. Express $\mathbf{d}^{(k)}$ in terms of $\mathbf{r}^{(j)}$ and $\beta^{(j)}$, $j = 0, 1, \ldots, k$. Using $\mathbf{x}^{(k+1)} = \sum_{j=0}^{k} \omega^{(j)} \mathbf{d}^{(j)}$, $\omega^{(j)} > 0$ and Theorem 2.10 show that the sequence $\|\mathbf{x}^{(j)}\|$, $j = 0, 1, \ldots, k+1$, increases monotonically.*

Exercise 2.15. *Use the standard form of the conjugate gradient method to solve*

$$\begin{pmatrix} 1 & 0 & 0 \\ 0 & 2 & 0 \\ 0 & 0 & 3 \end{pmatrix} \mathbf{x} = \begin{pmatrix} 1 \\ 1 \\ 1 \end{pmatrix}$$

starting with $\mathbf{x}^{(0)} = 0$. Show that the residuals $\mathbf{r}^{(0)}, \mathbf{r}^{(1)}$ and $\mathbf{r}^{(2)}$ are mutually orthogonal and that the search directions $\mathbf{d}^{(0)}, \mathbf{d}^{(1)}$ and $\mathbf{d}^{(2)}$ are mutually conjugate and that $\mathbf{x}^{(3)}$ is the solution.

2.16   Krylov Subspaces and Pre-Conditioning

Definition 2.9. *Let A be an $n \times n$ matrix, $\mathbf{b} \in \mathbb{R}^n$ a non-zero vector, then for a number m the space spanned by $A^j \mathbf{b}$, $j = 0, \ldots, m-1$ is the mth Krylov subspace of $\mathbb{R}^n$ and is denoted by $K_m(A, \mathbf{b})$.*

In our analysis of the conjugate gradient method we saw that in the kth iteration the space spanned by the search directions $\mathbf{d}^{(j)}$ and the space spanned by the gradients $\mathbf{g}^{(j)}$, $j = 0, \ldots, k$, are the same.

Lemma 2.3. *The space spanned by $\mathbf{g}^{(j)}$ (or $\mathbf{d}^{(j)}$), $j = 0, \ldots, k$, is the same as the $k+1^{\text{th}}$ Krylov subspace.*

Proof. For $k = 0$ we have $\mathbf{d}^{(0)} = -\mathbf{g}^{(0)} = \mathbf{b} \in K_1(A, \mathbf{b})$.

We assume that the space spanned by $\mathbf{g}^{(j)}$, $j = 0, \ldots, k$, is the same as the space spanned by $\mathbf{b}, A\mathbf{b}, \ldots, A^k \mathbf{b}$ and increase k by one.

In the formula $\mathbf{g}^{(k+1)} = \mathbf{g}^{(k)} + \omega^{(k)} A \mathbf{d}^{(k)}$ both $\mathbf{g}^{(k)}$ and $\mathbf{d}^{(k)}$ can be expressed as linear combinations of $\mathbf{b}, A\mathbf{b}, \ldots, A^k \mathbf{b}$ by the inductive hypothesis. Thus $\mathbf{g}^{(k+1)}$ can be expressed as a linear combination of $\mathbf{b}, A\mathbf{b}, \ldots, A^{k+1}\mathbf{b}$. Equally, using $\mathbf{d}^{(k+1)} = -\mathbf{g}^{(k+1)} + \beta^{(k)} \mathbf{d}^{(k)}$ we show that $\mathbf{d}^{(k+1)}$ lies in the space spanned by $\mathbf{b}, A\mathbf{b}, \ldots, A^{k+1}\mathbf{b}$, which is $K_{k+2}(A, \mathbf{b})$. This completes the proof. □

In the following lemma we show some properties of the Krylov subspaces.

Lemma 2.4. *Given A and $\mathbf{b}$. Then $K_m(A, \mathbf{b})$ is a subspace of $K_{m+1}(A, \mathbf{b})$ and there exists a positive integer $s \leq n$ such that for every $m \geq s$ $K_m(A, \mathbf{b}) = K_s(A, \mathbf{b})$. Furthermore, if we express $\mathbf{b}$ as $\mathbf{b} = \sum_{i=1}^{t} c_i \mathbf{v}_i$, where $\mathbf{v}_1, \ldots, \mathbf{v}_t$ are eigenvectors of A corresponding to distinct eigenvalues and all coefficients c_i, $i = 1, \ldots, t$, are non-zero, then $s = t$.*

Proof. Clearly, $K_m(A, \mathbf{b}) \subseteq K_{m+1}(A, \mathbf{b})$. The dimension of $K_m(A, \mathbf{b})$ is less than or equal to m, since it is spanned by m vectors. It is also at most n since

$K_m(A, \mathbf{b})$ is a subspace of $\mathbb{R}^n$. The first Krylov subspace has dimension 1. We let s be the greatest integer such that the dimension of $K_s(A, \mathbf{b})$ is s. Then the dimension of $K_{s+1}(A, \mathbf{b})$ cannot be $s+1$ by the choice of s. It has to be s, since $K_s(A, \mathbf{b}) \subseteq K_{s+1}(A, \mathbf{b})$. Hence the two spaces are the same. This means that $A^s\mathbf{b} \in K_s(A, \mathbf{b})$, i.e., $A^s\mathbf{b} = \sum_{j=0}^{s-1} a_j A^j\mathbf{b}$. But then

$$A^{s+r}\mathbf{b} = \sum_{j=0}^{s-1} a_j A^{j+r}\mathbf{b}$$

for any positive r. This means that also the spaces $K_{s+r+1}(A, \mathbf{b})$ and $K_{s+r}(A, \mathbf{b})$ are the same. Therefore, for every $m \geq s$ $K_m(A, \mathbf{b}) = K_s(A, \mathbf{b})$.

Suppose now that $\mathbf{b} = \sum_{i=1}^{t} c_i \mathbf{v}_i$, where $\mathbf{v}_1, \ldots, \mathbf{v}_t$ are eigenvectors of A corresponding to distinct eigenvalues λ_i. Then for every $j = 1, \ldots, s$

$$A^j\mathbf{b} = \sum_{i=1}^{t} c_i \lambda_i^j \mathbf{v}_i.$$

Thus $K_s(A, \mathbf{b})$ is a subspace of the space spanned by the eigenvectors, which has dimension t, since the eigenvectors are linearly independent. Hence $s \leq t$.

Now assume that $s < t$. The dimension of $K_t(A, \mathbf{b})$ is s. Thus there exists a linear combination of $\mathbf{b}, A\mathbf{b}, \ldots, A^{t-1}\mathbf{b}$ which equates to zero,

$$0 = \sum_{j=0}^{t-1} a_j A^j\mathbf{b} = \sum_{j=0}^{t-1} a_j A^j \sum_{i=1}^{t} c_i \mathbf{v}_i = \sum_{i=1}^{t} c_i \sum_{j=0}^{t-1} a_j \lambda_i^j \mathbf{v}_i.$$

Since the eigenvectors are linearly independent and all c_i are nonzero, we have

$$\sum_{j=0}^{t-1} a_j \lambda_i^j = 0$$

for the distinct eigenvalues λ_i, $i = 1, \ldots, t$. But a polynomial of degree $t - 1$ can have at most $t - 1$ roots. So we have a contradiction and $s = t$. $\quad\square$

From this lemma it follows that the number of iterations of the conjugate gradient method is at most the number of distinct eigenvalues of A. We can tighten this bound even more: if $\mathbf{b}$ is expressed as a linear combination of eigenvectors of A with distinct eigenvalues, then the number of iterations is at most the number of nonzero coefficients in this linear combination.

By changing variables, $\tilde{\mathbf{x}} = P^{-1}\mathbf{x}$, where P is a nonsingular $n \times n$ matrix, we can significantly reduce the work of the conjugate gradient method. Instead of solving $A\mathbf{x} = \mathbf{b}$ we solve the system $P^T A P \tilde{\mathbf{x}} = P^T \mathbf{b}$. $P^T A P$ is also symmetric and positive definite, since A is symmetric and positive definite and P is nonsingular. Hence we can apply the conjugate gradient method to obtain the solution $\tilde{\mathbf{x}}$, which then in turn gives $\mathbf{x}^* = P\tilde{\mathbf{x}}$. This procedure is called the *preconditioned conjugate gradient method* and P is called the *preconditioner*.

The speed of convergence of the conjugate gradient method depends on the condition number of A, which for symmetric, positive definite A is the ratio between the modulus of its largest and its least eigenvalue. Therefore, P should be chosen so that the condition number of $P^T AP$ is much smaller than the one of A.

This gives the *transformed preconditioned conjugate gradient method* with formulae:

$$\tilde{\mathbf{d}}^{(0)} = \tilde{\mathbf{r}}^{(0)} = P^T\mathbf{b} - P^T AP\tilde{\mathbf{x}}^{(0)},$$

$$\omega^{(k)} = \frac{\tilde{\mathbf{r}}^{(k)T}\tilde{\mathbf{r}}^{(k)}}{\tilde{\mathbf{d}}^{(k)T}P^T AP\tilde{\mathbf{d}}^{(k)}},$$

$$\tilde{\mathbf{x}}^{(k+1)} = \tilde{\mathbf{x}}^{(k)} + \omega^{(k)}\tilde{\mathbf{d}}^{(k)},$$

$$\tilde{\mathbf{r}}^{(k+1)} = \tilde{\mathbf{r}}^{(k)} - \omega^{(k)}P^T AP\tilde{\mathbf{d}}^{(k)},$$

$$\beta^{(k+1)} = \frac{\tilde{\mathbf{r}}^{(k+1)T}\tilde{\mathbf{r}}^{(k+1)}}{\tilde{\mathbf{r}}^{(k)T}\tilde{\mathbf{r}}^{(k)}},$$

$$\tilde{\mathbf{d}}^{(k+1)} = \tilde{\mathbf{r}}^{(k+1)} + \beta^{(k+1)}\tilde{\mathbf{d}}^{(k)}.$$

The number of iterations is at most the dimension of the Krylov subspace $K_n(P^T AP, P^T\mathbf{b})$. This space is spanned by $(P^T AP)^j P^T\mathbf{b} = P^T(APP^T)^j\mathbf{b}$, $j = 0, \ldots, n-1$. Since P is nonsingular, so is P^T. It follows that the dimension of $K_n(P^T AP, P^T\mathbf{b})$ is the same as the dimension of $K_n(APP^T, \mathbf{b})$.

It is undesirable in this method that P has to be computed. However, with a few careful changes of variables P can be eliminated. Instead, the matrix $S = PP^T$ is used. We see later why this is advantageous.

Firstly, we use $\tilde{\mathbf{x}}^{(k)} = P^{-1}\mathbf{x}^{(k)}$, or equivalently $\mathbf{x}^{(k)} = P\tilde{\mathbf{x}}^{(k)}$. Setting $\mathbf{r}^{(k)} = P^{-T}\tilde{\mathbf{r}}^{(k)}$ and $\mathbf{d}^{(k)} = P\tilde{\mathbf{d}}^{(k)}$, we derive the *untransformed preconditioned conjugate gradient method*:

$$\mathbf{r}^{(0)} = P^{-T}\tilde{\mathbf{r}}^{(0)} \qquad\qquad = \mathbf{b} - A\mathbf{x}^{(0)},$$

$$\mathbf{d}^{(0)} = P\tilde{\mathbf{d}}^{(0)} = PP^T\mathbf{b} - PP^T A\mathbf{x}^{(0)} \qquad = S\mathbf{r}^{(0)},$$

$$\omega^{(k)} = \frac{(P^T\mathbf{r}^{(k)})^T P^T\mathbf{r}^{(k)}}{(P^{-1}\mathbf{d}^{(k)})^T P^T APP^{-1}\mathbf{d}^{(k)}} \qquad = \frac{\mathbf{r}^{(k)^T}S\mathbf{r}^{(k)}}{\mathbf{d}^{(k)^T}A\mathbf{d}^{(k)}},$$

$$\mathbf{x}^{(k+1)} = P\tilde{\mathbf{x}}^{(k+1)} = P\tilde{\mathbf{x}}^{(k)} + \omega^{(k)}P\tilde{\mathbf{d}}^{(k)} \qquad = \mathbf{x}^{(k)} + \omega^{(k)}\mathbf{d}^{(k)},$$

$$\mathbf{r}^{(k+1)} = P^{-T}\tilde{\mathbf{r}}^{(k+1)} \qquad\qquad = \mathbf{r}^{(k)} - \omega^{(k)}A\mathbf{d}^{(k)},$$
$$= P^{-T}\tilde{\mathbf{r}}^{(k)} - \omega^{(k)}P^{-T}P^T AP\tilde{\mathbf{d}}^{(k)}$$

$$\beta^{(k+1)} = \frac{(P^T\mathbf{r}^{(k+1)})^T P^T\mathbf{r}^{(k+1)}}{(P^T\mathbf{r}^{(k)})^T P\mathbf{r}^{(k)}} \qquad = \frac{\mathbf{r}^{(k+1)^T}S\mathbf{r}^{(k+1)}}{\mathbf{r}^{(k)^T}S\mathbf{r}^{(k)}},$$

$$\mathbf{d}^{(k+1)} = P\tilde{\mathbf{d}}^{(k+1)} = P\tilde{\mathbf{r}}^{(k+1)} + \beta^{(k+1)}P\tilde{\mathbf{d}}^{(k)} \quad = S\mathbf{r}^{(k+1)} + \beta^{(k+1)}\mathbf{d}^{(k)},$$

The effectiveness of the preconditioner S is determined by the condition number of $AS = APP^T$ (and occasionally by its clustering of eigenvalues). The problem remains of finding S which is close enough to A to improve convergence, but the cost of computing $S\mathbf{r}^{(k)}$ is low.

The perfect preconditioner would be $S = A^{-1}$, since for this preconditioner $AS = I$ has the condition number 1. Unfortunately, finding this preconditioner is equivalent to solving $A\mathbf{x} = \mathbf{b}$, the original problem for which we seek a preconditioner. Hence this fails to be a useful preconditioner at all.

The simplest useful choice for the matrix S is a diagonal matrix whose inverse has the same diagonal entries as A. Making the diagonal entries of AS equal to one often causes the eigenvalues of $P^T A P$ to be close to one. This is known as *diagonal preconditioning* or *Jacobi preconditioning*. It is trivial to invert a diagonal matrix, but often this is a mediocre preconditioner.

Another possibility is to let P be the inverse of the lower triangular part of A, possibly changing the diagonal elements in the hope that S is close to the inverse of A then.

A more elaborate preconditioner is *incomplete Cholesky preconditioning*. We express A as $S^{-1} + E$, where S^{-1} is symmetric, positive definite, close to A (such that the error E is small) and can be factorized easily into a Cholesky factorization, $S^{-1} = LL^T$. For example, S might be a banded matrix with small band width. Once the Cholesky factorization of S^{-1} is known, the main expense is calculating $\mathbf{v} = S\mathbf{r}^{(k)}$. However, this is equivalent to solving $S^{-1}\mathbf{v} = LL^T\mathbf{v} = \mathbf{r}^{(k)}$, which can be done efficiently by back substitution. Unfortunately, incomplete Cholesky preconditioning is not always stable.

Many preconditioners, some quite sophisticated, have been developed. In general, conjugate gradients should nearly always be used with a preconditioner for large-scale applications. For moderate n the standard form of the conjugate gradients method usually converges with an acceptably small value of $\|\mathbf{r}^{(k)}\|$ in far fewer than n iterations and there is no need for preconditioning.

Exercise 2.16. *Let A be an $n \times n$ tridiagonal matrix of the form*

$$A = \begin{pmatrix} \alpha & \beta & & & \\ \beta & \alpha & \beta & & \\ & \ddots & \ddots & \ddots & \\ & & \beta & \alpha & \beta \\ & & & \beta & \alpha \end{pmatrix}.$$

Verify that $\alpha > 2\beta > 0$ implies that the matrix is positive definite. We now precondition the system with the lower triangular, bidiagonal matrix P being the inverse of

$$P^{-1} = \begin{pmatrix} \gamma & & & \\ \delta & \gamma & & \\ & \ddots & \ddots & \\ & & \delta & \gamma \end{pmatrix}.$$

Determine γ and δ such that the inverse of $S = PP^T$ is the same as A apart *from the (n, n) entry. Prove that the preconditioned gradient method converges in just two iterations then.*

As a closing remark on conjugate gradients, we have seen that the method of normal equations solves $A^T A\mathbf{x} = A^T\mathbf{b}$ for which conjugate gradients can be used as long as $A\mathbf{x} = \mathbf{b}$ is not underdetermined, because only then is $A^T A$ nonsingular. However, the condition number of $A^T A$ is the square of the condition number of A, so convergence can be significantly slower. An important technical point is that $A^T A$ is never computed explicitly, since it is less sparse. Instead when calculating $A^T A\mathbf{d}^{(k)}$, first $A\mathbf{d}^{(k)}$ is computed and this is then multiplied by A^T. It is also numerically more stable to calculate $\mathbf{d}^{(k)^T} A^T A\mathbf{d}^{(k)}$ as the inner product of $A\mathbf{d}^{(k)}$ with itself.

2.17 Eigenvalues and Eigenvectors

So far we only considered eigenvalues when analyzing the properties of numerical methods. In the following sections we look at how to determine eigenvalues and eigenvectors. Let A be a real $n \times n$ matrix. The *eigenvalue equation* is given by

$$A\mathbf{v} = \lambda\mathbf{v},$$

where λ is scalar. It may be complex if A is not symmetric. There exists $\mathbf{v} \in \mathbb{R}^n$ satisfying the eigenvalue equation if and only if the determinant $\det(A - \lambda I)$ is zero. The function $p(\lambda) = \det(A - \lambda I)$, $\lambda \in \mathbb{C}$, is a polynomial of degree n. However, calculating the eigenvalues by finding the roots of p is generally unsuitable because of loss of accuracy due to rounding errors. In Chapter 1, Fundamentals, we have seen how even finding the roots of a quadratic can be difficult due to loss of significance.

If the polynomial has some multiple roots and if A is not symmetric, then A might have less than n linearly independent eigenvalues. However, there are always n mutually orthogonal real eigenvectors when A is symmetric. In the following we assume that A has n linearly independent eigenvectors $\mathbf{v}_j$ for each eigenvalue λ_j, $j = 1, \ldots, n$. This can be achieved by perturbing A slightly if necessary. The task is now to find $\mathbf{v}_j$ and λ_j, $j = 1, \ldots, n$.

2.18 The Power Method

The *power method* forms the basic of many iterative algorithms for the calculation of eigenvalues and eigenvectors. It generates a single eigenvector and eigenvalue of A.

1. Pick a starting vector $\mathbf{x}^{(0)} \in \mathbb{R}^n$ satisfying $\|\mathbf{x}^{(0)}\| = 1$. Set $k = 0$ and choose a tolerance $\epsilon > 0$.

2. Calculate $\mathbf{x}^{(k+1)} = A\mathbf{x}^{(k)}$ and find the real number λ that minimizes

$\|\mathbf{x}^{(k+1)} - \lambda \mathbf{x}^{(k)}\|$. This is given by the *Rayleigh quotient*

$$\lambda = \frac{\mathbf{x}^{(k)T} A \mathbf{x}^{(k)}}{\mathbf{x}^{(k)T} \mathbf{x}^{(k)}}.$$

3. Accept λ as an eigenvalue and $\mathbf{x}^{(k)}$ as an eigenvector , if $\|\mathbf{x}^{(k+1)} - \lambda \mathbf{x}^{(k)}\| \leq \epsilon$.

4. Otherwise, replace $\mathbf{x}^{(k+1)}$ by $\mathbf{x}^{(k+1)}/\|\mathbf{x}^{(k+1)}\|$, increase k by one, and go back to step 2.

Theorem 2.11. *If there is one eigenvalue of A whose modulus is larger than the moduli of the other eigenvalues, then the power method terminates with an approximation to this eigenvalue and the corresponding eigenvector as long as the starting vector has a component of the largest eigenvector (in exact arithmetic).*

Proof. Let $|\lambda_1| \leq |\lambda_2| \leq \ldots \leq |\lambda_{n-1}| < |\lambda_n|$ be the eigenvalues ordered by modulus and $\mathbf{v}_1, \ldots, \mathbf{v}_n$ the corresponding eigenvectors of unit length. Let the starting vector $\mathbf{x}^{(0)} = \sum_{i=1}^{n} c_i \mathbf{v}_i$ be chosen such that c_n is nonzero. Then $\mathbf{x}^{(k)}$ is a multiple of

$$A^k \mathbf{x}^{(0)} = \sum_{i=1}^{n} c_i \lambda_i^k \mathbf{v}_i = c_n \lambda_n^k \left(\mathbf{v}_n + \sum_{i=1}^{n-1} \frac{c_i}{c_n} \left(\frac{\lambda_i}{\lambda_n} \right)^k \mathbf{v}_i \right),$$

since in every iteration the new approximation is scaled to have length 1. Since $\|\mathbf{x}^{(k)}\| = \|\mathbf{v}_n\| = 1$, it follows that $\mathbf{x}^{(k)} = \pm \mathbf{v}_n + O(|\lambda_{n-1}/\lambda_n|^k)$, where the sign is determined by $c_n \lambda_n^k$, since we scale by a positive factor in each iteration. The fraction $|\lambda_{n-1}/\lambda_n|$ characterizes the rate of convergence. Thus if λ_{n-1} is similar in size to λ_n convergence is slow. However, if λ_n is considerably larger than the other eigenvalues in modulus, convergence is fast.

Termination occurs, since

$$\begin{aligned} \|\mathbf{x}^{(k+1)} - \lambda \mathbf{x}^{(k)}\| &= \min_\lambda \|\mathbf{x}^{(k+1)} - \lambda \mathbf{x}^{(k)}\| \leq \|\mathbf{x}^{(k+1)} - \lambda_n \mathbf{x}^{(k)}\| \\ &= \|A\mathbf{x}^{(k)} - \lambda_n \mathbf{x}^{(k)}\| = \|A\mathbf{v}_n - \lambda_n \mathbf{v}_n\| + O(|\lambda_{n-1}/\lambda_n|^k) \\ &= O(|\lambda_{n-1}/\lambda_n|^k) \overset{k \to \infty}{\longrightarrow} 0. \end{aligned}$$

$\square$

Exercise 2.17. *Let A be the bidiagonal $n \times n$ matrix*

$$A = \begin{pmatrix} \lambda & 1 & & \\ & \ddots & \ddots & \\ & & \lambda & 1 \\ & & & \lambda \end{pmatrix}.$$

Find an explicit expression for A^k. Letting $n = 3$, the sequence $\mathbf{x}^{(k+1)}$, $k =$

$0, 1, 2, \ldots$, *is generated by the power method* $\mathbf{x}^{(k+1)} = A\mathbf{x}^{(k)}/\|\mathbf{x}^{(k)}\|$, *starting with some* $\mathbf{x}^{(0)} \in \mathbb{R}^3$. *From the expression for* A^k, *deduce that the second and third component of* $\mathbf{x}^{(k)}$ *tend to zero as* k *tend to infinity. Further show that this implies* $A\mathbf{x}^{(k)} - \lambda\mathbf{x}^{(k)}$ *tends to zero.*

The power method usually works well if the modulus of one of the eigenvalues is substantially larger than the moduli of the other $n-1$ eigenvalues. However, it can be unacceptably slow, if there are eigenvalues of similar size. In the case $c_n = 0$, the method should find the eigenvector $\mathbf{v}_m$ for the largest m for which c_m is nonzero in exact arithmetic. Computer rounding errors can, however, introduce a small nonzero component of $\mathbf{v}_n$, which will grow and the method converges to $\mathbf{v}_n$ eventually.

Two other methods are the *Arnoldi* and *Lancozs method*, which are Krylow subspace methods and exploit the advantages associated with Krylow subspaces; that is, convergence is guaranteed in a finite number of iterations (in exact arithmetic). They are described in [6] *Numerical Linear Algebra and Applications* by B. Datta. In the following we examine a variation of the power method dealing with complex conjugate pairs of eigenvalues.

When A is real and not symmetric, then some eigenvalues occur in complex conjugate pairs. In this case we might have two eigenvalues with largest modulus. The *two-stage power method* is used then.

1. Pick a starting vector $\mathbf{x}^{(0)} \in \mathbb{R}^n$ satisfying $\|\mathbf{x}^{(0)}\| = 1$. Set $k = 0$ and choose a tolerance $\epsilon > 0$. Calculate $\mathbf{x}^{(1)} = A\mathbf{x}^{(0)}$.

2. Calculate $\mathbf{x}^{(k+2)} = A\mathbf{x}^{(k+1)}$ and find the real numbers α and β that minimize $\|\mathbf{x}^{(k+2)} + \alpha\mathbf{x}^{(k+1)} + \beta\mathbf{x}^{(k)}\|$.

3. Let λ_+ and λ_- be the roots of the quadratic equation $\lambda^2 + \alpha\lambda + \beta = 0$. If $\|\mathbf{x}^{(k+2)} + \alpha\mathbf{x}^{(k+1)} + \beta\mathbf{x}^{(k)}\| \leq \epsilon$ holds, then accept λ_+ as eigenvalue with eigenvector $\mathbf{x}^{(k+1)} - \lambda_-\mathbf{x}^{(k)}$ and λ_- as eigenvalue with eigenvector $\mathbf{x}^{(k+1)} - \lambda_+\mathbf{x}^{(k)}$.

4. Otherwise, replace the vector $\mathbf{x}^{(k+1)}$ by $\mathbf{x}^{(k+1)}/\|\mathbf{x}^{(k+2)}\|$ and $\mathbf{x}^{(k+2)}$ by $\mathbf{x}^{(k+2)}/\|\mathbf{x}^{(k+2)}\|$, keeping the relationship $\mathbf{x}^{(k+2)} = A\mathbf{x}^{(k+1)}$, then increase k by one and go back to step 2.

Lemma 2.5. *If* $\|\mathbf{x}^{(k+2)} + \alpha\mathbf{x}^{(k+1)} + \beta\mathbf{x}^{(k)}\| = 0$ *in step 3 of the two-stage power method, then* $\mathbf{x}^{(k+1)} - \lambda_-\mathbf{x}^{(k)}$ *and* $\mathbf{x}^{(k+1)} - \lambda_+\mathbf{x}^{(k)}$ *satisfy the eigenvalue equations*

$$A(\mathbf{x}^{(k+1)} - \lambda_-\mathbf{x}^{(k)}) = \lambda_+(\mathbf{x}^{(k+1)} - \lambda_-\mathbf{x}^{(k)}),$$
$$A(\mathbf{x}^{(k+1)} - \lambda_+\mathbf{x}^{(k)}) = \lambda_-(\mathbf{x}^{(k+1)} - \lambda_+\mathbf{x}^{(k)}).$$

Proof. Using $A\mathbf{x}^{(k+1)} = \mathbf{x}^{(k+2)}$ and $A\mathbf{x}^{(k)} = \mathbf{x}^{(k+1)}$, we have

$$A(\mathbf{x}^{(k+1)} - \lambda_-\mathbf{x}^{(k)}) - \lambda_+(\mathbf{x}^{(k+1)} - \lambda_-\mathbf{x}^{(k)})$$

$$= \mathbf{x}^{(k+2)} - (\lambda_+ + \lambda_-)\mathbf{x}^{(k+1)} + \lambda_+\lambda_-\mathbf{x}^{(k)}$$

$$= \mathbf{x}^{(k+2)} + \alpha\mathbf{x}^{(k+1)} + \beta\mathbf{x}^{(k)} = 0.$$

This proves the first assertion, the second one follows similarly. □

If $\|\mathbf{x}^{(k+2)} + \alpha\mathbf{x}^{(k+1)} + \beta\mathbf{x}^{(k)}\| = 0$, then the vectors $\mathbf{x}^{(k)}$ and $\mathbf{x}^{(k+1)}$ span an *eigenspace* of A. This means that if A is applied to any linear combination of $\mathbf{x}^{(k)}$ and $\mathbf{x}^{(k+1)}$, then the result is again a linear combination of $\mathbf{x}^{(k)}$ and $\mathbf{x}^{(k+1)}$. Since $\|\mathbf{x}^{(k+2)} + \alpha\mathbf{x}^{(k+1)} + \beta\mathbf{x}^{(k)}\| = 0$, it is easy to see that

$$\mathbf{x}^{(k+2)} = A\mathbf{x}^{(k+1)} = -\alpha\mathbf{x}^{(k+1)} - \beta\mathbf{x}^{(k)}$$

is a linear combination of $\mathbf{x}^{(k)}$ and $\mathbf{x}^{(k+1)}$. Let $\mathbf{v} = a\mathbf{x}^{(k+1)} + b\mathbf{x}^{(k)}$ be any linear combination, then

$$A\mathbf{v} = aA\mathbf{x}^{(k+1)} + bA\mathbf{x}^{(k)} = a\mathbf{x}^{(k+2)} + b\mathbf{x}^{(k+1)} = a\left(-\alpha\mathbf{x}^{(k+1)} - \beta\mathbf{x}^{(k)}\right) + b\mathbf{x}^{(k+1)}$$

is a linear combination of $\mathbf{x}^{(k)}$ and $\mathbf{x}^{(k+1)}$.

In the two-stage power method a norm of the form $\|\mathbf{u} + \alpha\mathbf{v} + \beta\mathbf{w}\|$ has to be minimized. This is equivalent to minimizing

$$\left(\mathbf{u} + \alpha\mathbf{v} + \beta\mathbf{w}\right)^T \left(\mathbf{u} + \alpha\mathbf{v} + \beta\mathbf{w}\right) =$$
$$\mathbf{u}^T\mathbf{u} + \alpha^2\mathbf{v}^T\mathbf{v} + \beta^2\mathbf{w}^T\mathbf{w} + 2\alpha\mathbf{u}^T\mathbf{v} + 2\beta\mathbf{u}^T\mathbf{w} + 2\alpha\beta\mathbf{v}^T\mathbf{w}.$$

Differentiating with respect to α and β, we see that the gradient vanishes if α and β satisfy the following system of equations

$$\begin{pmatrix} \mathbf{v}^T\mathbf{v} & \mathbf{v}^T\mathbf{w} \\ \mathbf{v}^T\mathbf{w} & \mathbf{w}^T\mathbf{w} \end{pmatrix} \begin{pmatrix} \alpha \\ \beta \end{pmatrix} = \begin{pmatrix} -\mathbf{u}^T\mathbf{v} \\ -\mathbf{u}^T\mathbf{w} \end{pmatrix},$$

which can be solved easily.

The two-stage power method can find complex eigenvectors, although most of its operations are in real arithmetic. The complex values λ_+ and λ_- and their eigenvectors are only calculated if $\|\mathbf{x}^{(k+2)} + \alpha\mathbf{x}^{(k+1)} + \beta\mathbf{x}^{(k)}\|$ is acceptably small. The main task in each operation is a matrix by vector product. Therefore, both power methods benefit greatly from sparsity in A. However, in practice the following method is usually preferred, since it is more efficient.

When solving systems of linear equations iteratively, we have seen that convergence can be sped up by relaxation. Relaxation essentially shifts the eigenvalues of the iteration matrix so that they lie in an interval symmetrically about zero, which leads to faster convergence. The eigenvectors of $A - sI$ for $s \in \mathbb{R}$ are also the eigenvectors of A. The eigenvalues of $A - sI$ are the same as the eigenvalues of A shifted by s. We use this fact by choosing a *shift* s to achieve faster convergence. However, the shift is not chosen such that the interval containing all eigenvalues is symmetric about zero. The component of the jth eigenvector is reduced by the factor $|\lambda_j/\lambda_n|$ in each iteration. If $|\lambda_j|$ is close to λ_n the reduction is small. For example, if A has $n-1$ eigenvalues in the interval $[100, 101]$ and $\lambda_n = 102$, the reduction factors lie in the interval $[100/102, 101/102] \approx [0.98, 0.99]$. Thus convergence is slow. However, for $s =$

100.5, the first $n-1$ eigenvalues lie in $[-0.5, 0.5]$ and the largest eigenvalue is 1.5. Now the reduction factor is at least $0.5/1.5 = 1/3$. Occasionally some prior knowledge (Gerschgorin theorem) is available, which provides good choices for the shift.

This gives the *power method with shifts*.

1. Pick a starting vector $\mathbf{x}^{(0)} \in \mathbb{R}^n$ satisfying $\|\mathbf{x}^{(0)}\| = 1$. Set $k = 0$ and choose a tolerance $\epsilon > 0$ and a shift $s \in \mathbb{R}$.

2. Calculate $\mathbf{x}^{(k+1)} = (A - sI)\mathbf{x}^{(k)}$. Find the real number λ that minimizes $\|\mathbf{x}^{(k+1)} - \lambda\mathbf{x}^{(k)}\|$.

3. Accept $\lambda + s$ as an eigenvalue and $\mathbf{x}^{(k)}$ as an eigenvector, if we have $\|\mathbf{x}^{(k+1)} - \lambda\mathbf{x}^{(k)}\| \le \epsilon$.

4. Otherwise, replace $\mathbf{x}^{(k+1)}$ by $\mathbf{x}^{(k+1)}/\|\mathbf{x}^{(k+1)}\|$, increase k by one and go back to step 2.

The following method also uses shifts, but with a different intention.

2.19   Inverse Iteration

The method described in this section is called *inverse iteration* and is very effective in practice. It is similar to the power method with shifts, except that, instead of $\mathbf{x}^{(k+1)}$ being a multiple of $(A - sI)\mathbf{x}^{(k)}$, it is calculated as a scalar multiple of the solution to

$$(A - sI)\mathbf{x}^{(k+1)} = \mathbf{x}^{(k)}, \qquad k = 0, 1, \ldots,$$

where s is a scalar that may depend on k. Thus the inverse power method is the power method applied to the matrix $(A - sI)^{-1}$. If s is close to an eigenvalue, then the matrix $A - sI$ has an eigenvalue close to zero, but this implies that $(A - sI)^{-1}$ has a very large eigenvalue and we have seen that in this case the power method converges fast.

In every iteration $\mathbf{x}^{(k+1)}$ is scaled such that $\|\mathbf{x}^{(k+1)}\| = 1$. We see that the calculation of $\mathbf{x}^{(k+1)}$ requires the solution of an $n \times n$ system of equations.

If s is constant in every iteration such that $A - sI$ is nonsingular, then $\mathbf{x}^{(k+1)}$ is a multiple of $(A - sI)^{-k-1}\mathbf{x}^{(0)}$. As before we let $\mathbf{x}^{(0)} = \sum_{j=1}^{n} c_j \mathbf{v}_j$, where $\mathbf{v}_j$, $j = 1, \ldots, n$, are the linearly independent eigenvectors. The eigenvalue equation then implies $(A - sI)\mathbf{v}_j = (\lambda_j - s)\mathbf{v}_j$. For the inverse we have then $(A - sI)^{-1}\mathbf{v}_j = (\lambda_j - s)^{-1}\mathbf{v}_j$. It follows that $\mathbf{x}^{(k+1)}$ is a multiple of

$$(A - sI)^{-k-1}\mathbf{x}^{(0)} = \sum_{j=0}^{n} c_j (A - sI)^{-k-1}\mathbf{v}_j = \sum_{j=0}^{n} c_j (\lambda_j - s)^{-k-1}\mathbf{v}_j.$$

Let m be the index of the smallest number of $|\lambda_j - s|$. $j = 1, \ldots, n$. If c_m is nonzero then $\mathbf{x}^{(k+1)}$ tends to be multiple of $\mathbf{v}_m$ as k tends to infinity. The speed of convergence can be excellent if s is close to λ_m. It can be improved even more if s is adjusted during the iterations as the following implementation shows.

1. Set s to an estimate of an eigenvalue of A. Either pick a starting vector $\mathbf{x}^{(0)} \neq 0$ or let it be chosen automatically in step 4. Set $k = 0$ and choose a tolerance $\epsilon > 0$.

2. Calculate (with pivoting if necessary) the LU factorization of $(A - sI) = LU$.

3. Stop if U is singular (in this case one or more diagonal elements are zero), because then s is an eigenvalue of A, while the corresponding eigenvector lies in the null space of U and can be found easily by back substitution.

4. If $k = 0$ and $\mathbf{x}^{(0)}$ has not been chosen, let i be the index of the smallest diagonal element of U, i.e., $|U_{i,i}| \leq |U_{j,j}|$, $i \neq j$. We define $\mathbf{x}^{(1)}$ by $U\mathbf{x}^{(1)} = \mathbf{e}_i$ and let $\mathbf{x}^{(0)} = L\mathbf{e}_i$, where $\mathbf{e}_i$ is the ith standard unit vector. Then $(A - sI)\mathbf{x}^{(1)} = \mathbf{x}^{(0)}$. Otherwise, solve $(A - sI)\mathbf{x}^{(k+1)} = LU\mathbf{x}^{(k+1)} = \mathbf{x}^{(k)}$ by back substitution to obtain $\mathbf{x}^{(k+1)}$

5. Let λ be the number which minimizes $\|\mathbf{x}^{(k)} - \lambda\mathbf{x}^{(k+1)}\|$.

6. Stop if $\|\mathbf{x}^{(k)} - \lambda\mathbf{x}^{(k+1)}\| \leq \epsilon\|\mathbf{x}^{(k+1)}\|$. Since

$$\|\mathbf{x}^{(k)} - \lambda\mathbf{x}^{(k+1)}\| = \|(A - sI)\mathbf{x}^{(k+1)} - \lambda\mathbf{x}^{(k+1)}\| = \|A\mathbf{x}^{(k+1)} - (s + \lambda)\mathbf{x}^{(k+1)}\|,$$

we let $s + \lambda$ and $\mathbf{x}^{(k+1)}/\|\mathbf{x}^{(k+1)}\|$ be the approximation to the eigenvalue and its corresponding eigenvector.

7. Otherwise, replace $\mathbf{x}^{(k+1)}$ by $\mathbf{x}^{(k+1)}/\|\mathbf{x}^{(k+1)}\|$, increase k by one and either return to step 4 without changing s or to step 2 after replacing s by $s + \lambda$.

The order of convergence is illustrated in the following exercise.

Exercise 2.18. *Let A be a symmetric 2×2 matrix with distinct eigenvalues $\lambda_1 > \lambda_2$ with normalized corresponding eigenvectors $\mathbf{v}_1$ and $\mathbf{v}_2$. Starting with $\mathbf{x}^{(0)} \neq 0$, the sequence $\mathbf{x}^{(k+1)}$, $k = 0, 1, \ldots$, is generated by inverse iteration. In every iteration we let $s^{(k)}$ be the Rayleigh quotient $s^{(k)} = \mathbf{x}^{(k)T}A\mathbf{x}^{(k)}/\|\mathbf{x}^{(k)}\|^2$. Show that if $\mathbf{x}^{(k)} = (\mathbf{v}_1 + \epsilon^{(k)}\mathbf{v}_2)/\sqrt{1 + \epsilon^{(k)2}}$, where $|\epsilon^{(k)}|$ is small, then $|\epsilon^{(k+1)}|$ is of magnitude $|\epsilon^{(k)}|^3$. That is, the order of convergence is 3.*

Adjusting s and calculating an LU-factorization in every iteration seems excessive. The same s can be retained for a few iterations and adjusted when necessary However, if A is an *upper Hessenberg matrix*, that is, if every element of A below the first subdiagonal is zero (i.e., $A_{ij} = 0, j < i - 1$), inverse iterations are very efficient, because in this case the LU factorization requires $O(n^2)$ operations when A is nonsymmetric and $O(n)$ operations if A is symmetric.

We have seen earlier when examining how to arrive at a QR factorization, that Givens rotations and Householder reflections can be used to introduce zeros below the diagonal. These can be used to transform A into an upper Hessenberg matrix. They are also used in the next section.

2.20   Deflation

Suppose we have found one solution of the eigenvector equation $A\mathbf{v} = \lambda\mathbf{v}$ (or possibly a pair of complex conjugate eigenvalues with their corresponding eigenvectors), where A is an $n \times n$ matrix. *Deflation* constructs an $(n-1) \times (n-1)$ (or $(n-2)\times(n-2)$) matrix, say B, whose eigenvalues are the other $n-1$ (or $n-2$) eigenvalues of A. The concept is based on the following theorem.

Theorem 2.12. *Let A and S be $n \times n$ matrices, S being nonsingular. Then $\mathbf{v}$ is an eigenvector of A with eigenvalue λ if and only if $S\mathbf{v}$ is an eigenvector of SAS^{-1} with the same eigenvalue. S is called a* similarity transformation.

Proof.

$$A\mathbf{v} = \lambda\mathbf{v} \Leftrightarrow AS^{-1}(S\mathbf{v}) = \lambda\mathbf{v} \Leftrightarrow (SAS^{-1})(S\mathbf{v}) = \lambda(S\mathbf{v}).$$

□

Let's assume one eigenvalue λ and its corresponding eigenvector have been found. In deflation we apply a similarity transformation S to A such that the first column of SAS^{-1} is λ times the first standard unit vector $\mathbf{e}_1$,

$$(SAS^{-1})\mathbf{e}_1 = \lambda\mathbf{e}_1.$$

Then we can let B be the bottom right $(n-1) \times (n-1)$ submatrix of SAS^{-1}. We see from the above theorem that it is sufficient to let S have the property $S\mathbf{v} = a\mathbf{e}_1$, where a is any nonzero scalar.

If we know a complex conjugate pair of eigenvalues, then there is a two-dimensional eigenspace associated with them. Eigenspace means that if A is applied to any vector in the eigenspace, then the result will again lie in the eigenspace. Let $\mathbf{v}_1$ and $\mathbf{v}_2$ be vectors spanning the eigenspace. For example these could have been found by the two-stage power method. We need to find a similarity transformation S which maps the eigenspace to the space spanned by the first two standard basis vectors $\mathbf{e}_1$ and $\mathbf{e}_2$. Let S_1 such that $S_1\mathbf{v}_1 = a\mathbf{e}_1$. In addition let $\hat{\mathbf{v}}$ be the vector composed of the last $n-1$ components of $S_1\mathbf{v}_2$. We then let S_2 be of the form

$$S_2 = \begin{pmatrix} 1 & 0 & \cdots & 0 \\ 0 & & & \\ \vdots & & \hat{S} & \\ 0 & & & \end{pmatrix},$$

where $\hat{S}$ is a $(n-1) \times (n-1)$ matrix such that the last $n-2$ components of $\hat{S}\hat{\mathbf{v}}$ are zero. The matrix $S = S_2 S_1$ then maps the eigenspace to the space spanned by the first two standard basis vectors $\mathbf{e}_1$ and $\mathbf{e}_2$, since $S\mathbf{v}_1 = a\mathbf{e}_1$ and $S\mathbf{v}_2$ is a linear combination of $\mathbf{e}_1$ and $\mathbf{e}_2$. The matrix SAS^{-1} then has zeros in the last $n-2$ entries of the first two columns.

For symmetric A we want B to be symmetric also. This can be achieved if S is an orthogonal matrix, since then $S^{-1} = S^T$ and SAS^{-1} remains symmetric. A Householder reflection is the most suitable choice:

$$S = I - 2\frac{\mathbf{u}\mathbf{u}^T}{\|\mathbf{u}\|^2}.$$

As in the section on Householder reflections we let $u_i = v_i$ for $i = 2, \ldots, n$ and choose u_1 such that $2\mathbf{u}^T\mathbf{v} = \|\mathbf{u}\|^2$, if $\mathbf{v}$ is a known eigenvector. This gives $u_1 = v_1 \pm \|\mathbf{v}\|$. The sign is chosen such that loss of significance is avoided. With this choice we have

$$S\mathbf{v} = \left(I - 2\frac{\mathbf{u}\mathbf{u}^T}{\|\mathbf{u}\|^2}\right)\mathbf{v} = \pm\|\mathbf{v}\|\mathbf{e}_1.$$

Since the last $n-1$ components of $\mathbf{u}$ and $\mathbf{v}$ are the same, the calculation of $\mathbf{u}$ only requires $O(n)$ operations. Further using the fact that $S^{-1} = S^T = S$, since S is not only orthogonal, but also symmetric, SAS^{-1} can be calculated as

$$
\begin{aligned}
SAS^{-1} = SAS &= \left(I - 2\frac{\mathbf{u}\mathbf{u}^T}{\|\mathbf{u}\|^2}\right)A\left(I - 2\frac{\mathbf{u}\mathbf{u}^T}{\|\mathbf{u}\|^2}\right) \\
&= A - 2\frac{\mathbf{u}\mathbf{u}^T A}{\|\mathbf{u}\|^2} - 2\frac{A\mathbf{u}\mathbf{u}^T}{\|\mathbf{u}\|^2} + 4\frac{\mathbf{u}\mathbf{u}^T A\mathbf{u}\mathbf{u}^T}{\|\mathbf{u}\|^4}.
\end{aligned}
$$

The number of operations is $O(n^2)$ to form $A\mathbf{u}$ and then $A\mathbf{u}\mathbf{u}^T$. The matrix $\mathbf{u}\mathbf{u}^T A$ is just the transpose of $A\mathbf{u}\mathbf{u}^T$, since A is symmetric. Further $\mathbf{u}^T A\mathbf{u}$ is a scalar which can be calculated in $O(n)$ operations once $A\mathbf{u}$ is known. It remains to calculate $\mathbf{u}\mathbf{u}^T$ in $O(n^2)$ operations. Thus the overall number of iterations is $O(n^2)$.

Once an eigenvector $\hat{\mathbf{w}} \in \mathbb{R}^{n-1}$ of B is found, we extend it to a vector $\mathbf{w} \in \mathbb{R}^n$ by letting the first component equal zero and the last $n-1$ components equal $\hat{\mathbf{w}}$. This is an eigenvector of SAS^{-1}. The corresponding eigenvector of A is $S^{-1}\mathbf{w} = S\mathbf{w}$.

Further eigenvalue/eigenvector pairs can be found by deflating B and continuing the process.

Exercise 2.19. *The symmetric matrix*

$$A = \begin{pmatrix} 3 & 2 & 4 \\ 2 & 0 & 2 \\ 4 & 2 & 3 \end{pmatrix}$$

has the eigenvector $\mathbf{v} = (2, 1, 2)^T$. *Use a Householder reflection to find an* *orthogonal matrix S such that $S\mathbf{v}$ is a multiple of the first standard unit vector $\mathbf{e}_1$. Calculate SAS. The resultant matrix is suitable for deflation and hence identify the remaining eigenvalues and eigenvectors.*

We could achieve the same form of SAS^{-1} using successive Givens rotations instead of one Householder reflection. However, this makes sense only if there are already many zero elements in the first column of A.

The following algorithm for deflation can be used for nonsymmetric matrices as well as symmetric ones. Let v_i, $i = 1, \ldots, n$, be the components of the eigenvector $\mathbf{v}$. We can assume $v_1 \neq 0$, since otherwise the variables could be reordered. Let S be the $n \times n$ matrix which is identical to the $n \times n$ identity matrix except for the off-diagonal elements of the first column of S, which are $S_{i,1} = -v_i/v_1$, $i = 2, \ldots, n$.

$$
S = \begin{pmatrix}
1 & 0 & \cdots & \cdots & 0 \\
-v_2/v_1 & 1 & \ddots & & \vdots \\
\vdots & 0 & \ddots & \ddots & \vdots \\
\vdots & \vdots & \ddots & \ddots & 0 \\
-v_n/v_1 & 0 & \cdots & 0 & 1
\end{pmatrix}.
$$

Then S is nonsingular, has the property $S\mathbf{v} = v_1\mathbf{e}_1$, and thus is suitable for our purposes. The inverse S^{-1} is also identical to the identity matrix except for the off-diagonal elements of the first column of S^{-1} which are $(S^{-1})_{i,1} = +v_i/v_1$, $i = 2, \ldots, n$. Because of this form of S and S^{-1}, SAS^{-1} and hence B can be calculated in only $O(n^2)$ operations. Moreover, the last $n - 1$ columns of SAS^{-1} and SA are the same, since the last $n - 1$ columns of S^{-1} are taken from the identity matrix, and thus B is just the bottom $(n - 1) \times (n - 1)$ submatrix of SA. Therefore, for every integer $i = 1, \ldots, n-1$ we calculate the ith row of B by subtracting v_{i+1}/v_1 times the first row of A from the $(i+1)^{\text{th}}$ row of A and deleting the first component of the resultant row vector.

The following algorithm provides deflation in the form of block matrices. It is known as the *QR algorithm*, since QR factorizations are calculated again and again. Set $A_0 = A$. For $k = 0, 1, \ldots$ calculate the QR factorization $A_k = Q_k R_k$, where Q_k is orthogonal and R_k is upper triangular. Set $A_{k+1} = R_k Q_k$. The eigenvalues of A_{k+1} are the same as the eigenvalues of A_k, since

$$
A_{k+1} = R_k Q_k = Q_k^{-1} Q_k R_k Q_k = Q_k^{-1} A_k Q_k
$$

is a similarity transformation. Furthermore, $Q_k^{-1} = Q_k^T$, since Q_k is orthogonal. So if A_k is symmetric so is A_{k+1}. Surprisingly often the matrix A_{k+1} can be regarded as deflated. That is, it has the block structure

$$
A_{k+1} = \begin{pmatrix} B & C \\ D & E \end{pmatrix},
$$

where B and E are square $m \times m$ and $(n - m) \times (n - m)$ matrices and where all entries in D are close to zero. We can then calculate the eigenvalues of B and E separately, possibly again with the QR algorithm, except for 1×1 and 2×2 blocks where the eigenvalue problem is trivial. The space spanned by $\mathbf{e}_1, \ldots, \mathbf{e}_m$ can be regarded as an *eigenspace* of A_{k+1}, since, if $D = 0$, $A_{k+1}\mathbf{e}_i$, $i = 1, \ldots, m$, again lies in this space. Equally the space spanned by $\mathbf{e}_{m+1}, \ldots, \mathbf{e}_n$ can be regarded as an eigenspace of A_{k+1}.

The concept of eigenspaces is important when dealing with a complex conjugate pair of eigenvalues λ and $\bar{\lambda}$ with corresponding eigenvectors $\mathbf{v}$ and $\bar{\mathbf{v}}$ in $\mathbb{C}^n$. However, we are operating in $\mathbb{R}^n$. The real and imaginary parts $\mathrm{Re}(\mathbf{v})$ and $\mathrm{Im}(\mathbf{v})$ form an eigenspace of $\mathbb{R}^n$; that is, A applied to any linear combination of $\mathrm{Re}(\mathbf{v})$ and $\mathrm{Im}(\mathbf{v})$ will again be a linear combination of $\mathrm{Re}(\mathbf{v})$ and $\mathrm{Im}(\mathbf{v})$.

In this situation we choose S such that S applied to any vector in the space spanned by $\mathrm{Re}(\mathbf{v})$ and $\mathrm{Im}(\mathbf{v})$ is a linear combination $\mathbf{e}_1$ and $\mathbf{e}_2$. The matrix SAS^{-1}, then, consists of a 2×2 block in the top left corner and an $(n-2) \times (n-2)$ block B in the bottom right, and the last $(n-2)$ elements in the first two columns are zero. The search for eigenvalues can then continue with B. The following exercise illustrates this.

Exercise 2.20. *Show that the vectors* $\mathbf{x}$, $A\mathbf{x}$, *and* $A^2\mathbf{x}$ *are linearly dependent for*

$$
A = \begin{pmatrix} 1 & \frac{1}{4} & 0 & -\frac{3}{4} \\ 3 & \frac{1}{2} & 2 & -\frac{1}{2} \\ 0 & -\frac{3}{4} & 1 & \frac{1}{4} \\ 2 & -\frac{1}{2} & 3 & \frac{1}{2} \end{pmatrix} \quad \text{and } \mathbf{x} = \begin{pmatrix} 1 \\ 4 \\ 1 \\ 4 \end{pmatrix}.
$$

From this, calculate two eigenvalues of A. *Obtain by deflation a* 2×2 *matrix whose eigenvalues are the remaining eigenvalues of* A.

2.21 Revision Exercises

Exercise 2.21. *(a) Explain the technique of splitting for solving the linear system* $A\mathbf{x} = \mathbf{b}$ *iteratively where* A *is an* $n \times n$, *non-singular matrix. Define the iteration matrix* H *and state the property it has to satisfy to ensure convergence.*

(b) Define the Gauss–Seidel and Jacobi iterations and state their iteration matrices, respectively.

(c) Describe relaxation and briefly consider the cases when the relaxation parameter ω *equals 0 and 1.*

(d) Show how the iteration matrix H_ω *of the relaxed method is related to the iteration matrix* H *of the original method and thus how the eigenvalues are related. How should* ω *be chosen?*

(e) *We now consider the tridiagonal matrix A with diagonal elements $A_{i,i} = 1$ and off-diagonal elements $A_{i,i-1} = A_{i,i+1} = 1/4$. Calculate the iteration matrices H of the Jacobi method and H_ω of the relaxed Jacobi method.*

(f) *The eigenvectors of both H and H_ω are $\mathbf{v}_1, \ldots, \mathbf{v}_n$ where the ith component of $\mathbf{v}_k$ is given by $(\mathbf{v}_k)_i = \sin\frac{\pi i k}{n+1}$. Calculate the eigenvalues of H by evaluating $H\mathbf{v}_k$ (Hint: $\sin(x \pm y) = \sin x \cos y \pm \cos x \sin y$).*

(g) *Using the formula for the eigenvalues of H_ω derived earlier, state the eigenvalues of H_ω and show that the relaxed method converges for $0 < \omega \le 4/3$.*

Exercise 2.22. *Let A be an $n \times n$ matrix with n linearly independent eigenvectors. The eigenvectors are normalized to have unit length.*

(a) *Describe the power method to generate a single eigenvector and eigenvalue of A. Define the Rayleigh quotient in the process.*

(b) *Which assumption is crucial for the power method to converge? What characterizes the rate of convergence? By expressing the starting vector $\mathbf{x}^{(0)}$ as a linear combination of eigenvectors, give an expression for $\mathbf{x}^{(k)}$.*

(c) *Given*

$$A = \begin{pmatrix} 1 & 1 & 1 \\ 1 & 1 & 0 \\ 1 & 0 & 1 \end{pmatrix} \quad and \quad \mathbf{x}^{(0)} = \begin{pmatrix} 2 \\ 1 \\ -1 \end{pmatrix},$$

calculate $\mathbf{x}^{(1)}$ and $\mathbf{x}^{(2)}$ and evaluate the Rayleigh quotient $\lambda^{(k)}$ for $k = 0, 1, 2$.

(d) *Suppose now that for a different matrix A the eigenvalues of largest modulus are a complex conjugate pair of eigenvalues, λ and $\bar{\lambda}$. In this case the vectors $\mathbf{x}^{(k)}$, $A\mathbf{x}^{(k)}$ and $A^2\mathbf{x}^{(k)}$ tend to be linearly dependent. Assuming that they are linearly dependent, show how this can be used to calculate the eigenvalues λ and $\bar{\lambda}$.*

(e) *For large k, the iterations yielded the following vectors*

$$\mathbf{x}^{(k)} = \begin{pmatrix} 1 \\ 1 \\ 1 \end{pmatrix}, A\mathbf{x}^{(k)} = \begin{pmatrix} 2 \\ 3 \\ 4 \end{pmatrix} \quad and \quad A^2\mathbf{x}^{(k)} = \begin{pmatrix} 2 \\ 4 \\ 6 \end{pmatrix}.$$

Find the coefficients of the linear combination of these vectors which equates to zero. Thus deduce two eigenvalues of A.

Exercise 2.23. *(a) Given an $n \times n$ matrix A, define the concept of LU factorization and how it can be used to solve the system of equations $A\mathbf{x} = \mathbf{b}$.*

(b) *State two other applications of the LU factorization.*

(c) *Describe the algorithm to obtain an LU factorization. How many operations does this generally require?*

(d) *Describe the concept of pivoting in the context of solving the system of equations $A\mathbf{x} = \mathbf{b}$ by LU factorization.*

(e) *How does the algorithm need to be adjusted if in the process we encounter a column with all entries equal to zero? What does it mean if there is a column consisting entirely of zeros in the process?*

(f) *How can sparsity be exploited in the LU factorization?*

(g) *Calculate the LU factorization with pivoting of the matrix*

$$A = \begin{pmatrix} 2 & 1 & 1 & 0 \\ 4 & 3 & 3 & 1 \\ 8 & 7 & 9 & 5 \\ 6 & 7 & 9 & 8 \end{pmatrix}.$$

Exercise 2.24. (a) *Define the QR factorization of an $n \times n$ matrix A explaining what Q and R are. Show how the QR factorization can be used to solve the system of equations $A\mathbf{x} = \mathbf{b}$.*

(b) *How is the QR factorization defined, if A is an $m \times n$ ($m \neq n$) rectangular matrix? What does it mean if R is said to be in standard form?*

(c) *For an $m \times n$ matrix A and $\mathbf{b} \in \mathbb{R}^m$, explain how the QR factorization can be used to solve the least squares problem of finding $\mathbf{x}^* \in \mathbb{R}^n$ such that*

$$\|A\mathbf{x}^* - \mathbf{b}\| = \min_{\mathbf{x} \in \mathbb{R}^n} \|A\mathbf{x}^* - \mathbf{b}\|,$$

where the norm is the Euclidean distance $\|\mathbf{v}\| = \sqrt{\sum_{i=1}^m |v_i|^2}$.

(d) *Define a Householder reflection H in general and prove that H is an orthogonal matrix.*

(e) *Find a Householder reflection H, such that for*

$$A = \begin{pmatrix} 2 & 4 \\ 1 & -1 \\ 2 & 1 \end{pmatrix}$$

the first column of HA is a multiple of the first standard unit vector and calculate HA.

(f) *Having found H in the previous part, calculate $H\mathbf{b}$ for*

$$b = \begin{pmatrix} 1 \\ 5 \\ 1 \end{pmatrix}.$$

(g) *Using the results of the previous two parts, find the* $\mathbf{x} \in \mathbb{R}^2$ *which minimizes* $\|Ax - b\|$ *and calculate the minimum.*

Exercise 2.25. (a) *Explain the technique of splitting for solving the linear system* $A\mathbf{x} = \mathbf{b}$ *iteratively where* A *is an* $n \times n$, *non-singular matrix. Define the iteration matrix* H *and state the property it has to satisfy to ensure convergence.*

(b) *Define what it means for a matrix to be positive definite. Show that all diagonal elements of a positive definite matrix are positive.*

(c) *State the Householder-John theorem and explain how it can be used to design iterative methods for solving* $A\mathbf{x} = \mathbf{b}$.

(d) *Let the iteration matrix* H *have a* real *eigenvector* $\mathbf{v}$ *with* real *eigenvalue* λ. *Show that the condition of the Householder-John theorem implies that* $|\lambda| < 1$.

(e) *We write* A *in the form* $A = L + D + U$, *where* L *is the subdiagonal (or strictly lower triangular),* D *is the diagonal and* U *is the superdiagonal (or strictly upper triangular) portion of* A. *The following iterative scheme is suggested*

$$(L + \omega D)\mathbf{x}^{(k+1)} = -[(1 - \omega)D + U]\mathbf{x}^{(k)} + \mathbf{b}.$$

Using the Householder-John theorem, for which values of ω *does the scheme converge in the case when* A *is symmetric and positive definite?*

Exercise 2.26. *Let* A *be an* $n \times n$ *matrix which is symmetric and positive definite and let* $\mathbf{b} \in \mathbb{R}^n$.

(a) *Explain why solving* $A\mathbf{x} = \mathbf{b}$ *is equivalent to minimizing the quadratic function*

$$F(\mathbf{x}) = \frac{1}{2}\mathbf{x}^T A\mathbf{x} - \mathbf{x}^T\mathbf{b}.$$

By considering $\mathbf{x}^* + \mathbf{v}$ *where* $\mathbf{x}^*$ *denotes the vector where* F *takes it extremum, show why the extremum has to be a minimum.*

(b) *Having calculated* $\mathbf{x}^{(k)}$ *in the* k^{th} *iteration, the descent methods pick a search direction* $\mathbf{d}^{(k)}$ *in the next iteration which satisfy the descent condition. Define the descent condition.*

(c) *The next approximation* $\mathbf{x}^{(k+1)}$ *is calculated as* $\mathbf{x}^{(k+1)} = \mathbf{x}^{(k)} + \omega^{(k)}\mathbf{d}^{(k)}$. *Derive an expression for* $\omega^{(k)}$ *using the gradient* $\mathbf{g}^{(k)} = \nabla F(\mathbf{x}^{(k)}) = A\mathbf{x}^{(k)} - \mathbf{b}$.

(d) *Derive an expression for the new gradient* $\mathbf{g}^{(k+1)}$ *and a relationship between it and the search direction* $\mathbf{d}^{(k)}$.

(e) Explain how the search direction $\mathbf{d}^{(k)}$ is chosen in the steepest descent method and give a motivation for this choice.

(f) Define the concept of conjugacy.

(g) How are the search directions chosen in the conjugate gradient method? Derive an explicit formula for the search direction $\mathbf{d}^{(k)}$ stating the conjugacy condition in the process. What is $\mathbf{d}^{(0)}$?

(h) Which property do the gradients in the conjugate gradient method satisfy?

(i) Perform the conjugate gradient method to solve the system

$$\begin{pmatrix} 2 & -1 \\ -1 & 2 \end{pmatrix} \mathbf{x} = \begin{pmatrix} 1 \\ 0 \end{pmatrix}$$

starting from $\mathbf{x}^{(0)} = 0$.

Exercise 2.27. *(a) Use Gaussian elimination with backwards substitution to solve the linear system:*

$$\begin{array}{rrrcl}
5x_1+ & 10x_2+ & 9x_3 & = & 4 \\
10x_1+ & 26x_2+ & 26x_3 & = & 10 \\
15x_1+ & 54x_2+ & 66x_3 & = & 27
\end{array}$$

(b) How is the LU factorization defined, if A is an $n \times n$ square matrix, and how can it be used to solve the system of equations $A\mathbf{x} = \mathbf{b}$?

(c) Describe the algorithm to obtain an LU factorization.

(d) By which factor does the number of operations increase to obtain an LU factorization if n is increased by a factor of 10?

(e) What needs to be done if during Gaussian elimination or LU factorization a zero entry is encountered on the diagonal? Distinguish two different cases.

(f) Describe scaled and total pivoting. Explain why it is necessary under certain circumstances.

(g) Perform an LU factorization on the matrix arising from the system of equations given in (a).

Exercise 2.28. *(a) Define an $n \times n$ Householder reflection H in general and prove that H is a symmetric and orthogonal matrix.*

(b) For a general, non-zero vector $\mathbf{v} \in \mathbb{R}^n$, describe the construction of a Householder reflection which transforms $\mathbf{v}$ into a multiple of the first unit vector $\mathbf{e}_1 = (1, 0, \dots, 0)^T$.

(c) For $\mathbf{v} = (0, 1, -1)^T$ calculate H and $H\mathbf{v}$ such that the last two components of $H\mathbf{v}$ are zero.

(d) Let A be an $n \times n$ real matrix with a real eigenvector $\mathbf{v} \in \mathbb{R}^n$ with real eigenvector λ. Explain how a similarity transformation can be used to obtain an $(n-1) \times (n-1)$ matrix B whose $n-1$ eigenvalues are the other $n-1$ eigenvalues of A.

(e) The matrix

$$A = \begin{pmatrix} 1 & 1 & 1 \\ 1 & 1 & 2 \\ 1 & 2 & 1 \end{pmatrix}$$

has the eigenvector $\mathbf{v} = (0, 1, -1)^T$ with eigenvalue -1. Using the H obtained in (c) calculate HAH^T and thus calculate two more eigenvalues.

Exercise 2.29. (a) Explain the technique of splitting for solving the linear system $A\mathbf{x} = \mathbf{b}$ iteratively where A is an $n \times n$, non-singular matrix. Define the iteration matrix H and state the property it has to satisfy to ensure convergence.

(b) Define the Gauss–Seidel and Jacobi iterations and state their iteration matrices, respectively.

(c) Let

$$A = \begin{pmatrix} 2 & \frac{\sqrt{3}}{2} & \frac{1}{2} \\ \frac{\sqrt{3}}{2} & 2 & \frac{\sqrt{3}}{2} \\ \frac{1}{2} & \frac{\sqrt{3}}{2} & 2 \end{pmatrix}.$$

Derive the iteration matrix for the Jacobi iterations and state the eigenvalue equation. Check that the numbers $-3/4, 1/4, 1/2$ satisfy the eigenvalue equation and thus are the eigenvalues of the iteration matrix.

(d) The matrix given in (c) is positive definite. State the Householder–John theorem and apply it to show that the Gauss–Seidel iterations for this matrix converge.

(e) Describe relaxation and show how the the iteration matrix H_ω of the relaxed method is related to the iteration matrix H of the original method and thus how the eigenvalues are related. How should ω be chosen?

(f) For the eigenvalues given in (c) calculate the best choice of ω and the eigenvalues of the relaxed method.

CHAPTER 3

Interpolation and Approximation Theory

Interpolation describes the problem of finding a curve (called the *interpolant*) that passes through a given set of real values $f_0, f_1, \ldots, f_n$ at real data points $x_0, x_1, \ldots, x_n$, which are sometimes called *abscissae* or *nodes*. Different forms of interpolant exist. The theory of interpolation is important as a basis for numerical integration known also as *quadrature*. Approximation theory on the other hand seeks an approximation such that an *error norm* is minimized.

3.1 Lagrange Form of Polynomial Interpolation

The simplest case is to find a straight line

$$p(x) = a_1 x + a_0$$

through a pair of points given by (x_0, f_0) and (x_1, f_1). This means solving 2 equations, one for each data point. Thus we have 2 *degrees of freedom*. For a quadratic curve there are 3 degrees of freedom, fitting a cubic curve we have 4 degrees of freedom, etc.

Let $\mathbb{P}_n[x]$ denote the linear space of all real polynomials of degree at most n. Each $p \in \mathbb{P}_n[x]$ is uniquely defined by its $n + 1$ coefficients. This gives $n + 1$ degrees of freedom, while interpolating $x_0, x_1, \ldots, x_n$ gives rise to $n + 1$ conditions.

As we have mentioned above, in determining the polynomial interpolant we can solve a linear system of equations. However, this can be done more easily.

Definition 3.1 (Lagrange cardinal polynomials). *These are given by*

$$L_k(x) := \prod_{\substack{l=0 \\ l \neq k}}^{n} \frac{x - x_l}{x_k - x_l}, \qquad x \in \mathbb{R}.$$

Note that the Lagrange cardinal polynomials lie in $\mathbb{P}_n[x]$. It is easy to verify that $L_k(x_k) = 1$ and $L_k(x_j) = 0$ for $j \neq k$. The interpolant is then given by the *Lagrange formula*

$$p(x) = \sum_{k=0}^{n} f_k L_k(x) = \sum_{k=0}^{n} f_k \prod_{\substack{l=0 \\ l \neq k}}^{n} \frac{x - x_l}{x_k - x_l}.$$

Exercise 3.1. *Let the function values $f(0), f(1), f(2)$, and $f(3)$ be given. We want to estimate*

$$f(-1), f'(1) \ and \ \int_0^3 f(x)dx.$$

To this end, we let p be the cubic polynomial that interpolates these function values, and then approximate by

$$p(-1), p'(1) \ and \ \int_0^3 p(x)dx.$$

Using the Lagrange formula, show that every approximation is a linear combination of the function values with constant coefficients and calculate these coefficients. Show that the approximations are exact if f is any cubic polynomial.

Lemma 3.1 (Uniqueness). *The polynomial interpolant is unique.*

Proof. Suppose that two polynomials $p, q \in \mathbb{P}_n[x]$ satisfy $p(x_i) = q(x_i) = f_i$, $i = 0, \dots, n$. Then the nth degree polynomial $p - q$ vanishes at $n + 1$ distinct points. However, the only n^{th} degree polynomial with $n + 1$ or more zeros is the zero polynomial. Therefore $p = q$. $\qquad\square$

Exercise 3.2 (Birkhoff–Hermite interpolation). *Let a, b, and c be distinct real numbers, and let $f(a), f(b), f'(a), f'(b)$, and $f'(c)$ be given. Because there are five function values, a possibility is to approximate f by a polynomial of degree at most four that interpolates the function values. Show by a general argument that this interpolation problem has a solution and that the solution is unique if and only if there is no nonzero polynomial $p \in \mathbb{P}_4[x]$ that satisfies $p(a) = p(b) = p'(a) = p'(b) = p'(c) = 0$. Hence, given a and b, show that there exists a possible value of $c \neq a, b$ such that there is no unique solution.*

Let $[a, b]$ be a closed interval of $\mathbb{R}$. $C[a, b]$ is the space of continuous functions from $[a, b]$ to $\mathbb{R}$ and we denote by $C^k[a, b]$ the set of such functions which have continuous k^{th} derivatives.

Theorem 3.1 (The error of polynomial interpolation). *Given $f \in C^{n+1}[a, b]$ and $f(x_i) = f_i$, where $x_0, \dots, x_n$ are pairwise distinct, let $p \in \mathbb{P}_n[x]$ be the interpolating polynomial. Then for every $x \in [a, b]$, there exists $\xi \in [a, b]$ such that*

$$f(x) - p(x) = \frac{1}{(n+1)!} f^{(n+1)}(\xi) \prod_{i=0}^{n} (x - x_i). \tag{3.1}$$

Proof. Obviously, the formula (3.1) is true when $x = x_j$ for $j = 0, \ldots, n$, since both sides of the equation vanish. Let x be any other fixed point in the interval and for $t \in [a, b]$ let

$$\phi(t) := [f(t) - p(t)] \prod_{i=0}^{n} (x - x_i) - [f(x) - p(x)] \prod_{i=0}^{n} (t - x_i).$$

For $t = x_j$ the first term vanishes, since $f(x_j) = f_j = p(x_j)$, and by construction the product in the second term vanishes. We also have $\phi(x) = 0$, since then the two terms cancel. Hence ϕ has at least $n + 2$ distinct zeros in $[a, b]$. By *Rolle's theorem* if a function with continuous derivative vanishes at two distinct points, then its derivative vanishes at an intermediate point. Since $\phi \in C^{n+1}[a, b]$, we can deduce that ϕ' vanishes at $n + 1$ distinct points in $[a, b]$. Applying Rolle again, we see that ϕ'' vanishes at n distinct points in $[a, b]$. By induction, $\phi^{(n+1)}$ vanishes once, say at $\xi \in [a, b]$. Since p is an nth degree polynomial, we have $p^{(n+1)} \equiv 0$. On the other hand,

$$\frac{d^{n+1}}{dt^{n+1}} \prod_{i=0}^{n} (t - x_i) = (n + 1)!.$$

Hence

$$0 = \phi^{(n+1)}(\xi) = f^{(n+1)}(\xi) \prod_{i=0}^{n} (x - x_i) - [f(x) - p(x)](n + 1)!$$

and the result follows. □

This gives us an expression to estimate the error. Runge gives an example where a polynomial interpolation to an apparently well-behaved function is not suitable. *Runge's example* is

$$R(x) = \frac{1}{1 + 25x^2} \quad \text{on } [-1, 1].$$

It behaves like a polynomial near the centre $x = 0$,

$$R(x) = 1 - 25x^2 + O(x^4),$$

but it behaves like a quadratic hyperbola near the endpoints $x = \pm 1$,

$$R(x) = \frac{1}{25x^2} + O(\frac{1}{(25x^2)^2}).$$

Thus a polynomial interpolant will perform badly at the endpoints, where the behaviour is not like a polynomial. Figure 3.1 illustrates this when using equally spaced interpolation points. Adding more interpolation points actually makes the largest error worse. The growth in the error is explained by the product in (3.1). One can help matters by clustering points towards the ends

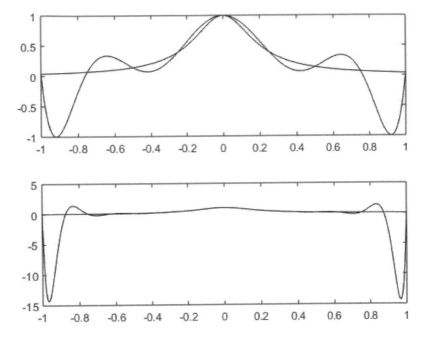

Figure 3.1 Interpolation of Runge's example with polynomials of degree 8 (top) and degree 16 (bottom)

of the interval by letting $x_k = \cos(\frac{\pi}{2} \frac{2k+1}{n+1})$, $k = 0, \ldots, n$. These are the so-called *Chebyshev points* or *nodes*.

These Chebyshev points are the zeros of the $(n+1)^{\text{th}}$ *Chebyshev polynomial*, which is defined on the interval $[-1, 1]$ by the *trigonometric definition*

$$T_{n+1}(x) = \cos((n+1)\arccos x), \qquad n \geq -1.$$

From this definition it is easy to see that the roots of T_{n+1} are $x_k = \cos(\frac{\pi}{2} \frac{2k+1}{n+1}) \in (-1, 1)$, $k = 0, \ldots, n$. Moreover, all the extrema of Chebyshev polynomials have values of either -1 or 1, two of which lie at the endpoints

$$\begin{aligned} T_{n+1}(1) &= 1 \\ T_{n+1}(-1) &= (-1)^{n+1}. \end{aligned}$$

In between we have extrema at $x_j = \cos \frac{j\pi}{n+1}$, $j = 1, \ldots, n$. Figure 3.2 shows the Chebyshev polynomial for $n = 9$. It is trivial to show that the Chebyshev polynomials satisfy the recurrence relation

$$\begin{aligned} T_0(x) &= 1 \\ T_1(x) &= x \\ T_{n+1}(x) &= 2xT_n(x) - T_{n-1}(x). \end{aligned}$$

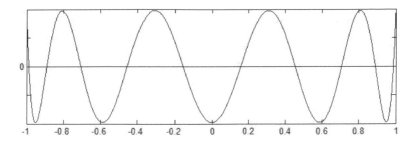

Figure 3.2 The 10th Chebyshev Polynomial

Lemma 3.2. *The maximum absolute value of $\prod_{i=0}^{n}(x - x_i)$ on $[-1, 1]$ is minimal if it is the normalized $(n+1)^{\text{th}}$ Chebyshev polynomial, i.e., $\prod_{i=0}^{n}(x - x_i) = 2^{-n}T_{n+1}(x)$. The maximum absolute value is then 2^{-n}.*

Proof. $\prod_{i=0}^{n}(x - x_i)$ describes an $(n + 1)^{\text{th}}$ degree polynomial with leading coefficient 1. From the recurrence relation we see that the leading coefficient of T_{n+1} is 2^n and thus $2^{-n}T_{n+1}$ has leading coefficient one. Let p be a polynomial of degree $n + 1$ with leading coefficient 1 with maximum absolute value $m < 2^{-n}$ on $[-1, 1]$. T_{n+1} has $n+2$ extreme points. At these points we have $|p(x)| \leq m < |2^{-n}T_{n+1}(x)|$. Moreover, for $x = \cos\frac{2k\pi}{n+1}$, $0 \leq 2k \leq n + 1$, where T_{n+1} has a maximum, we have

$$2^{-n}T_{n+1}(x) - p(x) \geq 2^{-n}T_{n+1}(x) - m > 0,$$

while for $x = \cos\frac{(2k+1)\pi}{n+1}$, $0 \leq 2k + 1 \leq n + 1$, where T_{n+1} has a minimum,

$$2^{-n}T_{n+1}(x) - p(x) \leq 2^{-n}T_{n+1}(x) + m < 0.$$

Thus the function $2^{-n}T_{n+1}(x) - p(x)$ changes sign between the points $\cos\frac{k\pi}{n+1}$, $0 \leq k \leq n+1$. From the intermediate value theorem $2^{-n}T_{n+1} - p$ has at least $n + 1$ roots. However, this is impossible, since $2^{-n}T_{n+1} - p$ is a polynomial of degree n, since the leading coefficients cancel. Thus we have a contradiction and $2^{-n}T_{n+1}(x)$ gives the minimal value. □

For a general interval $[a, b]$ the Chebyshev points are $x_k = (a + b + (b - a)\cos(\frac{\pi}{2}\frac{2k+1}{n+1}))/2$.

Another interesting fact about Chebyshev polynomials is that they form a set of polynomials which are orthogonal with respect to the weight function $(1 - x^2)^{-1/2}$ on $(-1, 1)$. More specifically we have

$$\int_{-1}^{1} T_m(x)T_n(x)\frac{dx}{\sqrt{1 - x^2}} = \begin{cases} \pi, & m = n = 0, \\ \frac{\pi}{2}, & m = n \geq 1, \\ 0 & m \neq n. \end{cases}$$

We will study orthogonal polynomials in much more detail when considering polynomial best approximations. Because of the orthogonality every polynomial p of degree n can be expressed as

$$p(x) = \sum_{k=0}^{n} \check{p}_k T_k(x),$$

where the coefficients $\check{p}_k$ are given by

$$\check{p}_0 = \frac{1}{\pi} \int_{-1}^{1} p(x) \frac{dx}{\sqrt{1-x^2}}, \quad \check{p}_k = \frac{2}{\pi} \int_{-1}^{1} p(x) T_k(x) \frac{dx}{\sqrt{1-x^2}}, \quad k = 1, \ldots, n.$$

We will encounter Chebyshev polynomials also again in spectral methods for the solution of partial differential equations.

3.2   Newton Form of Polynomial Interpolation

Another way to describe the polynomial interpolant was introduced by Newton. First we need to introduce some concepts, however.

Definition 3.2 (Divided difference). *Given pairwise distinct points $x_0, x_1, \ldots, x_n \in [a, b]$, let $p \in \mathbb{P}_n[x]$ interpolate $f \in C^n[a, b]$ at these points. The coefficient of x^n in p is called the divided difference of degree n and denoted by $f[x_0, x_1, \ldots, x_n]$.*

From the Lagrange formula we see that

$$f[x_0, x_1, \ldots, x_n] = \sum_{k=0}^{n} f(x_k) \prod_{\substack{l=0 \\ l \neq k}}^{n} \frac{1}{x_k - x_l}.$$

Theorem 3.2. *There exists $\xi \in [a, b]$ such that*

$$f[x_0, x_1, \ldots, x_n] = \frac{1}{n!} f^{(n)}(\xi).$$

Proof. Let p be the polynomial interpolant. The difference $f - p$ has at least $n + 1$ zeros. By applying Rolle's theorem n times, it follows that the nth derivative $f^{(n)} - p^{(n)}$ is zero at some $\xi \in [a, b]$. Since p is of degree n, $p^{(n)}$ is constant, say c, and we have $f^{(n)}(\xi) = c$. On the other hand the coefficient of x^n in p is given by $\frac{1}{n!}c$, since the nth derivative of x^n is $n!$. Hence we have

$$f[x_0, x_1, \ldots, x_n] = \frac{1}{n!} c = \frac{1}{n!} f^{(n)}(\xi).$$

$\square$

Thus, divided differences can be used to approximate derivatives.

Exercise 3.3. *Let f be a real valued function and let p be the polynomial of degree at most n that interpolates f at the pairwise distinct points $x_0, x_1, \ldots, x_n$. Furthermore, let x be any real number that is not an interpolation point. Deduce for the error at x*

$$f(x) - p(x) = f[x_0, \ldots, x_n, x] \prod_{j=0}^{n} (x - x_j).$$

(Hint: Use the definition for the divided difference $f[x_0, \ldots, x_n, x]$.)

We might ask what the divided difference of degree zero is. It is the coefficient of the zero degree interpolating polynomial, i.e., a constant. Hence $f[x_i] = f(x_i)$. Using the formula for linear interpolation between two points $(x_i, f(x_i))$ and $(x_j, f(x_j))$, the interpolating polynomial is given by

$$p(x) = \left(\frac{x - x_j}{x_i - x_j} \right) f(x_i) + \left(\frac{x - x_i}{x_j - x_i} \right) f(x_j)$$

and thus we obtain

$$f[x_i, x_j] = \frac{f[x_i] - f[x_j]}{x_i - x_j}.$$

More generally, we have the following theorem

Theorem 3.3. *Let $x_0, x_1, \ldots, x_{k+1}$ be pairwise distinct points, where $k \geq 0$. Then*

$$f[x_0, x_1, \ldots, x_k, x_{k+1}] = \frac{f[x_1, \ldots, x_{k+1}] - f[x_0, \ldots, x_k]}{x_{k+1} - x_0}.$$

Proof. Let $p, q \in \mathbb{P}_k[x]$ be the polynomials that interpolate f at $x_0, \ldots, x_k$ and $x_1, \ldots, x_{k+1}$ respectively. Let

$$r(x) := \frac{(x - x_0)q(x) + (x_{k+1} - x)p(x)}{x_{k+1} - x_0} \in \mathbb{P}_{k+1}[x].$$

It can be easily seen that $r(x_i) = f(x_i)$ for $i = 0, \ldots, k + 1$. Hence r is the unique interpolating polynomial of degree $k + 1$ and the coefficient of x^{k+1} in r is given by the formula in the theorem. □

This recursive formula gives a fast way to calculate the *divided difference table* shown in Figure 3.3. This requires $O(n^2)$ operations and calculates the numbers $f[x_0, \ldots x_l]$ for $l = 0, \ldots, n$. These are needed for the alternative representation of the interpolating polynomial.

Exercise 3.4. *Implement a routine calculating the divided difference table.*

Theorem 3.4 (Newton Interpolation Formula). *Let $x_0, x_1, \ldots, x_n$ be pairwise*

$$f[x_0] \searrow$$
$$\qquad \nearrow \quad f[x_0, x_1] \searrow$$
$$f[x_1] \searrow \qquad \qquad \nearrow \quad f[x_0, x_1, x_2]$$
$$\qquad \nearrow \quad f[x_1, x_2] \searrow$$
$$f[x_2] \searrow \qquad \qquad \qquad \qquad \ddots$$
$$\qquad \qquad \qquad \qquad \qquad \qquad \qquad \ddots$$
$$\qquad \qquad \qquad \qquad \qquad \vdots \qquad \qquad \qquad f[x_0, \ldots, x_n].$$
$$\vdots \qquad \qquad \qquad \qquad \qquad \qquad \qquad \cdot\cdot$$
$$\qquad \qquad \qquad \qquad f[x_{n-2}, x_{n-1}, x_n] \qquad \cdot\cdot$$
$$\qquad \qquad \nearrow$$
$$\qquad \quad f[x_{n-1}, x_n] \nearrow$$
$$f[x_n] \nearrow$$

Figure 3.3 Divided difference table

distinct. The polynomial

$$p_n(x) \quad := \quad f[x_0] + f[x_0, x_1](x - x_0) + \cdots + f[x_0, \ldots, x_{n-1}] \prod_{i=0}^{n-2}(x - x_i)$$
$$+ f[x_0, \ldots, x_n] \prod_{i=0}^{n-1}(x - x_i) \in \mathbb{P}_n[x]$$

satisfies $p_n(x_i) = f(x_i)$, $i = 0, 1, \ldots, n$.

Proof. The proof is done by induction on n. The statement is obvious for $n = 0$ since the interpolating polynomial is the constant $f(x_0)$. Suppose that the assertion is true for n. Let $p \in \mathbb{P}_{n+1}[x]$ be the interpolating polynomial at $x_0, \ldots, x_{n+1}$. The difference $p - p_n$ vanishes at x_i for $i = 0, \ldots, n$ and hence it is a multiple of $\prod_{i=0}^{n}(x - x_i)$. By definition of divided differences the coefficient of x^{n+1} in p is $f[x_0, \ldots, x_{n+1}]$. Since $p_n \in \mathbb{P}_n[x]$, it follows that $p(x) - p_n(x) = f[x_0, \ldots, x_{n+1}] \prod_{i=0}^{n}(x - x_i)$. The explicit form of p_{n+1} follows by adding p_n to $p - p_n$. $\qquad \square$

Here is a MATLAB implementation of evaluating the interpolating polynomial given the interpolation points and divided difference table.

```
function [y] = NewtonEval(inter, d, x )
% Calculates the values of the polynomial in Newton form at the
% points given in x
% inter input parameter, the interpolation points
% d input parameter, divided differences prescribing the polynomial

[n,m]=size(inter); % finding the size of inter
[k,l] = size(d); % finding the size of d
if m≠1 || l≠1
    disp('input need to be column vectors');
```

```
        return;
end
if n≠k
    disp('input dimensions do not agree')
    return;
end
m = size(x); % number of evaluation points
y = d(1) * ones(m); % first term of the sum in the Newton form
temp = x - inter(1) * ones(m); % temp holds the product
% Note that y and temp are vectors
for i=2:n % add the terms of the sum in the Newton form one after
            % the other
    y = y + d(i) * temp;
    temp = temp .* (x - inter(i)); % Note: *. is element—wise
                                   % multiplication
end
end
```

Exercise 3.5. *Given $f, g \in C[a, b]$, let $h := fg$. Prove by induction that the divided differences of h satisfy the relation*

$$h[x_0, \ldots, x_n] = \sum_{j=0}^{n} f[x_0, \ldots, x_j] g[x_j, \ldots, x_n].$$

By using the representation as derivatives of the differences and by letting the points $x_0, \ldots, x_n$ coincide, deduce the Leibniz formula for the nth derivative of a product of two functions.

The Newton interpolation formula has several advantages over the Lagrange formula. Provided that the divided differences are known, it can be evaluated at a given point x in just $O(n)$ operations as long as we employ the nested multiplication as in the *Horner scheme*

$$p_n(x) = \{\cdots\{\{f[x_0, \ldots, x_n](x - x_{n-1}) + f[x_0, \ldots, x_{n-1}]\} \times (x - x_{n-2})$$
$$+ f[x_0, \ldots, x_{n-2}]\} \times (x - x_{n-3}) + \cdots\} + f[x_0].$$

The MATLAB implementation using the Horner scheme is as follows

```
function [y] = NewtonHorner(inter, d, x )
```

However, both representations of the interpolating polynomials have their advantages. For example the Lagrange formula is often better when the interpolating polynomial is part of a larger mathematical expression such as in Gaussian quadrature.

If $x_0, \ldots, x_n$ are equally spaced and arranged consecutively, letting $h = x_{i+1} - x_i$ for each $i = 0, \ldots, n - 1$, we can rewrite the Newton formula for

$x = x_0 + sh$. Then $x - x_i = (s - i)h$, and we have

$$p_n(x_0 + sh) = f[x_0] + f[x_0, x_1]sh + \cdots + f[x_0, \ldots, x_{n-1}] \prod_{i=0}^{n-2}(s - i)h^{n-1}$$

$$+ f[x_0, \ldots, x_n] \prod_{i=0}^{n-1}(s - i)h^n$$

This is called the *Newton Forward Divided Difference Formula*.
Defining the *finite difference*

$$\Delta f(x_0) := f(x_1) - f(x_0)$$

we see that $f[x_0, x_1] = h^{-1}\Delta f(x_0)$. We will encounter finite differences again when deriving solutions to partial differential equations. Since

$$\Delta^2 f(x_0) = \Delta f(x_1) - \Delta f(x_0) = hf[x_1, x_2] - hf[x_0, x_1] = 2h^2 f[x_0, x_1, x_2],$$

where we use the recurrence formula for divided differences, we can deduce by induction that

$$\Delta^j f(x_0) = j!h^j f[x_0, \ldots, x_j].$$

Hence we can rewrite the Newton formula as

$$p_n(x_0 + sh) = f(x_0) + \sum_{j=1}^{n} \prod_{i=0}^{j-1}(s - i)\frac{1}{j!}\Delta^j f(x_0) = f(x_0) + \sum_{j=1}^{n} \binom{s}{j} \Delta^j f(x_0).$$

In this form the formula looks suspiciously like a finite analog of the Taylor expansion. The Taylor expansion tells where a function will go based on the values of the function and its derivatives (its rate of change and the rate of change of its rate of change, etc.) at one given point x. Newtons formula is based on finite differences instead of instantaneous rates of change.

If the points are reordered as $x_n, \ldots, x_0$, we can again rewrite the Newton formula for $x = x_n + sh$. Then $x - x_i = (s + n - i)h$, since $x_i = x_n - (n - i)h$, and we have

$$p_n(x_n + sh) = f[x_n] + f[x_n, x_{n-1}]sh + \cdots + f[x_n, \ldots, x_1] \prod_{i=0}^{n-2}(s + i)h^{n-1}$$

$$+ f[x_n, \ldots, x_0] \prod_{i=0}^{n-1}(s + i)h^n.$$

This is called the *Newton Backward Divided Difference Formula*.
The degree of an interpolating polynomial can be increased by adding more points and more terms. In Newton's form points can simply be added at one end. Newton's forward formula can add new points to the right, and Newton's backward formula can add new points to the left. The accuracy of

an interpolation polynomial depends on how far the point of interest is from the middle of the interpolation points used. Since points are only added at the same end, the accuracy only increases at that end. There are formulae by Gauss, Stirling, and Bessel to remedy this problem (see for example [17] A. Ralston, P. Rabinowitz *First Course in Numerical Analysis*).

The Newton form of the interpolating polynomial can be viewed as one of a class of methods for generating successively higher order interpolation polynomials. These are known as *iterated linear interpolation* and are iterative. They are all based on the following lemma.

Lemma 3.3. *Let* $x_{i_1}, x_{i_2}, \ldots, x_{i_m}$ *be* m *distinct points and denote by* $p_{i_1,i_2,\ldots,i_m}$ *the polynomial of degree* $m-1$ *satisfying*

$$p_{i_1,i_2,\ldots,i_m}(x_{i_\nu}) = f(x_{i_\nu}), \qquad \nu = 1, \ldots, m.$$

Then for $n \geq 0$, *if* x_j, x_k *and* x_{i_ν}, $\nu - 1, \ldots, n$, *are any* $n+2$ *distinct points, then*

$$p_{i_1,i_2,\ldots,i_n,j,k}(x) = \frac{(x - x_k)p_{i_1,i_2,\ldots,i_n,j}(x) - (x - x_j)p_{i_1,i_2,\ldots,i_n,k}(x)}{x_j - x_k}.$$

Proof. For $n = 0$, there are no additional points x_{i_ν} and $p_j(x) \equiv f(x_j)$ and $p_k(x) \equiv f(x_k)$. It can be easily seen that the right-hand side defines a polynomial of degree at most $n + 1$, which takes the values $f(x_{i_\nu})$ at x_{i_ν} for $\nu = 1, \ldots, n$, $f(x_j)$ at x_j and $f(x_k)$ at x_k. Hence the right-hand side is the unique interpolation polynomial. □

We see that with iterated linear interpolation, points can be added anywhere. The variety of methods differ in the order in which the values $(x_j, f(x_j))$ are employed. For many applications, additional function values are generated on the fly and thus their number are not known in advance. For such cases we may always employ the latest pair of values as in

$$
\begin{array}{cc|cccc}
x_0 & f(x_0) & & & & \\
x_1 & f(x_1) & p_{0,1}(x) & & & \\
x_2 & f(x_2) & p_{1,2}(x) & p_{0,1,2}(x) & & \\
\vdots & \vdots & \vdots & \vdots & & \\
x_n & f(x_n) & p_{n-1,n}(x) & p_{n-2,n-1,n}(x) & \cdots & p_{0,1,\ldots,n}(x).
\end{array}
$$

The rows are computed sequentially. Any $p_{\ldots}(x)$ is calculated using the two quantities to its left and diagonally above it. To determine the $(n+1)^{\text{th}}$ row only the nth row needs to be known. As more points are generated, rows of greater lengths need to be calculated and stored. The algorithm stops whenever

$$|p_{0,1,\ldots,n}(x) - p_{0,1,\ldots,n-1}(x)| < \epsilon.$$

This scheme is known as *Neville's iterated interpolation*.

An alternative scheme is *Aitken's iterated interpolation* given by

$$
\begin{array}{cc|cccc}
x_0 & f(x_0) & & & & \\
x_1 & f(x_1) & p_{0,1}(x) & & & \\
x_2 & f(x_2) & p_{0,2}(x) & p_{0,1,2}(x) & & \\
\vdots & \vdots & \vdots & \vdots & & \\
x_n & f(x_n) & p_{0,n}(x) & p_{0,1,n}(x) & \cdots & p_{0,1,\ldots,n}(x).
\end{array}
$$

The basic difference between the two methods is that in Aitken's method the interpolants on the row with x_k use points with subscripts nearest 0, while in Neville's they use points with subscripts nearest to k.

3.3 Polynomial Best Approximations

We now turn our attention to *best approximations* where best is defined by a *norm* (possibly introduced by a scalar product) which we try to minimize. Recall that a *scalar* or *inner product* is any function $\mathbb{V} \times \mathbb{V} \to \mathbb{R}$, where $\mathbb{V}$ is a real vector space, subject to the three axioms

Symmetry:
$$\langle x, y \rangle = \langle y, x \rangle \text{ for all } x, y \in \mathbb{V},$$

Non-negativity:
$$\langle x, x \rangle \geq 0 \text{ for all } x \in \mathbb{V} \text{ and } \langle x, x \rangle = 0 \text{ if and only if } x = 0, \text{ and}$$

Linearity:
$$\langle ax + by, z \rangle = a\langle x, z \rangle + b\langle y, z \rangle \text{ for all } x, y, z \in \mathbb{V}, \ a, b \in \mathbb{R}.$$

We already encountered the vector space $\mathbb{R}^n$ and its scalar product with the QR factorization of matrices. Another example of a vector space is the space of polynomials of degree n, $\mathbb{P}_n[x]$, but no scalar product has been defined for it so far.

Once a scalar product is defined, we can define *orthogonality*: $x, y \in \mathbb{V}$ are *orthogonal* if $\langle x, y \rangle = 0$. A *norm* can be defined by

$$\|x\| = \sqrt{\langle x, x \rangle} \qquad x \in \mathbb{V}.$$

For $\mathbb{V} = C[a, b]$, the space of all continuous functions on the interval $[a, b]$, we can define a scalar product using a fixed positive function $w \in C[a, b]$, the *weight function*, in the following way

$$\langle f, g \rangle := \int_a^b w(x)f(x)g(x)dx \text{ for all } f, g \in C[a, b].$$

All three axioms are easily verified for this scalar product. The associated norm is

$$\|f\|_2 = \sqrt{\langle f, f \rangle} = \sqrt{\int_a^b w(x)f^2(x)dx}.$$

For $w(x) \equiv 1$ this is known as the L_2-norm. Note that $\mathbb{P}_n$ is a subspace of $C[a, b]$.

Generally L_p *norms* are defined by

$$\|f\|_p = \left(\int_a^b |f(x)|^p dx \right)^{1/p}.$$

The vector space of functions for which this integral exists is denoted by L_p. Unless $p = 2$, this is a normed space, but not an inner product space, because this norm does not satisfy the parallelogram equality given by

$$2\|f\|_p^2 + 2\|g\|_p^2 = \|f + g\|_p^2 + \|f - g\|_p^2$$

required for a norm to have an associated inner product.

Let g be an approximation to the function f. Often g is chosen to lie in a certain subspace, for example the space of polynomials of a certain degree. The best approximation chooses g such that the norm of the error $\|f - g\|$ is minimized. Different choices of norm give different approximations.

For $p \to \infty$ the norm becomes

$$\|f\|_\infty = \max_{x \in [a,b]} |f(x)|.$$

We actually have already seen the *best L_∞ approximation* from $\mathbb{P}_n[x]$. It is the interpolating polynomial where the interpolation points are chosen to be the Chebyshev points. This is why the best approximation with regards to the L_∞ norm is sometimes called the *Chebyshev approximation*. It is also known as *Minimax approximation*, since the problem can be rephrased as finding g such that

$$\min_g \max_{x \in [a,b]} |f(x) - g(x)|.$$

3.4   Orthogonal polynomials

Given an inner product, we say that $p_n \in \mathbb{P}_n[x]$ is the nth *orthogonal polynomial* if $\langle p_n, p \rangle = 0$ for all $p \in \mathbb{P}_{n-1}[x]$. We have already seen the orthogonal polynomials with regards to the weight function $(1 - x^2)^{-1/2}$. These are the Chebyshev polynomials. Different scalar products lead to different orthogonal polynomials. Orthogonal polynomials stay orthogonal if multiplied by a constant. We therefore introduce a *normalization* by requiring the leading coefficient to equal one. These polynomials are called *monic*, the noun being *monicity*.

Theorem 3.5. *For every $n \geq 0$ there exists a unique monic orthogonal polynomial p_n of degree n. Any $p \in \mathbb{P}_n[x]$ can be expressed as a linear combination of $p_0, p_1, \ldots p_n$.*

Proof. We first consider uniqueness. Assume that there are two monic orthogonal polynomials $p_n, \tilde{p}_n \in \mathbb{P}_n[x]$. Let $p = p_n - \tilde{p}_n$ which is of degree $n - 1$, since the x^n terms cancel, because in both polynomials the leading coefficient is 1. By definition of orthogonal polynomials, $\langle p_n, p \rangle = 0 = \langle \tilde{p}_n, p \rangle$. Thus we can write

$$0 = \langle p_n, p \rangle - \langle \tilde{p}_n, p \rangle = \langle p_n - \tilde{p}_n, p \rangle = \langle p, p \rangle,$$

and hence $p \equiv 0$.

We provide a constructive proof for the existence by induction on n. We let $p_0(x) \equiv 1$. Assume that $p_0, p_1, \ldots, p_n$ have already been constructed, consistent with both statements of the theorem. Let $q(x) := x^{n+1} \in \mathbb{P}_{n+1}[x]$. Following the *Gram–Schmidt algorithm*, we construct

$$p_{n+1}(x) = q(x) - \sum_{k=0}^{n} \frac{\langle q, p_k \rangle}{\langle p_k, p_k \rangle} p_k(x).$$

It is of degree $n + 1$ and it is monic, since all the terms in the sum are of degree less than or equal to n. Let $m \in 0, 1, \ldots, n$.

$$\langle p_{n+1}, p_m \rangle = \langle q, p_m \rangle - \sum_{k=0}^{n} \frac{\langle q, p_k \rangle}{\langle p_k, p_k \rangle} \langle p_k, p_m \rangle = \langle q, p_m \rangle - \frac{\langle q, p_m \rangle}{\langle p_m, p_m \rangle} \langle p_m, p_m \rangle = 0.$$

Hence p_{n+1} is orthogonal to $p_0, \ldots, p_n$ and consequently to all $p \in \mathbb{P}_n[x]$ due to the second statement of the theorem.

Finally, to prove that any $p \in \mathbb{P}_{n+1}[x]$ is a linear combination of $p_0, \ldots, p_n, p_{n+1}$, we note that p can be written as $p = cp_{n+1} + \tilde{p}$, where c is the coefficient of x^{n+1} in p and where $\tilde{p} \in \mathbb{P}_n[x]$. Due to the induction hypothesis, $\tilde{p}$ is a linear combination of $p_0, \ldots, p_n$ and hence p is a linear combination of $p_0, \ldots, p_n, p_{n+1}$. □

Well-known examples of orthogonal polynomials are ($\alpha, \beta > -1$):

Name	Notation	Interval	Weight function
Legendre	P_n	$[-1, 1]$	$w(x) \equiv 1$
Jacobi	$P_n^{(\alpha, \beta)}$	$(-1, 1)$	$w(x) = (1 - x)^\alpha (1 + x)^\beta$
Chebyshev (first kind)	T_n	$(-1, 1)$	$w(x) = (1 - x^2)^{-1/2}$
Chebyshev (second kind)	U_n	$[-1, 1]$	$w(x) = (1 - x^2)^{1/2}$
Laguerre	L_n	$[0, \infty)$	$w(x) = e^{-x}$
Hermite	H_n	$(-\infty, \infty)$	$w(x) = e^{-x^2}$

Polynomials can be divided by each other. For example, the polynomial $x^2 - 1$ has zeros at ± 1. Thus $x - 1$ and $x + 1$ are factors of $x^2 - 1$ or written as a division

$$(x^2 + 0x - 1)/(x - 1) = x + 1$$
$$x^2 - x$$
$$ x - 1.$$

Exercise 3.6. *The polynomial $p(x) = x^4 - x^3 - x^2 - x - 2$ is zero for $x = -1$ and $x = 2$. Find two factors $q_1(x)$ and $q_2(x)$ and divide $p(x)$ by those to obtain a quadratic polynomial. Find two further zeros of $p(x)$ as the zeros of that quadratic polynomial.*

Exercise 3.7. *The functions $p_0, p_1, p_2, \ldots$ are generated by the Rodrigues formula*

$$p_n(x) = e^x \frac{d^n}{dx^n}(x^n e^{-x}), \qquad x \in \mathbb{R}^+.$$

Show that these functions are polynomials and prove by integration by parts that for every $p \in \mathbb{P}_{n-1}[x]$ we have the orthogonality condition $\langle p_n, p \rangle = 0$ with respect to the scalar product given by

$$\langle f, g \rangle := \int_0^\infty e^{-x} f(x) g(x) dx.$$

Thus these polynomials are the Laguerre polynomials. Calculate p_3, p_4, and p_5 from the Rodrigues formula.

The proof of Theorem 3.5 is constructive, but it suffers from loss of accuracy in the calculation of the inner products. With Chebyshev polynomials we have seen that they satisfy a three-term recurrence relation. This is also true for monic orthogonal polynomials of other weight functions, as the following theorem shows.

Theorem 3.6. *Monic orthogonal polynomials are given by the recurrence relation*

$$\begin{aligned} p_{-1}(x) &\equiv 0, \\ p_0(x) &\equiv 1, \\ p_{n+1}(x) &= (x - \alpha_n)p_n(x) - \beta_n p_{n-1}(x), \end{aligned}$$

where

$$\alpha_n = \frac{\langle p_n, x p_n \rangle}{\langle p_n, p_n \rangle}, \qquad \beta_n = \frac{\langle p_n, p_n \rangle}{\langle p_{n-1}, p_{n-1} \rangle} > 0.$$

Proof. For $n \geq 0$ let $p(x) := p_{n+1}(x) - (x - \alpha_n)p_n(x) + \beta_n p_{n-1}(x)$. We will show that p is actually zero. Since p_{n+1} and xp_n are monic, it follows that $p \in \mathbb{P}_n[x]$. Furthermore,

$$\langle p, p_l \rangle = \langle p_{n+1}, p_l \rangle - \langle (x - \alpha_n)p_n, p_l \rangle + \beta_n \langle p_{n-1}, p_l \rangle = 0, \qquad l = 0, \ldots, n-2,$$

since from the definition of the inner product, $\langle (x - \alpha_n)p_n, p_l \rangle = \langle p_n, (x - \alpha_n)p_l \rangle$, and since p_{n-1}, p_n and p_{n+1} are orthogonal polynomials. Moreover,

$$\begin{aligned} \langle p, p_{n-1} \rangle &= \langle p_{n+1}, p_{n-1} \rangle - \langle (x - \alpha_n)p_n, p_{n-1} \rangle + \beta_n \langle p_{n-1}, p_{n-1} \rangle \\ &= -\langle p_n, x p_{n-1} \rangle + \beta_n \langle p_{n-1}, p_{n-1} \rangle. \end{aligned}$$

Now because of monicity, $x p_{n-1} = p_n + q$, where $q \in \mathbb{P}_{n-1}[x]$, we have

$$\langle p, p_{n-1} \rangle = -\langle p_n, p_n \rangle + \beta_n \langle p_{n-1}, p_{n-1} \rangle = 0$$

due to the definition of β_n. Finally,

$$\langle p, p_n \rangle = \langle p_{n+1}, p_n \rangle - \langle (x-\alpha_n)p_n, p_n \rangle + \beta_n \langle p_{n-1}, p_n \rangle = -\langle xp_n, p_n \rangle + \alpha_n \langle p_n, p_n \rangle = 0$$

from the definition of α_n. It follows that p is orthogonal to $p_0, \ldots, p_n$ which form a basis of $\mathbb{P}_n[x]$ which is only possible for the zero polynomial. Hence $p \equiv 0$ and the assertion is true. □

Exercise 3.8. *Continuing from the previous exercise, show that the coefficients of p_3, p_4 and p_5 are compatible with a three-term recurrence relation of the form*

$$p_5(x) = (\gamma x - \alpha)p_4(x) - \beta p_3(x), \qquad x \in \mathbb{R}^+.$$

3.5 Least-Squares Polynomial Fitting

We now have the tools at our hands to find the best polynomial approximation $p \in \mathbb{P}_n[x]$, which minimizes $\langle f - p, f - p \rangle$ subject to a given inner product $\langle g, h \rangle = \int_a^b w(x)g(x)h(x)dx$ where $w(x) > 0$ is the weight function. Let $p_0, p_1, \ldots, p_n$ be the orthogonal polynomials, $p_l \in \mathbb{P}_l[x]$, which form a basis of $\mathbb{P}_n[x]$. Therefore there exist constants $c_0, c_1, \ldots, c_n$ such that $p = \sum_{j=0}^{n} c_j p_j$. Because of orthogonality the inner product simplifies to

$$\begin{aligned}
\langle f - p, f - p \rangle &= \left\langle f - \sum_{j=0}^{n} c_j p_j, f - \sum_{j=0}^{n} c_j p_j \right\rangle \\
&= \langle f, f \rangle - 2\sum_{j=0}^{n} c_j \langle f, p_j \rangle + \sum_{j=0}^{n} c_j^2 \langle p_j, p_j \rangle.
\end{aligned}$$

This is a quadratic function in the c_js and we can minimize it to find optimal values for the c_js. Differentiating with respect to c_j gives

$$\frac{\partial}{\partial c_j} \langle f - p, f - p \rangle = -2\langle p_j, f \rangle + 2c_j \langle p_j, p_j \rangle, \qquad j = 0, \ldots, n.$$

Setting the gradient to zero, we obtain

$$c_j = \frac{\langle p_j, f \rangle}{\langle p_j, p_j \rangle}$$

and thus

$$p = \sum_{j=0}^{n} \frac{\langle p_j, f \rangle}{\langle p_j, p_j \rangle} p_j.$$

The value of the error norm is then

$$\langle f - p, f - p \rangle = \langle f, f \rangle - \sum_{j=0}^{n} \frac{\langle f, p_j \rangle^2}{\langle p_j, p_j \rangle}.$$

The coefficients c_j, $j = 0, \ldots, n$, are independent of n. Thus, more and more terms can be added until $\langle f - p, f - p \rangle$ is below a given tolerance ϵ or in other words, until $\sum_{j=0}^{n} \frac{\langle f, p_j \rangle^2}{\langle p_j, p_j \rangle}$ becomes close to $\langle f, f \rangle$.

So far our analysis was concerned with $f \in C[a, b]$ and the inner product was defined by an integral over $[a, b]$. However, what if we only have discrete function values $f_1, \ldots, f_m$, $m > n$ at pairwise distinct points $x_1, \ldots, x_m$ available?

An inner product can be defined in the following way

$$\langle g, h \rangle := \sum_{j=1}^{m} g(x_j) h(x_j).$$

We then seek $p \in \mathbb{P}_n[x]$ that minimizes $\langle f - p, f - p \rangle$. A straightforward approach is to express p as $\sum_{k=0}^{n} c_k x^k$ and to find optimal values for $c_0, \ldots, c_n$. An alternative approach, however, is to construct orthogonal polynomials $p_0, p_1, \ldots$ with regards to this inner product. Letting $p_0(x) \equiv 1$, we construct

$$p_1(x) = x - \frac{\langle x, p_0 \rangle}{\langle p_0, p_0 \rangle} p_0(x) = x - \frac{1}{m} \sum_{j=1}^{m} x_j.$$

We continue to construct orthogonal polynomials according to

$$p_k(x) = x^k - \sum_{i=0}^{k-1} \frac{\langle x^k, p_i \rangle}{\langle p_i, p_i \rangle} p_i(x) = x^k - \sum_{i=0}^{k-1} \frac{\sum_{j=1}^{m} x_j^k p_i(x_j)}{\sum_{j=1}^{m} p_i(x_j)^2} p_i(x).$$

In the fraction we see the usual inner product of the vector $(p_i(x_1), \ldots, p_i(x_m))^T$ with itself and with the vector $(x_1^k, \ldots, x_m^k)^T$.

Once the orthogonal polynomials are constructed, we find p by

$$p(x) = \sum_{k=0}^{n} \frac{\langle p_k, f \rangle}{\langle p_k p_k \rangle} p_k(x).$$

For each k the work to find p_k is bounded by a multiple of m and thus the complete cost is $O(mn)$. The only difference to the continuous case is that we cannot keep adding terms, since we only have enough data to construct $p_0, p_1, \ldots, p_{m-1}$.

3.6   The Peano Kernel Theorem

In the following we study a general abstract result by Guiseppe Peano, which applies to the error of a wide class of numerical approximations. Suppose we are given an approximation, e.g., to a function, a derivative at a given point, or a quadrature as an approximation to an integral. For $f \in C^{k+1}[a, b]$ let $L(f)$ be the approximation error. For example, when approximating the function this could be the maximum of the absolute difference between the function and the approximation over the interval $[a, b]$. In the case of a quadrature it is the absolute difference between the value given by the quadrature and the value of the integral. Thus L maps the space of functions $C^{k+1}[a, b]$ to $\mathbb{R}$. We assume that L is a *linear functional*, i.e., $L(\alpha f + \beta g) = \alpha L(f) + \beta L(g)$ for all $\alpha, \beta \in \mathbb{R}$. We also assume that the approximation is constructed in such a way that it is correct for all polynomials of degree at most k, i.e., $L(p) = 0$ for all $p \in \mathbb{P}_k[x]$.

At a given point $x \in [a, b]$, $f(x)$ can be written as its *Taylor polynomial with integral remainder term*

$$
\begin{aligned}
f(x) \quad = \quad & f(a) + (x - a)f'(a) + \tfrac{(x-a)^2}{2!}f''(a) + \cdots + \tfrac{(x-a)^k}{k!}f^{(k)}(a) \\
& + \tfrac{1}{k!}\int_a^x (x - \theta)^k f^{(k+1)}(\theta)d\theta.
\end{aligned}
$$

This can be verified by integration by parts. Applying the functional L to both sides of the equation, we obtain

$$
L(f) = \frac{1}{k!}L\left(\int_a^x (x - \theta)^k f^{(k+1)}(\theta)d\theta \right), \qquad x \in [a, b],
$$

since the first terms of the Taylor expansion are polynomials of degree at most k. To make the integration independent of x, we can use the notation

$$
(x - \theta)_+^k := \begin{cases} (x - \theta)^k, & x \geq \theta, \\ 0, & x \leq \theta. \end{cases}
$$

Then

$$
L(f) = \frac{1}{k!}L\left(\int_a^b (x - \theta)_+^k f^{(k+1)}(\theta)d\theta \right).
$$

We now make the important assumption that the order of the integral and functional can be exchanged. For most approximations, calculating L involves differentiation, integration and linear combination of function values. In the case of quadratures (which we will encounter later and which are a form of numerical integration), L is the difference of the quadrature rule which is a linear combination of function values and the integral. Both these operations can be exchanged with the integral.

Definition 3.3 (Peano kernel). *The* Peano kernel K *of* L *is the function defined by*

$$
K(\theta) := L[(x - \theta)_+^k] \ \text{for } x \in [a, b].
$$

Theorem 3.7 (Peano kernel theorem). *Let L be a linear functional such that $L(p) = 0$ for all $p \in \mathbb{P}_k[x]$. Provided that the exchange of L with the integration is valid, then for $f \in C^{k+1}[a, b]$*

$$
L(f) = \frac{1}{k!}\int_a^b K(\theta)f^{(k+1)}(\theta)d\theta.
$$

Theorem 3.8. *If K does not change sign in (a, b), then for $f \in C^{k+1}[a, b]$*

$$L(f) = \frac{1}{k!} \left[\int_a^b K(\theta) d\theta \right] f^{(k+1)}(\xi)$$

for some $\xi \in (a, b)$.

Proof. Without loss of generality K is positive on $[a, b]$. Then

$$L(f) \geq \frac{1}{k!} \int_a^b K(\theta) \min_{x \in [a,b]} f^{(k+1)}(x) d\theta = \frac{1}{k!} \left(\int_a^b K(\theta) d\theta \right) \min_{x \in [a,b]} f^{(k+1)}(x).$$

Similarly we can obtain an upper bound employing the maximum of $f^{(k+1)}$ over $[a, b]$. Hence

$$\min_{x \in [a,b]} f^{(k+1)}(x) \leq \frac{L(f)}{\frac{1}{k!} \int_a^b K(\theta) d\theta} \leq \max_{x \in [a,b]} f^{(k+1)}(x).$$

The result of the theorem follows by the intermediate value theorem. · □

As an example for the application of the Peano kernel theorem we consider the approximation of a derivative by a linear combination of function values. More specifically, $f'(0) = -\frac{3}{2} f(0) + 2f(1) - \frac{1}{2} f(2)$. Hence, $L(f) := f'(0) - [-\frac{3}{2} f(0) + 2f(1) - \frac{1}{2} f(2)]$. It can be easily verified by letting $p(x) = 1, x, x^2$ and using linearity that $L(p) = 0$ for all $p \in \mathbb{P}_2[x]$. Therefore, for $f \in C^3[a, b]$,

$$L(f) = \frac{1}{2} \int_0^2 K(\theta) f'''(\theta) d\theta.$$

The Peano kernel is

$$K(\theta) = L[(x - \theta)_+^2] = 2(0 - \theta)_+ - [-\frac{3}{2}(0 - \theta)_+^2 + 2(1 - \theta)_+^2 - \frac{1}{2}(2 - \theta)_+^2].$$

Using the definition of $(x - \theta)_+^k$, we obtain

$$K(\theta) = \begin{cases} -2\theta + \frac{3}{2}\theta^2 - 2(1 - \theta)^2 + \frac{1}{2}(2 - \theta)^2 \equiv 0 & \theta \leq 0 \\ -2(1 - \theta)^2 + \frac{1}{2}(2 - \theta)^2 = \theta(2 - \frac{3}{2}\theta) & 0 \leq \theta \leq 1 \\ \frac{1}{2}(2 - \theta)^2 & 1 \leq \theta \leq 2 \\ 0 & \theta \geq 2. \end{cases}$$

We see that $K \geq 0$. Moreover,

$$\int_0^2 K(\theta) d\theta = \int_0^1 \theta(2 - \frac{3}{2}\theta) d\theta + \int_1^2 \frac{1}{2}(2 - \theta)^2 d\theta = \frac{2}{3}.$$

Hence $L(f) = \frac{1}{2} \frac{2}{3} f'''(\xi) = \frac{1}{3} f'''(\xi)$ for some $\xi \in (0, 2)$. Thus the error in approximating the derivative at zero is $\frac{1}{3} f'''(\xi)$ for some $\xi \in (0, 2)$.

Exercise 3.9. *Express the divided difference* $f[0, 1, 2, 3]$ *in the form*

$$L(f) = f[0, 1, 2, 3] = \frac{1}{2} \int_0^3 K(\theta) f'''(\theta) d\theta,$$

assuming that $f \in C^3[0, 3]$. *Sketch the kernel function* $K(\theta)$ *for* $\theta \in [0, 3]$. *By integrating* $K(\theta)$ *and using the mean value theorem, show that*

$$f[0, 1, 2, 3] = \frac{1}{6} f'''(\xi)$$

for some point $\xi \in [0, 3]$.

Exercise 3.10. *We approximate the function value of* $f \in C^2[0, 1]$ *at* $\frac{1}{2}$ *by* $p(\frac{1}{2}) = \frac{1}{2}[f(0) + f(1)]$. *Find the least constants* c_0, c_1 *and* c_2 *such that*

$$\left| f\left(\frac{1}{2}\right) - p\left(\frac{1}{2}\right) \right| \leq c_k \|f^{(k)}\|_\infty, \qquad k = 0, 1, 2.$$

For $k = 0, 1$ *work from first principles and for* $k = 2$ *apply the Peano kernel theorem.*

3.7   Splines

The problem with polynomial interpolation is that with increasing degree the polynomial 'wiggles' from data point to data point. Low-order polynomials do not display this behaviour. Let's interpolate data by fitting two cubic polynomials, $p_1(x)$ and $p_2(x)$, to different parts of the data meeting at the point x^*. Each cubic polynomial has four coefficients and thus we have 8 degrees of freedom and hence can fit 8 data points. However, the two polynomial pieces are unlikely to meet at x^*. We need to ensure some continuity. If we let $p_1(x^*) = p_2(x^*)$, then the curve is at least continuous, but we are losing one degree of freedom. The fit

$$\begin{array}{rcl} p_1'(x^*) & = & p_2'(x^*) \\ p_1''(x^*) & = & p_2''(x^*) \end{array}$$

take up one degree of freedom each. This gives a smooth curve, but we are only left with five degrees of freedom and thus can only fit 5 data points. If we also require the third derivative to be continuous, the two cubics become the same cubic which is uniquely specified by the four data points.

The point x^* is called a *knot point*. To fit more data we specify $n+1$ such knots and fit a curve consisting of n separate cubics between them, which is continuous and also has continuity of the first and second derivative. This has $n + 3$ degrees of freedom. This curve is called a *cubic spline*.

The two-dimensional equivalent of the cubic spline is called *thin-plate spline*. A linear combination of thin-plate splines passes through the data points exactly while minimizing the so-called bending energy, which is defined as the integral over the squares of the second derivatives

$$\int \int (f_{xx}^2 + 2f_{xy}^2 + f_{yy}^2) dx dy$$

where $f_x = \frac{\partial f}{\partial x}$. The name thin-plate spline refers to bending a thin sheet of metal being held in place by bolts as in the building of a ship's hull. However, here we will consider splines in one dimension.

Definition 3.4 (Spline function). *The function s is called a spline function of degree k if there exist points $a = x_0 < x_1 < \ldots < x_n = b$ such that s is a polynomial of degree at most k on each of the intervals $[x_{j-1}, x_j]$ for $j = 1, \ldots, n$ and such that s has continuous $k-1$ derivatives. In other words, $s \in C^{k-1}[a, b]$. We call s a linear, quadratic, cubic, or quartic spline for $k = 1, 2, 3,$ or 4. The points $x_0, \ldots, x_n$ are called knots and the points $x_1, \ldots, x_{n-1}$ are called interior knots.*

A spline of degree k can be written as

$$s(x) = \sum_{i=0}^{k} c_i x^i + \sum_{j=1}^{n-1} d_j (x - x_j)_+^k \qquad (3.2)$$

for $x \in [a, b]$. Recall the notation

$$(x - x_j)_+^k := \begin{cases} (x - x_j)^k, & x \geq x_j, \\ 0, & x \leq x_j. \end{cases}$$

introduced for the Peano kernel. The first sum defines a general polynomial on $[x_0, x_1]$ of degree k. Each of the functions $(x - x_j)_+^k$ is a spline itself with only one knot at x_j and continuous $k - 1$ derivatives. These derivatives all vanish at x_j. Thus (3.2) describes all possible spline functions.

Theorem 3.9. *Let S be the set of splines of degree k with fixed knots $x_0, \ldots, x_n$, then S is a linear space of dimension $n + k$.*

Proof. Linearity is implied since differentiation and continuity are linear. The notation in (3.2) implies that S has dimension at most $n + k$. Hence it is sufficient to show that if $s \equiv 0$, then all the coefficients c_i, $i = 0, \ldots, k$, and d_j $j = 1, \ldots, n - 1$, vanish. Considering the interval $[x_0, x_1]$, then $s(x)$ is equal to $\sum_{i=0}^{k} c_i x^i$ on this interval and has an infinite number of zeros. Hence the coefficients c_i, $i = 0, \ldots, k$ have to be zero. To deduce $d_j = 0$, $j = 1, \ldots, n-1$, we consider each interval $[x_j, x_{j+1}]$ in turn. The polynomial there has again infinite zeros from which $d_j = 0$ follows. $\square$

Definition 3.5 (Spline interpolation). *Let $f \in C[a, b]$ be given. The spline interpolant to f is obtained by constructing the spline s that satisfies $s(x_i) = f(x_i)$ for $i = 0, \ldots, n$.*

For $k = 1$ this gives us piecewise linear interpolation. The spline s is a linear (or constant) polynomial between adjacent function values. These conditions define s uniquely.

For $k = 3$ we have cubic spline interpolation. Here the dimension of S is $n + 3$, but we have only $n + 1$ conditions given by the function values. Thus the data cannot define s uniquely.

In the following we show how the cubic spline is constructed and how uniqueness is achieved. If we have not only the values of s, but also of s' available at all the knots, then on $[x_{i-1}, x_i]$ for $i = 1, \ldots, n$ the polynomial piece of s can be written as

$$s(x) = s(x_{i-1}) + s'(x_{i-1})(x - x_{i-1}) + c_2(x - x_{i-1})^2 + c_3(x - x_{i-1})^3. \quad (3.3)$$

The derivative of s is

$$s'(x) = s'(x_{i-1}) + 2c_2(x - x_{i-1}) + 3c_3(x - x_{i-1})^2.$$

Obviously s and s' take the right value at $x = x_{i-1}$. Considering s and s' at $x = x_i$, we get two equations for the coefficients c_2 and c_3. Solving these we obtain

$$
\begin{aligned}
c_2 &= 3\frac{s(x_i) - s(x_{i-1})}{(x_i - x_{i-1})^2} - \frac{2s'(x_{i-1}) + s'(x_i)}{x_i - x_{i-1}}, \\
c_3 &= 2\frac{s(x_{i-1}) - s(x_i)}{(x_i - x_{i-1})^3} + \frac{s'(x_{i-1}) + s'(x_i)}{(x_i - x_{i-1})^2}.
\end{aligned}
\quad (3.4)
$$

The second derivative of s is

$$s''(x) = 2c_2 + 6c_3(x - x_{i-1}).$$

Considering two polynomial pieces in adjacent intervals $[x_{i-1}, x_i]$ and $[x_i, x_{i+1}]$, we have to ensure second derivative continuity at x_i. Calculating $s''(x_i)$ from the polynomial piece on the left interval and on the right interval gives two expressions which must be equal. This implies for $i = 1, \ldots, n-1$ the equation

$$\frac{s'(x_{i-1}) + 2s'(x_i)}{x_i - x_{i-1}} + \frac{2s'(x_i) + s'(x_{i+1})}{x_{i+1} - x_i} = \frac{3s(x_i) - 3s(x_{i-1})}{(x_i - x_{i-1})^2} + \frac{3s(x_{i+1}) - 3s(x_i)}{(x_{i+1} - x_i)^2}. \quad (3.5)$$

Assume we are given the function values $f(x_i)$, $i = 0, \ldots, n$, and the derivatives $f'(a)$ and $f'(b)$ at the endpoints $a = x_0$ and $b = x_n$. Thus we know $s'(x_0)$ and $s'(x_n)$. We seek the cubic spline that interpolates the augmented data. Note that now we have $n + 3$ conditions consistent with the dimension of the space of cubic splines. Equation (3.5) describes a system of $n - 1$ equations in the unknowns $s'(x_i)$, $i = 1, \ldots, n-1$, specified by a tridiagonal matrix S where the diagonal elements are

$$S_{i,i} = \frac{2}{x_i - x_{i-1}} + \frac{2}{x_{i+1} - x_i}$$

and the off-diagonal elements are

$$S_{i,i-1} = \frac{1}{x_i - x_{i-1}} \text{ and } S_{i,i+1} = \frac{1}{x_{i+1} - x_i}.$$

This matrix is nonsingular, since it is diagonally dominant. That means that for each row the absolute value of the diagonal element is larger than the sum of the absolute values of the off-diagonal elements. Here the diagonal and elements on the first subdiagonal and superdiagonal are all positive, since the knots are ordered from smallest to largest. Thus the diagonal elements are twice the sum of the off-diagonal elements. Nonsingularity implies that the system has a unique solution and thus the cubic interpolation spline exists and is unique.

The right-hand side of the system of equations is given by the right-hand side of Equation (3.5) for $i = 2, \ldots, n - 2$ with $s(x_i) = f(x_i)$. For $i = 1$ the right-hand side is

$$\frac{3f(x_1) - 3f(x_0)}{(x_1 - x_0)^2} + \frac{3f(x_2) - 3f(x_1)}{(x_2 - x_1)^2} - \frac{f'(a)}{x_1 - x_0}.$$

For $i = n - 1$ the right-hand side is

$$\frac{3f(x_{n-1}) - 3f(x_{n-2})}{(x_{n-1} - x_{i-2})^2} + \frac{3f(x_n) - 3f(x_{n-1})}{(x_n - x_{n-1})^2} - \frac{f'(b)}{x_n - x_{n-1}}.$$

Once the values $s'(x_i)$, $i = 2, \ldots, n - 2$, are obtained from the system of equations, they can be inserted in (3.4) to calculate c_2 and c_3. We then have all the values to calculate $s(x)$ on $[x_{i-1}, x_i]$ given by Equation (3.3).

So far we have taken up the extra two degrees of freedom by requiring $s'(a) = f'(a)$ and $s'(b) = f'(b)$. In the following we examine how to take up the extra degrees of freedom further.

Let S be the space of cubic splines with knots $a = x_0 < x_1 < \ldots < x_n = b$ and let $\hat{s}$ and $\check{s}$ be two different cubic splines that satisfy $\hat{s}(x_i) = \check{s}(x_i) = f(x_i)$. The difference $\hat{s} - \check{s}$ is a non-zero element of S that is zero at all the knots. These elements of S form a subspace of S, say S_0. How to take up the extra degree of freedom is equivalent to which element from S_0 to choose.

To simplify the argument we let the knots be equally spaced, i.e., $x_i - x_{i-1} = h$ for $i = 1, \ldots, n$. Equation (3.5) then becomes

$$s'(x_{i-1}) + 4s'(x_i) + s'(x_{i+1}) = \frac{3}{h}[s(x_{i+1}) - s(x_{i-1})] \qquad (3.6)$$

for $i = 1 \ldots, n - 1$. For $s \in S_0$, the right-hand side is zero. Thus (3.6) is a recurrence relation where the solutions are given by $\alpha \lambda_1^i + \beta \lambda_2^i$, where λ_1, λ_2 are the roots of $\lambda^2 + 4\lambda + 1 = 0$. These are

$$\lambda_1 = \sqrt{3} - 2$$
$$\lambda_2 = -\sqrt{3} - 2 = -\frac{(\sqrt{3} + 2)(\sqrt{3} - 2)}{\sqrt{3} - 2} = 1/\lambda_1.$$

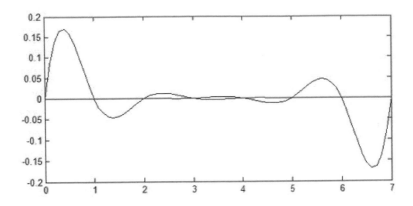

Figure 3.4 Basis of the subspace S_0 with 8 knots

If the coefficients α and β are chosen as $\alpha = 1, \beta = 0$ or $\alpha = 0, \beta = (\sqrt{3} - 2)^n$ we obtain two splines with values of the derivative at $x_0, \ldots, x_n$ given by

$$s'(x_i) = (\sqrt{3} - 2)^i \text{ and } s'(x_i) = (\sqrt{3} - 2)^{n-i} \qquad i = 0, \ldots, n.$$

These two splines are a convenient basis of S_0. One decays rapidly across the interval $[a, b]$ when moving from left to right, while the other decays rapidly when moving from right to left as can be seen in Figure 3.4. This basis implies that, for equally spaced knots, the freedom in a cubic spline interpolant is greatest near the end points of the interval $[a, b]$. Therefore it is important to take up this freedom by imposing an extra condition at each end of the interval as we have done by letting $s'(a) = f'(a)$ and $s'(b) = f'(b)$. The following exercise illustrates how the solution deteriorates if this is not done.

Exercise 3.11. *Let S be the set of cubic splines with knots $x_i = ih$ for $i = 0, \ldots, n$, where $h = 1/n$. An inexperienced user obtains an approximation to a twice-differentiable function f by satisfying the conditions $s'(0) = f'(0)$, $s''(0) = f''(0)$, and $s(x_i) = f(x_i)$, $i = 0, \ldots, n$. Show how the changes in the first derivatives $s'(x_i)$ propagate if $s'(0)$ is increased by a small perturbation ϵ, i.e., $s'(0) = f'(0) + \epsilon$, but the remaining data remain the same.*

Another possibility to take up this freedom is known as the *not-a-knot technique*. Here we require the third derivative s''' to be continuous at x_1 and x_{n-1}. It is called not-a-knot since there is no break between the two polynomial pieces at these points. Hence you can think of these knots as not being knots at all.

Definition 3.6 (Lagrange form of spline interpolation). *For $j = 0, \ldots, n$, let s_j be an element of S that satisfies*

$$s_j(x_i) = \delta_{ij}, \qquad i = 0, \ldots, n, \qquad\qquad (3.7)$$

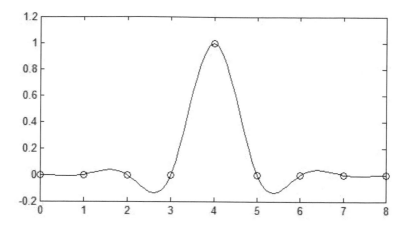

Figure 3.5 Example of a Lagrange cubic spline with 9 knots

where δ_{ij} is the Kronecker delta (i.e., $\delta_{jj} = 1$ and $\delta ij = 0$ for $i \neq j$). Figure 3.5 shows an example of a Lagrange cubic spline. Note that the splines s_j are not unique, since any element from S_0 can be added. Any spline interpolant to the data $f_0, \ldots, f_n$ can then be written as

$$s(x) = \sum_{j=0}^{n} f_j s_j(x) + \hat{s}(x),$$

where $\hat{s} \in S_0$. This expression is the Lagrange form of spline interpolation.

Theorem 3.10. Let S be the space of cubic splines with equally spaced knots $a = x_0 < x_1 < \ldots < x_n = b$ where $x_i - x_{i-1} = h$ for $i = 1, \ldots, n$. Then for each integer $j = 0, \ldots, n$ there is a cubic spline s_j that satisfies the Lagrange conditions (3.7) and that has the first derivative values

$$s_j'(x_i) = \begin{cases} -\frac{3}{h}(\sqrt{3} - 2)^{j-i}, & i = 0, \ldots, j-1, \\ 0, & i = j \\ \frac{3}{h}(\sqrt{3} - 2)^{i-j}, & i = j+1, \ldots, n. \end{cases}$$

Proof. We know that to achieve second derivative continuity, the derivatives have to satisfy (3.6). For s_j this means

$$s_j'(x_{i-1}) + 4s_j'(x_i) + s_j'(x_{i+1}) = 0 \qquad i = 1, \ldots, n-1, i \neq j-1, j+1$$

$$s_j'(x_{j-2}) + 4s_j'(x_{j-1}) + s_j'(x_j) = \frac{3}{h} \qquad i = j-1,$$

$$s_j'(x_j) + 4s_j'(x_{j+1}) + s_j'(x_{j+2}) = -\frac{3}{h} \qquad i = j+1.$$

It is easily verified that the values for $s_j'(x_i)$ given in the theorem satisfy these. For example, considering the last two equations, we have

$$
\begin{aligned}
s_j'(x_{j-2}) + 4s_j'(x_{j-1}) + s_j'(x_j) &= -\frac{3}{h}(\sqrt{3}-2)^2 - 4\frac{3}{h}(\sqrt{3}-2) \\
&= -\frac{3}{h}(7 - 4\sqrt{3} + 4\sqrt{3} - 8] = \frac{3}{h}
\end{aligned}
$$

and

$$
\begin{aligned}
s_j'(x_j) + 4s_j'(x_{j+1}) + s_j'(x_{j+2}) &= 4\frac{3}{h}(\sqrt{3}-2) + \frac{3}{h}(\sqrt{3}-2)^2 \\
&= \frac{3}{h}(-8 + 4\sqrt{3} - 4\sqrt{3} + 7) = -\frac{3}{h}.
\end{aligned}
$$

$\square$

Since $\sqrt{3} - 2 \approx -0.268$, it can be deduced that $s_j(x)$ decays rapidly as $|x - x_j|$ increases. Thus the contribution of f_j to the cubic spline interpolant s decays rapidly as $|x - x_j|$ increases. Hence for $x \in [a, b]$, the value of $s(x)$ is determined mainly by the data f_j for which $|x - x_j|$ is relatively small.

However, the Lagrange functions of quadratic spline interpolation do not enjoy these decay properties if the knots coincide with the interpolation points. Generally, on the interval $[x_i, x_{i+1}]$, the quadratic polynomial piece can be derived from the fact that the derivative is a linear polynomial and thus given by

$$
s'(x) = \frac{s'(x_{i+1}) - s'(x_i)}{x_{i+1} - x_i}(x - x_i) + s'(x_i).
$$

Integrating over x and using $s(x_i) = f_i$ we get

$$
s(x) = \frac{s'(x_{i+1}) - s'(x_i)}{2(x_{i+1} - x_i)}(x - x_i)^2 + s'(x_i)(x - x_i) + f_i.
$$

Using $s(x_{i+1}) = f_{i+1}$ we can solve for $s'(x_{i+1})$

$$
s'(x_{i+1}) = 2\frac{f_{i+1} - f_i}{x_{i+1} - x_i} - s'(x_i),
$$

which is a recurrence relation. The extra degree of freedom in quadratic splines can be taken up by letting $s'(x_0) = 0$ which gives the natural quadratic spline. Another possibility is letting $s'(x_0) = s'(x_1)$, then $s'(x_0) = \frac{f_2 - f_1}{x_2 - x_1}$.

For the Lagrange functions s_j, $j = 1, \ldots, n$, of quadratic spline interpolation the recurrence relation implies $s_j'(x_{i+1}) = -s_j'(x_i)$ for $i = 1, \ldots, j - 2$ and $i = j + 1, \ldots, n$. Thus the function will continue to oscillate. This can be seen in Figure 3.6.

Exercise 3.12. *Another strategy to use quadratic splines to interpolate equally spaced function values $f(jh)$, $j = 0, \ldots, n$, is to let s have the interior knots*

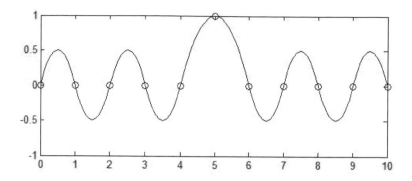

Figure 3.6 Lagrange function for quadratic spline interpolation

$(j + \frac{1}{2})h$, $j = 1, \dots, n - 2$. Verify that the values $s((j + \frac{1}{2})h)$, $j = 1, \dots, n - 2$, and the interpolation conditions $s(jh) = f(jh)$, $j = 0, \dots, n$, define the quadratic polynomial pieces of s. Further, prove that the continuity of s' at the interior knots implies the equations

$$s((j-\frac{1}{2})h)+6s((j+\frac{1}{2})h)+s((j+\frac{3}{2})h) = 4[f(jh)+f(jh+h)], \quad j = 2, \dots, n-3.$$

3.8 B-Spline

In this section we generalize the concept of splines.

Definition 3.7 (B-splines). *Let S be the linear space of splines of degree k that have the knots $a = x_0 < x_1 < \dots < x_n = b$, and let $0 \le p \le n - k - 1$ be an integer. The spline*

$$s(x) = \sum_{j=p}^{p+k+1} \lambda_j (x - x_j)_+^k, \qquad x \in \mathbb{R} \tag{3.8}$$

is called a B-spline if $s(x) \equiv 0$ for $x \le x_p$ and $s(x) \equiv 0$ for $x \ge x_{p+k+1}$, but the real coefficients λ_j, $j = p, \dots, p + k + 1$ do not vanish.

Recall that the notation $(\cdot)_+$ means that the expression is zero, if the term in the brackets becomes negative. Thus the spline defined by (3.8) satisfies $s(x) \equiv 0$ for $x \le x_p$ as required. For $x \ge x_{p+k+1}$, $s(x)$ is given by

$$s(x) = \sum_{j=p}^{p+k+1} \lambda_j (x - x_j)^k = \sum_{j=p}^{p+k+1} \lambda_j \sum_{i=0}^{k} \binom{k}{i} x^{k-i} (-x_j)^i.$$

This being identical to 0, means that there are infinitely many zeros, and this in turn means that the coefficients of x^{k-i} have to vanish for all $i = 0, \dots, k$.

Therefore

$$\sum_{j=p}^{p+k+1} \lambda_j x_j^i = 0, \qquad i = 0, \dots, k.$$

A solution to this problem exists, since $k + 2$ coefficients have to satisfy only $k + 1$ conditions. The matrix describing the system of equations is

$$A = \begin{pmatrix} 1 & \cdots & \cdots & 1 \\ x_p & x_{p+1} & \cdots & x_{p+k+1} \\ \vdots & \vdots & \ddots & \vdots \\ x_p^k & x_{p+1}^k & \cdots & x_{p+k+1}^k \end{pmatrix}.$$

If λ_{p+k+1} is zero, then the system reduces to a $(k + 1) \times (k + 1)$ matrix where the last column of A is removed. This however is a *Vandermonde matrix*, which is non-singular, since all x_i, $i = p \dots, p + k + 1$ are distinct. It follows then that all the other coefficients $\lambda_p, \dots, \lambda_{p+k}$ are also zero.

Therefore λ_{p+k+1} has to be nonzero. This can be chosen and the system can be solved uniquely for the remaining $k+1$ coefficients, since A with the last column removed is nonsingular. Therefore, apart from scaling by a constant, the B-spline of degree k with knots $x_p < x_{p+1} < \dots < x_{p+k+1}$ that vanishes outside (x_p, x_{p+k+1}) is unique.

Theorem 3.11. *Apart from scaling, the coefficients λ_j, $j = p, \dots, p + k + 1$, in (3.8) are given by*

$$\lambda_j = \left[\prod_{i=p, i \neq j}^{p+k+1} (x_j - x_i) \right]^{-1}.$$

Therefore the B-spline is

$$B_p^k(x) = \sum_{j=p}^{p+k+1} \left[\prod_{i=p, i \neq j}^{p+k+1} (x_j - x_i) \right]^{-1} (x - x_j)_+^k, \qquad x \in \mathbb{R}. \qquad (3.9)$$

Proof. The function $B_p^k(x)$ is a polynomial of degree at most k for $x \geq x_{p+k+1}$

$$\begin{aligned} B_p^k(x) &= \sum_{j=p}^{p+k+1} \left[\prod_{i=p, i \neq j}^{p+k+1} (x_j - x_i) \right]^{-1} \sum_{l=0}^{k} \binom{k}{l} x^{k-l}(-x_j)^l \\ &= \sum_{l=0}^{k} \binom{k}{l} (-1)^l \left(\sum_{j=p}^{p+k+1} \left[\prod_{i=p, i \neq j}^{p+k+1} (x_j - x_i) \right]^{-1} x_j^l \right) x^{k-l}. \end{aligned}$$

On the other hand, let $0 \leq l \leq k$ be an integer. We use the Lagrange interpolation formula to interpolate x^l, at the points $x_p, \dots, x_{p+k+1}$

$$x^l = \sum_{j=p}^{p+k+1} \left[\prod_{i=p, i \neq j}^{p+k+1} \frac{x - x_i}{x_j - x_i} \right] x_j^l.$$

Comparing the coefficient of x^{k+1} on both sides of the equation gives

$$0 = \sum_{j=p}^{p+k+1} \left[\prod_{i=p, i \neq j}^{p+k+1} \frac{1}{x_j - x_i} \right] x_j^l \text{ for } l = 0, \ldots, k.$$

The right-hand side is a factor of the coefficient of x^{k-l} in $B_p^k(x)$ for $x \geq x_{p+k+1}$. It follows that the coefficient of x^{k-l} in $B_p^k(x)$, $x \geq x_{p+k+1}$, is zero for $l = 0, 1, \ldots, k$. Thus $B_p^k(x)$ vanishes for $x \geq x_{p+k+1}$ and is the required B-spline. □

The advantage of B-splines is that the nonzero part is confined to an interval which contains only $k + 2$ consecutive knots. This is also known as the spline having *finite support*.

As an example we consider the $(n + 1)$-dimensional space of linear splines with the usual knots $a = x_0 < \ldots < x_n = b$. We introduce extra knots x_{-1} and x_{n+1} outside the interval and we let B_{j-1}^1 be the linear spline which satisfies the conditions $B_{j-1}^1(x_i) = \delta_{ij}$, $i = 0, \ldots, n$, where δ_{ij} denotes the *Kronecker delta*. Then every s in the space of linear splines can be written as

$$s(x) = \sum_{j=0}^{n} s(x_j) B_{j-1}^1(x), \qquad a \leq x \leq b.$$

These basis functions are often called *hat functions* because of their shape, as Figure 3.7 shows.

For general k we can generate a set of B-splines for the $(n+k)$- dimensional space S of splines of degree k with the knots $a = x_0 < x_1 < \ldots < x_n = b$. We add k additional knots both to the left of a and to the right of b. Thus the full list of knots is $x_{-k} < x_{-k+1} < \ldots < x_{n+k}$. We let B_p^k be the function as defined in (3.9) for $p = -k, -k+1, \ldots, n-1$, where we restrict the range of x to the interval $[a, b]$ instead of $x \in \mathbb{R}$. Therefore for each $p = -k, -k+1, \ldots, n-1$ the function $B_p^k(x)$, $a \leq x \leq b$, lies in S and vanishes outside the interval (x_p, x_{p+k+1}). Figure 3.8 shows these splines for $k = 3$.

Theorem 3.12. *The B-splines B_p^k, $p = -k, -k + 1, \ldots, n - 1$, form a basis of S.*

Proof. The number of B-splines is $n + k$ and this equals the dimension of S. To show that the B-splines form a basis, it is sufficient to show that a nontrivial linear combination of them cannot vanish identically in the interval $[a, b]$. Assume otherwise and let

$$s(x) = \sum_{j=-k}^{n-1} s_j B_j^k(x), \qquad x \in \mathbb{R},$$

where $s(x) \equiv 0$ for $a \leq x \leq b$. We know that $s(x)$ also has to be zero for

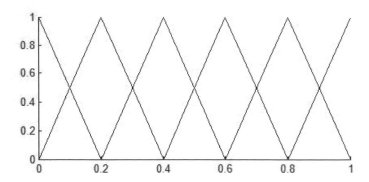

Figure 3.7 Hat functions generated by linear B-splines on equally spaced nodes

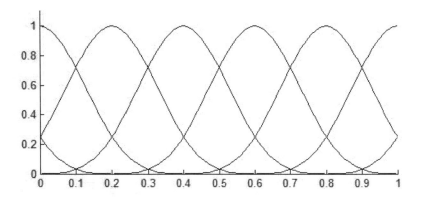

Figure 3.8 Basis of cubic B-splines

$x \leq x_{-k}$ and $x \geq x_{n+k}$. Considering $x \leq b$, then $s(x)$ is a B-spline which is identical to zero outside the interval (x_{-k}, x_0). However, this interval only contains $k + 1$ knots and we have seen that it is not possible to construct a B-spline with fewer than $k + 2$ knots. Thus $s(x)$ does not only vanish for $x \leq x_{-k}$ but also on the intervals $[x_j, x_{j+1}]$ for $j = -k, \ldots, n - 1$. By considering these in sequence we can deduce that $s_j = 0$ for $j = -k, \ldots, n - 1$ and the linear combination is the trivial linear combination. □

To use B-splines for interpolation at general points, let the distinct points $\xi_j \in [a, b]$ and data $f(\xi_j)$ be given for $j = 1, \ldots, n + k$. If s is expressed as

$$s(x) = \sum_{j=-k}^{n-1} s_j B_j^k(x), \qquad x \in [a, b],$$

then the coefficients s_j, $j = -k, \ldots, n-1$ have to satisfy the $(n+k) \times (n+k)$ system of equations given by

$$\sum_{j=-k}^{n-1} s_j B_j^k(\xi_i) = f(\xi_i), \qquad i = 1, \ldots, n+k.$$

The advantage of this form is that $B_j^k(\xi_i)$ is nonzero if and only if ξ_i is in the interval (x_j, x_{j+k+1}). Therefore there are at most $k+1$ nonzero elements in each row of the matrix. Thus the matrix is sparse.

The explicit form of a B-spline given in (3.9) is not very suitable for evaluation if x is close to x_{p+k+1}, since all $(\cdot)_+$ terms will be nonzero. However, a different representation of B-splines can be used for evaluation purposes. The following definition is motivated by formula (3.9) for $k = 0$.

Definition 3.8. *The* B-spline of degree 0 B_p^0 *is defined as*

$$B_p^0(x) = \begin{cases} 0, & x < x_p \text{ or } x > x_{p+1}, \\ (x_{p+1} - x_p)^{-1}, & x_p < x < x_{p+1}, \\ \frac{1}{2}(x_{p+1} - x_p)^{-1}, & x = x_p \text{ or } x = x_{p+1}. \end{cases}$$

Theorem 3.13 (B-spline recurrence relation). *For $k \geq 1$ the B-splines satisfy the following recurrence relation*

$$B_p^k(x) = \frac{(x - x_p)B_p^{k-1}(x) + (x_{p+k+1} - x)B_{p+1}^{k-1}(x)}{x_{p+k+1} - x_p}, \qquad x \in \mathbb{R}. \qquad (3.10)$$

Proof. In the proof one needs to show that the resulting function has the required properties to be a B-spline and this is left to the reader. $\qquad \square$

Let $s \in S$ be a linear combination of B-splines. If s needs to be evaluated for $x \in [a, b]$, we pick j between 0 and $n-1$ such that $x \in [x_j, x_{j+1}]$. It is only necessary to calculate $B_p^k(x)$ for $p = j - k, j - k + 1, \ldots, j$, since the other B-splines of order k vanish on this interval. The calculation starts with the one nonzero value of $B_p^0(x)$, $p = 0, 1, \ldots, n$. (It could be two nonzero values if x happens to coincide with a knot.) Then for $l = 1, 2, \ldots, k$ the values $B_p^l(x)$ for $p = j - l, j - l + 1, \ldots, j$ are generated by the recurrence relation given in (3.10), keeping in mind that $B_{j-l-1}^l(x)$ and $B_{j+1}^l(x)$ are zero. Note that as the order of the B-splines increases, the number of B-splines not zero on the interval $[x_j, x_{j+1}]$ increases.

$$B_j^0(x) \nearrow \; B_{j-1}^1(x) \; \nearrow \; B_{j-2}^2(x) \;\cdots\; B_{j-k}^k(x)$$
$$\searrow \; B_j^1(x) \; \searrow \; B_{j-1}^2(x) \;\cdots$$
$$\searrow \; B_j^2(x) \;\cdots\; B_j^k(x)$$

This means that s can be evaluated at $x \in [a, b]$ in $O(k^2)$ operations.

3.9 Revision Exercises

Exercise 3.13. *(a) Let Q_k, $k = 0, 1, \ldots$, be a set of polynomials orthogonal with respect to some inner product $\langle \cdot, \cdot \rangle$ in the interval $[a, b]$. Let f be a continuous function in $[a, b]$. Write explicitly the least-squares polynomial approximation to f by a polynomial of degree n in terms of the polynomials Q_k, $k = 0, 1, \ldots$.*

(b) Let an inner product be defined by the formula

$$\langle g, h \rangle = \int_{-1}^{1} (1 - x^2)^{-1/2} g(x) h(x) dx.$$

The orthogonal polynomials are the Chebyshev polynomials of the first kind given by $Q_k(x) = \cos(k \arccos x)$, $k \geq 0$. Using the substitution $x = \cos \theta$, calculate the inner products $\langle Q_k, Q_k \rangle$ for $k \geq 0$. (Hint: $2 \cos^2 x = 1 + \cos 2x$.)

(c) For the inner product given above and the Chebyshev polynomials, calculate the inner products $\langle Q_k, f \rangle$ for $k \geq 0$, $k \neq 1$, where f is given by $f(x) = (1 - x^2)^{1/2}$. (Hint: $\cos x \sin y = \frac{1}{2}[\sin(x + y) - \sin(x - y)]$.)

(d) Now for $k = 1$, calculate the inner product $\langle Q_1, f \rangle$.

(e) Thus for even n write the least squares polynomial approximation to f as linear combination of the Chebyshev polynomials with the correct coefficients.

Exercise 3.14. *(a) Given a set of real values $f_0, f_1, \ldots, f_n$ at real data points $x_0, x_1, \ldots, x_n$, give a formula for the Lagrange cardinal polynomials and state their properties. Write the polynomial interpolant in the Lagrange form.*

(b) Define the divided difference of degree n, $f[x_0, x_1, \ldots, x_n]$, and give a formula for it derived from the Lagrange form of the interpolant. What is the divided difference of degree zero?

(c) State the recursive formula for the divided differences and proof it.

(d) State the Newton form of polynomial interpolation and show how it can be evaluated in just $O(n)$ operations.

(e) Let $x_0 = 4, x_1 = 6, x_2 = 8$, and $x_3 = 10$ with data values $f_0 = 1, f_1 = 3, f_2 = 8$, and $f_3 = 20$. Give the Lagrange form of the polynomial interpolant.

(f) Calculate the divided difference table for $x_0 = 4, x_1 = 6, x_2 = 8$, and $x_3 = 10$ with data values $f_0 = 1, f_1 = 3, f_2 = 8$, and $f_3 = 20$.

(g) Thus give the Newton form of the polynomial interpolant.

Exercise 3.15. (a) Define the divided difference of degree n, $f[x_0, x_1, \ldots, x_n]$. What is the divided difference of degree zero?

(b) Prove the recursive formula for divided differences

$$f[x_0, x_1, \ldots, x_k, x_{n+1}] = \frac{f[x_1, \ldots, x_{n+1}] - f[x_0, \ldots, x_n]}{x_{n+1} - x_0}.$$

(c) By considering the polynomials $p, q \in \mathbb{P}_k[x]$ that interpolate f at $x_0, \ldots, x_{i-1}, x_{i+1}, \ldots, x_n$ and $x_0, \ldots, x_{j-1}, x_{j+1}, \ldots, x_n$, respectively, where $i \neq j$, construct a polynomial r, which interpolates f at $x_0, \ldots, x_n$. For the constructed r show that $r(x_k) = f(x_k)$ for $k = 0, \ldots, n$.

(d) Deduce that, for any $i \neq j$, we have

$$f[x_0, \ldots, x_n] = \frac{f[x_0, \ldots, x_{i-1}, x_{i+1}, \ldots, x_n] - f[x_0, \ldots, x_{j-1}, x_{j+1}, \ldots, x_n]}{x_j - x_i}.$$

(e) Calculate the divided difference table for $x_0 = 0, x_1 = 1, x_2 = 2$, and $x_3 = 3$ with data values $f_0 = 0, f_1 = 1, f_2 = 8$, and $f_3 = 27$.

(f) Using the above formula, calculate the divided differences $f[x_0, x_2]$, $f[x_0, x_2, x_3]$, and $f[x_0, x_1, x_3]$.

Exercise 3.16. (a) Give the definition of a spline function of degree k.

(b) Proof that the set of splines of degree k with fixed knots $x_0, \ldots, x_n$ is a linear space of dimension $n + k$.

(c) Turning to cubic splines. Let $s(x)$, $1 \leq x < \infty$, be a cubic spline that has the knots $x_j = \mu^j : j = 0, 1, 2, 3, \ldots$, where μ is a constant greater than 1. Prove that if s is zero at every knot, then its first derivatives satisfy the recurrence relation

$$\mu s'(x_{j-1}) + 2(\mu + 1)s'(x_j) + s'(x_{j+1}) = 0, \qquad j = 1, 2, 3, \ldots.$$

(d) Using

$$s(x) = s(x_{i-1}) + s'(x_{i-1})(x - x_{i-1}) + c_2(x - x_{i-1})^2 + c_3(x - x_{i-1})^3.$$

on the interval $[x_{i-1}, x_i]$, where

$$c_2 = 3\frac{s(x_i) - s(x_{i-1})}{(x_i - x_{i-1})^2} - \frac{2s'(x_{i-1}) + s'(x_i)}{x_i - x_{i-1}},$$

$$c_3 = 2\frac{s(x_{i-1}) - s(x_i)}{(x_i - x_{i-1})^3} + \frac{s'(x_{i-1}) + s'(x_i)}{(x_i - x_{i-1})^2},$$

deduce that s can be nonzero with a bounded first derivative.

(e) *Further show that such an s is bounded, if μ is at most $\frac{1}{2}(3 + \sqrt{5})$.*

Exercise 3.17. (a) *Given a set of real values $f_0, f_1, \ldots, f_n$ at real data points $x_0, x_1, \ldots, x_n$, give a formula for the Lagrange cardinal polynomials and state their properties. Write the polynomial interpolant in the Lagrange form.*

(b) *How many operations are necessary to evaluate the polynomial interpolant in the Lagrange form at x?*

(c) *Prove that the polynomial interpolant is unique.*

(d) *Using the Lagrange form of interpolation, compute the polynomial $p(x)$ that interpolates the data $x_0 = 0$, $x_1 = 1$, $x_2 = 2$ and $f_0 = 1$, $f_1 = 2$, $f_2 = 3$. What is the degree of $p(x)$?*

(e) *What is a divided difference and a divided difference table and for which form of interpolant is it used? Give the formula for the interpolant, how many operations are necessary to evaluate the polynomial in this form?*

(f) *Prove the relation used in a divided difference table.*

(g) *Write down the divided difference table for the interpolation problem given in (d). How does it change with the additional data $f_3 = 5$ at $x_3 = 3$?*

CHAPTER 4

Non-Linear Systems

4.1 Bisection, Regula Falsi, and Secant Method

We consider the solution of the equation $f(x) = 0$ for a suitably smooth function $f : \mathbb{R} \to \mathbb{R}$, i.e., we want to find a root of the function f. Any non-linear system can be expressed in this form.

If for a given interval $[a, b]$ $f(a)$ and $f(b)$ have opposite signs, then f must have at least one zero in the interval by the intermediate value theorem, if f is continuous. The *method of bisection* can be used to find the zero. It is *robust*, i.e., it is guaranteed to converge although at a possibly slow rate. It is also known as *binary search method*.

We repeatedly bisect the interval and select the interval in which the root must lie. At each step we calculate the midpoint $m = (a+b)/2$ and the function value $f(m)$. Unless m is itself a root (improbable, but not impossible), there are two cases: If $f(a)$ and $f(m)$ have opposite signs, then the method sets m as the new value for b. Otherwise if $f(m)$ and $f(b)$ have opposite signs, then the method sets m as the new a. The algorithm terminates, when $b - a$ is sufficiently small.

Suppose the calculation is performed in binary. In every step the width of the interval containing a zero is reduced by 50% and thus the distance of the end points to the zero is also at least halved. Therefore at worst the method will add one binary digit of accuracy in each step. So the iterations are linearly convergent.

The bisection method always chooses the mid-point of the current interval. It can be improved by considering the straight line between $f(a)$ and $f(b)$ which is given by

$$\frac{x - b}{a - b} f(a) + \frac{x - a}{b - a} f(b)$$

and where it crosses the axis. Then m is calculated by

$$m = \frac{f(b)a - f(a)b}{f(b) - f(a)}$$

and the interval containing the root is updated as before. The method is

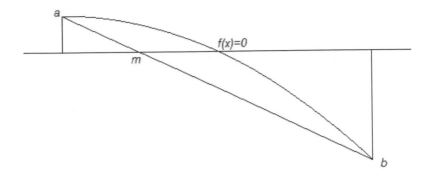

Figure 4.1 The rule of false position

illustrated in Figure 4.1. Loss of significance is unlikely since $f(a)$ and $f(b)$ have opposite signs. This method is sometimes called *regula falsi* or *rule of false position*, since we take a guess as to the position of the root.

At first glance it seems superior to the bisection method, since an approximation to the root is used. However, asymptotically one of the end-points will converge to the root, while the other one remains fixed. Thus only one end-point of the interval gets updated. We illustrate this behaviour with the function

$$f(x) = 2x^3 - 4x^2 + 3x,$$

which has a root for $x = 0$. We start with the initial interval $[-1, 1]$. The left end-point -1 is never replaced while the right end-point moves towards zero. Thus the length of the interval is always at least 1. In the first iteration the right end-point becomes $3/4$ and in the next iteration it is $159/233$. It converges to zero at a linear rate similar to the bisection method.

The value m is a weighted average of the function values. The method can be modified by adjusting the weights of the function values in the case where the same endpoint is retained twice in a row. The value m is then for example calculated according to

$$m = \frac{\frac{1}{2}f(b)a - f(a)b}{\frac{1}{2}f(b) - f(a)} \quad \text{or} \quad \frac{f(b)a - \frac{1}{2}f(a)b}{f(b) - \frac{1}{2}f(a)}.$$

This adjustment guarantees superlinear convergence. It and similar modifications to the regula falsi method are known as the *Illinois algorithm*. For more information on this modification of the regula falsi method see for example [5] G. Dahlquist, Â. Björck, *Numerical Methods*.

So far we always chose the interval containing the root, but a further development is to abandon the use of intervals. The *secant method* always retains the last two values instead of making sure to keep one point on either side of the root. Initializing $x^{(0)} = a$ and $x^{(1)} = b$ the method calculates

$$
\begin{aligned}
x^{(n+1)} &= \frac{f(x^{(n)})x^{(n-1)} - f(x^{(n-1)})x^{(n)}}{f(x^{(n)}) - f(x^{(n-1)})} \\
&= x^{(n)} - f(x^{(n)})\frac{x^{(n)} - x^{(n-1)}}{f(x^{(n)}) - f(x^{(n-1)})}.
\end{aligned}
\tag{4.1}
$$

This is the intersection of the secant through the points $(x^{(n-1)}, f(x^{(n-1)}))$ and $(x^{(n)}, f(x^{(n)}))$ with the axis. There is now the possibility of loss of significance, since $f(x^{(n-1)})$ and $f(x^{(n)})$ can have the same sign. Hence the denominator can become very small leading to large values, possibly leading away from the root.

Also, convergence may be lost, since the root is not bracketed by an interval anymore. If, for example, f is differentiable and the derivative vanishes at some point in the initial interval, the algorithm may not converge, since the secant can become close to horizontal and the intersection point will be far away from the root.

To analyze the properties of the secant method further, in particular to find the order of convergence, we assume that f is twice differentiable and that the derivative of f is bounded away from zero in a neighbourhood of the root. Let x^* be the root and let $e^{(n)}$ denote the error $e^{(n)} = x^{(n)} - x^*$ at the nth iteration. We can then express the error at the $(n+1)^{\text{th}}$ iteration in terms of the error in the previous two iterations:

$$
\begin{aligned}
e^{(n+1)} &= e^{(n)} - f(x^* + e^{(n)})\frac{e^{(n)} - e^{(n-1)}}{f(x^* + e^{(n)}) - f(x^* + e^{(n-1)})} \\
&= \frac{f(x^* + e^{(n)})e^{(n-1)} - f(x^* + e^{(n-1)})e^{(n)}}{f(x^* + e^{(n)}) - f(x^* + e^{(n-1)})} \\
&= e^{(n)}e^{(n-1)}\frac{\left(\frac{f(x^* + e^{(n)})}{e^{(n)}} - \frac{f(x^* + e^{(n-1)})}{e^{(n-1)}}\right)}{f(x^* + e^{(n)}) - f(x^* + e^{(n-1)})}.
\end{aligned}
$$

Using the Taylor expansion of $f(x^* + e^{(n)})$ we can write

$$
\frac{f(x^* + e^{(n)})}{e^{(n)}} = f'(x^*) + \frac{1}{2}f''(x^*)e^{(n)} + O([e^{(n)}]^2).
$$

Doing the same for $f(x^* + e^{(n-1)})/e^{(n-1)}$ and assuming that $x^{(n)}$ and $x^{(n-1)}$ are close enough to the root such that the terms $O([e^{(n)}]^2)$ and $O([e^{(n-1)}]^2)$ can be neglected, the expression for $e^{(n+1)}$ becomes

$$
\begin{aligned}
e^{(n+1)} &\approx e^{(n)}e^{(n-1)}\frac{1}{2}f''(x^*)\frac{e^{(n)} - e^{(n-1)}}{f(x^* + e^{(n)}) - f(x^* + e^{(n-1)})} \\
&\approx e^{(n)}e^{(n-1)}\frac{f''(x^*)}{2f'(x^*)},
\end{aligned}
$$

where we again used the Taylor expansion in the last step. Letting $C = |\frac{f''(x^*)}{2f'(x^*)}|$, the modulus of the error in the next iteration is then approximately

$$|e^{(n+1)}| \approx C|e^{(n)}||e^{(n-1)}|$$

or in other words $|e^{(n+1)}| = O(|e^{(n)}||e^{(n-1)}|)$. By definition the method is of order p if $|e^{(n+1)}| = O(|e^{(n)}|^p)$. We can write this also as $|e^{(n+1)}| = c|e^{(n)}|^p$ for some constant c which is equivalent to

$$\frac{|e^{(n+1)}|}{|e^{(n)}|^p} = c = O(1).$$

The O-notation helps us avoid noting down the constants C and c.

On the other hand,

$$\frac{|e^{(n+1)}|}{|e^{(n)}|^p} = \frac{O(|e^{(n)}||e^{(n-1)}|)}{|e^{(n)}|^p} = O(|e^{(n)}|^{1-p}|e^{(n-1)}|).$$

Using the fact that also $e^{(n)} = O(|e^{(n-1)}|^p)$, it follows that

$$O(|e^{(n-1)}|^{p-p^2+1}) = O(1).$$

This is only possible if $p - p^2 + 1 = 0$. The positive solution of this quadratic is $p = (\sqrt{5}+1)/2 \approx 1.618$. Thus we have calculated the order of convergence and it is superlinear.

4.2   Newton's Method

Looking closer at Equation (4.1), $f(x^{(n)})$ is divided by $\frac{f(x^{(n)})-f(x^{(n-1)})}{x^{(n)}-x^{(n-1)}}$. Taking the theoretical limit $x^{(n-1)} \to x^{(n)}$ this becomes the derivative $f'(x^{(n)})$ at $x^{(n)}$. This is known as *Newton's* or *Newton–Raphson method*,

$$x^{(n+1)} = x^{(n)} - \frac{f(x^{(n)})}{f'(x^{(n)})}. \tag{4.2}$$

Geometrically the secant through two points of a curve becomes the tangent to the curve in the limit of the points coinciding. The tangent to the curve f at the point $(x^{(n)}, f(x^{(n)}))$ has the equation

$$y = f(x^{(n)}) + f'(x^{(n)})(x - x^{(n)}).$$

The point $x^{(n+1)}$ is the point of intersection of this tangent with the x-axis. This is illustrated in Figure 4.2.

Let x^* be the root. The Taylor expansion of $f(x^*)$ about $x^{(n)}$ is

$$f(x^*) = f(x^{(n)}) + f'(x^{(n)})(x^* - x^{(n)}) + \frac{1}{2!}f''(\xi^{(n)})(x^* - x^{(n)})^2,$$

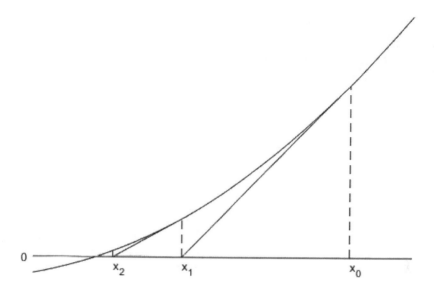

Figure 4.2 Newton's method

where $\xi^{(n)}$ lies between $x^{(n)}$ and x^*. Since x^* is the root, this equates to zero:

$$f(x^{(n)}) + f'(x^{(n)})(x^* - x^{(n)}) + \frac{1}{2!}f''(\xi^{(n)})(x^* - x^{(n)})^2 = 0.$$

Let's assume that f' is bounded away from zero in a neighbourhood of x^* and $x^{(n)}$ lies in this neighbourhood. We can then divide the above equation by $f'(x^{(n)})$. After rearranging, this becomes

$$\frac{f(x^{(n)})}{f'(x^{(n)})} + (x^* - x^{(n)}) = -\frac{1}{2}\frac{f''(\xi^{(n)})}{f'(x^{(n)})}(x^* - x^{(n)})^2.$$

Using (4.2) we can relate the error in the next iteration to the error in the current iteration

$$x^* - x^{(n+1)} = -\frac{1}{2}\frac{f''(\xi^{(n)})}{f'(x^{(n)})}(x^* - x^{(n)})^2.$$

This shows that under certain conditions the convergence of Newton's method is quadratic. The conditions are that there exists a neighbourhood U of the root where f' is bounded away from zero and where f'' is finite and that the starting point lies sufficiently close to x^*. More specifically, let

$$C = \frac{\sup_{x \in U}|f''(x)|}{\inf_{x \in U}|f'(x)|}.$$

Here the notation $\sup_{x \in U}|f''(x)|$ is the smallest upper bound of the values

$|f''|$ takes in U, while $\inf_{x \in U} |f'(x)|$ stands for the largest lower bound of the values $|f'|$ takes in U. Then

$$|x^* - x^{(n+1)}| \leq \frac{1}{2} C (x^* - x^{(n)})^2.$$

Here we can see that sufficiently close to x^* means that $\frac{1}{2} C |x^* - x^{(0)}| < 1$, otherwise the distance to the root may not decrease.

We take a closer look at the situations where the method fails. In some cases the function in question satisfies the conditions for convergence, but the point chosen as starting point is not sufficiently close to the root. In this case other methods such as bisection are used to obtain a better starting point.

The method fails if any of the iteration points happens to be a stationary point, i.e., a point where the first derivative vanishes. In this case the next iteration step is undefined, since the tangent there will be parallel to the x-axis and not intersect it. Even if the derivative is nonzero, but small, the next approximation may be far away from the root.

For some functions it can happen that the iteration points enter an infinite cycle. Take for example the polynomial $f(x) = x^3 - 2x + 2$. If 0 is chosen as the starting point, the first iteration produces 1, while the next iteration produces 0 again and so forth. The behaviour of the sequence produced by Newton's method is illustrated by *Newton fractals* in the complex plane $\mathbb{C}$. If $z \in \mathbb{C}$ is chosen as a starting point and if the sequence produced converges to a specific root then z is associated with this root. The set of initial points z that converge to the same root is called *basin of attraction* for that root. The fractals illustrate beautifully that a slight perturbation of the starting value can result into the algorithm converging to a different root.

Exercise 4.1. *Write a program which takes a polynomial of degree between 2 and 7 as input and colours the basins of attraction for each root a different colour. Try it out for the polynomial $z^n - 1$ for $n = 2, \ldots, 7$.*

A simple example where Newton's method diverges is the cube root $f(x) = \sqrt[3]{x}$ which is continuous and infinitely differentiable except for the root $x = 0$, where the derivative is undefined. For any approximation $x^{(n)} \neq 0$ the next approximation will be

$$x^{(n+1)} = x^{(n)} - \frac{x^{(n)1/3}}{\frac{1}{3} x^{(n)1/3-1}} = x^{(n)} - 3x^{(n)} = -2x^{(n)}.$$

In every iteration the algorithm overshoots the solution onto the other side further away than it was initially. The distance to the solution is doubled in each iteration.

There are also cases where Newton's method converges, but the rate of convergence is not quadratic. For example take $f(x) = x^2$. Then for every approximation $x^{(n)}$ the next approximation is

$$x^{(n+1)} = x^{(n)} - \frac{x^{(n)2}}{2x^{(n)}} = \frac{1}{2} x^{(n)}.$$

Thus the distance to the root is halved in every iteration comparable to the bisection method.

Newton's method readily generalizes to higher dimensional problems. Given a function $\mathbf{f} : \mathbb{R}^m \to \mathbb{R}^m$, we consider $\mathbf{f}$ as a vector of m functions

$$\mathbf{f}(\mathbf{x}) = \begin{pmatrix} f_1(\mathbf{x}) \\ \vdots \\ f_m(\mathbf{x}) \end{pmatrix},$$

where $\mathbf{x} \in \mathbb{R}^m$. Let $\mathbf{h} = (h_1, \ldots h_m)^T$ be a small perturbation vector. The multidimensional Taylor expansion of each function component f_i, $i = 1, \ldots, m$ is

$$f_i(\mathbf{x} + \mathbf{h}) = f_i(\mathbf{x}) + \sum_{k=1}^{m} \frac{\partial f_i(\mathbf{x})}{\partial x_k} h_k + O(\|\mathbf{h}\|^2).$$

The *Jacobian matrix* $J_f(\mathbf{x})$ has the entries $[J_f(\mathbf{x})]_{i,k} = \frac{\partial f_i(\mathbf{x})}{\partial x_k}$ and thus we can write in matrix notation

$$\mathbf{f}(\mathbf{x} + \mathbf{h}) = \mathbf{f}(\mathbf{x}) + J_f(\mathbf{x})\mathbf{h} + O(\|\mathbf{h}\|^2). \tag{4.3}$$

Assuming that $\mathbf{x}$ is an approximation to the root, we want to improve the approximation by choosing $\mathbf{h}$. Ignoring higher-order terms gathered in $O(\|\mathbf{h}\|^2)$ and setting $f(\mathbf{x} + \mathbf{h}) = 0$, we can solve (4.3) for $\mathbf{h}$. The new approximation is set to $\mathbf{x} + \mathbf{h}$. More formally the Newton iteration in higher dimensions is

$$\mathbf{x}^{(n+1)} = \mathbf{x}^{(n)} - J_f(\mathbf{x}^{(n)})^{-1}\mathbf{f}(\mathbf{x}^{(n)}).$$

However, computing the Jacobian can be a difficult and expensive operation. The next method computes an approximation to the Jacobian in each iteration step.

4.3   Broyden's Method

We return to the one-dimensional case for a comparison. Newton's method and the secant method are related in that the secant method replaces the derivative in Newton's method by a finite difference.

$$f'(x^{(n)}) \approx \frac{f(x^{(n)}) - f(x^{(n-1)})}{x^{(n)} - x^{(n-1)}}.$$

Rearranging gives

$$f'(x^{(n)})(x^{(n)} - x^{(n-1)}) \approx f(x^{(n)}) - f(x^{(n-1)}).$$

This can be generalized to

$$J_f(\mathbf{x}^{(n)})(\mathbf{x}^{(n)} - \mathbf{x}^{(n-1)}) \approx \mathbf{f}(\mathbf{x}^{(n)}) - \mathbf{f}(\mathbf{x}^{(n-1)}). \tag{4.4}$$

In *Broyden's method* the Jacobian is only calculated once in the first iteration as $A^{(0)} = J_f(\mathbf{x}^{(0)})$. In all the subsequent iterations the matrix is updated in such a way that it satisfies (4.4). That is, if the matrix multiplies $\mathbf{x}^{(n)} - \mathbf{x}^{(n-1)}$, the result is $\mathbf{f}(\mathbf{x}^{(n)}) - \mathbf{f}(\mathbf{x}^{(n-1)})$. This determines how the matrix acts on the one-dimensional subspace of $\mathbb{R}^m$ spanned by the vector $\mathbf{x}^{(n)} - \mathbf{x}^{(n-1)}$. However, this does not determine how the matrix acts on the $(m-1)$-dimensional complementary subspace. Or in other words (4.4) provides only m equations to specify an $m \times m$ matrix. The remaining degrees of freedom are taken up by letting $A^{(n)}$ be a minimal modification to $A^{(n-1)}$, minimal in the sense that $A^{(n)}$ acts the same as $A^{(n-1)}$ on all vectors orthogonal to $\mathbf{x}^{(n)} - \mathbf{x}^{(n-1)}$. These vectors are the $(m-1)$-dimensional complementary subspace. The matrix $A^{(n)}$ is then given by

$$A^{(n)} = A^{(n-1)} + \frac{\mathbf{f}(\mathbf{x}^{(n)}) - \mathbf{f}(\mathbf{x}^{(n-1)}) - A^{(n-1)}(\mathbf{x}^{(n)} - \mathbf{x}^{(n-1)})}{(\mathbf{x}^{(n)} - \mathbf{x}^{(n-1)})^T(\mathbf{x}^{(n)} - \mathbf{x}^{(n-1)})}(\mathbf{x}^{(n)} - \mathbf{x}^{(n-1)})^T.$$

If $A^{(n)}$ is applied to $\mathbf{x}^{(n)} - \mathbf{x}^{(n-1)}$, most terms vanish and the desired result $\mathbf{f}(\mathbf{x}^{(n)}) - \mathbf{f}(\mathbf{x}^{(n-1)})$ remains. The second term vanishes whenever $A^{(n)}$ is applied to a vector $\mathbf{v}$ orthogonal to $\mathbf{x}^{(n)} - \mathbf{x}^{(n-1)}$, since in this case $(\mathbf{x}^{(n)} - \mathbf{x}^{(n-1)})^T\mathbf{v} = 0$, and $A^{(n)}$ acts on these vectors exactly as $A^{(n-1)}$.

The next approximation is then given by

$$\mathbf{x}^{(n+1)} = \mathbf{x}^{(n)} - (A^{(n)})^{-1}\mathbf{f}(\mathbf{x}^{(n)}).$$

Just as in Newton's method, we do not calculate the inverse directly. Instead we solve

$$A^{(n)}\mathbf{h} = -\mathbf{f}(\mathbf{x}^{(n)})$$

for some perturbation $\mathbf{h} \in \mathbb{R}^m$ and let $\mathbf{x}^{(n+1)} = \mathbf{x}^{(n)} + \mathbf{h}$.

However, if the inverse of the initial Jacobian $A^{(0)}$ has been calculated the inverse can be updated in only $O(m^2)$ operations using the *Sherman–Morrison formula*, which states that for a non-singular matrix A and vectors $\mathbf{u}$ and $\mathbf{v}$ such that $\mathbf{v}^T A^{-1}\mathbf{u} \neq -1$ we have

$$(A + \mathbf{u}\mathbf{v}^T)^{-1} = A^{-1} - \frac{A^{-1}\mathbf{u}\mathbf{v}^T A^{-1}}{1 + \mathbf{v}^T A^{-1}\mathbf{u}}.$$

Letting

$$A = A^{(n-1)},$$

$$\mathbf{u} = \frac{\mathbf{f}(\mathbf{x}^{(n)}) - \mathbf{f}(\mathbf{x}^{(n-1)}) - A^{(n-1)}(\mathbf{x}^{(n)} - \mathbf{x}^{(n-1)})}{(\mathbf{x}^{(n)} - \mathbf{x}^{(n-1)})^T(\mathbf{x}^{(n)} - \mathbf{x}^{(n-1)})},$$

$$\mathbf{v} = \mathbf{x}^{(n)} - \mathbf{x}^{(n-1)},$$

we have a fast update formula.

Broyden's method is not as fast as the quadratic convergence of Newton's method. But the smaller operation count per iteration is often worth it. In the next section we return to the one-dimensional case.

4.4   Householder Methods

Householder methods are Newton-type methods with higher order of convergence. The first of these is *Halley's method* with cubic order of convergence if the starting point is close enough to the root x^*. As with Newton's method it can be applied in the case where the root x^* is simple; that is, the derivative of f is bounded away from zero in a neighbourhood of x^*. In addition f needs to be three times continuously differentiable.

The method is derived by noting that the function defined by

$$g(x) = \frac{f(x)}{\sqrt{|f'(x)|}}$$

also has a root at x^*. Applying Newton's method to g gives

$$x^{(n+1)} = x^{(n)} - \frac{g(x^{(n)})}{g'(x^{(n)})}.$$

The derivative of $g(x)$ is given by

$$g'(x) = \frac{2[f'(x)]^2 - f(x)f''(x)}{2f'(x)\sqrt{|f'(x)|}}.$$

With this the update formula is

$$x^{(n+1)} = x^{(n)} - \frac{2f(x^{(n)})f'(x^{(n)})}{2[f'(x^{(n)})]^2 - f(x^{(n)})f''(x^{(n)})}.$$

The formula can be rearranged to show the similarity between Halley's method and Newton's method

$$x^{(n+1)} = x^{(n)} - \frac{f(x^{(n)})}{f'(x^{(n)})}\left[1 - \frac{f(x^{(n)})}{f'(x^{(n)})}\frac{f''(x^{(n)})}{2f'(x^{(n)})}\right]^{-1}.$$

We see that when the second derivative is close to zero near x^* then the iteration is nearly the same as Newton's method. The expression $f(x^{(n)})/f'(x^{(n)})$ is only calculated once. This form is particularly useful when $f''(x^{(n)})/f'(x^{(n)})$ can be simplified.

The technique is also known as *Bailey's method* when written in the following form:

$$x^{(n+1)} = x^{(n)} - f(x^{(n)})\left[f'(x^{(n)}) - \frac{f(x^{(n)})f''(x^{(n)})}{2f'(x^{(n)})}\right]^{-1}.$$

As with Newton's method the convergence behaviour of Halley's method depends strongly on the position of the starting point as the following exercise shows.

Exercise 4.2. *Write a program which takes a polynomial of degree between 2 and 7 as input, applies Halley's root finding method, and colours the basins of attraction for each root a different colour. Try it out for the polynomial $z^n - 1$ for $n = 2, \ldots, 7$.*

More generally, the Householder methods of order $k \in \mathbb{N}$ are given by the formula

$$x^{(n+1)} = x^{(n)} + k\frac{(1/f)^{(k-1)}(x^{(n)})}{(1/f)^{(k)}(x^{(n)})},$$

where $(1/f)^{(k)}$ denotes the kth derivative of $1/f$. If f is $k+1$ times continuously differentiable and the root at x^* is a simple root then the rate of convergence is $k + 1$, provided that the starting point is sufficiently close to x^*.

Of course all these methods depend on the ability to calculate the derivatives of f. The following method however does not.

4.5   Müller's Method

The methods in this section and the following are motivated by the secant method. The secant method interpolates two function values linearly and takes the intersection of this interpolant with the x-axis as the next approximation to the root. *Müller's method* uses three function values and quadratic interpolation instead. Figure 4.3 illustrates this.

Having calculated $x^{(n-2)}$, $x^{(n-1)}$ and $x^{(n)}$, a polynomial $p(x) = a(x - x^{(n)})^2 + b(x - x^{(n)}) + c$ is fitted to the data:

$$\begin{aligned}
f(x^{(n-2)}) &= a(x^{(n-2)} - x^{(n)})^2 + b(x^{(n-2)} - x^{(n)}) + c, \\
f(x^{(n-1)}) &= a(x^{(n-1)} - x^{(n)})^2 + b(x^{(n-1)} - x^{(n)}) + c, \\
f(x^{(n)}) &= c.
\end{aligned}$$

Solving for a and b, we have

$$a = \frac{(x^{(n-1)} - x^{(n)})[f(x^{(n-2)}) - f(x^{(n)})] - (x^{(n-2)} - x^{(n)})[f(x^{(n-1)}) - f(x^{(n)})]}{(x^{(n-2)} - x^{(n)})(x^{(n-1)} - x^{(n)})(x^{(n-2)} - x^{(n-1)})}$$

$$b = \frac{(x^{(n-2)} - x^{(n)})^2[f(x^{(n-1)}) - f(x^{(n)})] - (x^{(n-1)} - x^{(n)})^2[f(x^{(n-2)}) - f(x^{(n)})]}{(x^{(n-2)} - x^{(n)})(x^{(n-1)} - x^{(n)})(x^{(n-2)} - x^{(n-1)})}$$

The next approximation $x^{(n+1)}$ is one of the roots of p and the one closer to $x^{(n)}$ is chosen. To avoid errors due to loss of significance we use the alternative formula for the roots derived in 1.4,

$$x^{(n+1)} - x^{(n)} = \frac{-2c}{b + \text{sgn}(b)\sqrt{b^2 - 4ac}}, \tag{4.5}$$

where $\text{sgn}(b)$ denotes the sign of b. This way the root which gives the largest denominator and thus is closest to $x^{(n)}$ is chosen.

Note that $x^{(n+1)}$ can be complex even if all previous approximations have been real. This is in contrast to previous root-finding methods where the

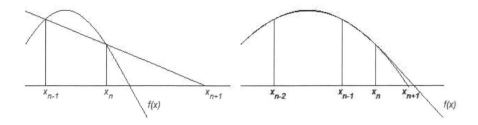

Figure 4.3 Illustration of the secant method on the left and Müller's method on the right

iterates remain real if the starting value is real. This behaviour can be an advantage, if complex roots are to be found or a disadvantage if the roots are known to be real.

An alternative representation uses the Newton form of the interpolating polynomial

$$p(x) \quad = \quad f(x^{(n)}) + (x - x^{(n)})f[x^{(n)}, x^{(n-1)}]$$
$$+ (x - x^{(n)})(x - x^{(n-1)})f[x^{(n)}, x^{(n-1)}, x^{(n-2)}],$$

where $f[x^{(n)}, x^{(n-1)}]$ and $f[x^{(n)}, x^{(n-1)}, x^{(n-2)}]$ denote divided differences. After some manipulation using the recurrence relation for divided differences, one can see that

$$a \quad = \quad f[x^{(n)}, x^{(n-1)}, x^{(n-2)}],$$
$$b \quad = \quad f[x^{(n)}, x^{(n-1)}] + f[x^{(n)}, x^{(n-2)}] - f[x^{(n-1)}, x^{(n-2)}].$$

Thus (4.5) can be evaluated fast using divided differences.

The order of convergence is approximately 1.84 if the initial approximations $x^{(0)}$, $x^{(1)}$, and $x^{(2)}$ are close enough to the simple root x^*. This is better than the secant method with 1.62 and worse than Newton's method with 2.

4.6   Inverse Quadratic Interpolation

The occurrence of complex approximants can be avoided by interpolating the inverse of f. This method is known as the *inverse quadratic interpolation method*. To interpolate the inverse one uses the Lagrange interpolation formula swapping the roles of f and x

$$
\begin{aligned}
p(y) \;=\; & \frac{(y - f(x^{(n-1)}))(y - f(x^{(n)}))}{(f(x^{(n-2)}) - f(x^{(n-1)}))(f(x^{(n-2)}) - f(x^{(n)}))} x^{(n-2)} \\
& + \frac{(y - f(x^{(n-2)}))(y - f(x^{(n)}))}{(f(x^{(n-1)}) - f(x^{(n-2)}))(f(x^{(n-1)}) - f(x^{(n)}))} x^{(n-1)} \\
& + \frac{(y - f(x^{(n-2)}))(y - f(x^{(n-1)}))}{(f(x^{(n)}) - f(x^{(n-2)}))(f(x^{(n)}) - f(x^{(n-1)}))} x^{(n)}.
\end{aligned}
$$

Now we are interested in a root of f and thus substitute $y = 0$ and use $p(0) = x^{(n+1)}$. Since the inverse of f is interpolated this immediately gives a formula for $x^{(n+1)}$

$$
\begin{aligned}
x^{(n+1)} \;=\; & \frac{f(x^{(n-1)})f(x^{(n)})}{(f(x^{(n-2)}) - f(x^{(n-1)}))(f(x^{(n-2)}) - f(x^{(n)}))} x^{(n-2)} \\
& + \frac{f(x^{(n-2)})f(x^{(n)})}{(f(x^{(n-1)}) - f(x^{(n-2)}))(f(x^{(n-1)}) - f(x^{(n)}))} x^{(n-1)} \\
& + \frac{f(x^{(n-2)})f(x^{(n-1)})}{(f(x^{(n)}) - f(x^{(n-2)}))(f(x^{(n)}) - f(x^{(n-1)}))} x^{(n)}.
\end{aligned}
$$

If the starting point is close to the root then the order of convergence is approximately 1.8. However, the algorithm can fail completely if any two of the function values $f(x^{(n-2)})$, $f(x^{(n-1)})$, and $f(x^{(n)})$ coincide. However, it plays an important role in mixed algorithms, which are considered in the last section.

4.7 Fixed Point Iteration Theory

All the algorithms we have encountered so far are examples of *fixed point iterative methods*. In these methods each iteration has the form

$$x^{(n+1)} = g(x^{(n)}). \tag{4.6}$$

If we have convergence to some limit x^*, then x^* is a *fixed point* of the function g, i.e., $x^* = g(x^*)$, hence the name.

Theorem 4.1. *Suppose for the iteration given by (4.6), we have*

$$|g(x) - g(x')| \le \lambda |x - x'| \tag{4.7}$$

for all x and x' in an interval $I = [x^{(0)} - \delta, x^{(0)} + \delta]$ for some constant λ. This means g is Lipschitz continuous on I. Suppose further that $\lambda < 1$ and $|x^{(0)} - g(x^{(0)})| < (1 - \lambda)\delta$. Then

1. *all iterates lie in I,*

2. *the iterates converge and*

3. the solution is unique.

Proof. We have

$$
\begin{aligned}
|x^{(n+1)} - x^{(n)}| &= |g(x^{(n)}) - g(x^{(n-1)})| \\
&\leq \lambda |x^{(n)} - x^{(n-1)}| \\
&\leq \lambda^n |x^{(1)} - x^{(0)}| \\
&\leq \lambda^n (1-\lambda)\delta,
\end{aligned}
$$

since $x^{(1)} - x^{(0)} = g(x^{(0)}) - x^{(0)}$. Thus for each new iterate

$$
\begin{aligned}
|x^{(n+1)} - x^{(0)}| &\leq \sum_{k=0}^{n} |x^{(k+1)} - x^{(k)}| \\
&\leq \sum_{k=0}^{n} \lambda^k (1-\lambda)\delta \\
&\leq (1 - \lambda^{n+1})\delta \leq \delta.
\end{aligned}
$$

Hence all iterates lie in the interval I. This also means that the sequence of iterates is bounded. Moreover,

$$
\begin{aligned}
|x^{(n+p)} - x^{(n)}| &\leq \sum_{k=n}^{n+p-1} |x^{(k+1)} - x^{(k)}| \\
&\leq \sum_{k=n}^{n+p-1} \lambda^k (1-\lambda)\delta \\
&\leq (1 - \lambda^p)\lambda^n \delta \xrightarrow{n \to \infty} 0.
\end{aligned}
$$

Since $\lambda < 1$, the sequence is a *Cauchy sequence* and hence converges.

Suppose the solution is not unique, i.e., there exist $x^* \neq \tilde{x}^*$ such that $x^* = g(x^*)$ and $\tilde{x}^* = g(\tilde{x}^*)$. Then

$$
\begin{aligned}
|x^* - \tilde{x}^*| &= |g(x^*) - g(\tilde{x}^*)| \\
&\leq \lambda |x^* - \tilde{x}^*| \\
&< |x^* - \tilde{x}^*|,
\end{aligned}
$$

which is a contradiction. □

If in (4.7) x is close to x' and $g(x)$ is differentiable, then it is reasonable to assume that $\lambda \approx |g'(x)|$. Indeed it can be shown that if $|g'(x^*)| < 1$ for a fixed point x^* then there exists an interval of convergence. Such a fixed point x^* is called a *point of attraction* for the iterative method.

A good way to visualize fixed point iteration is through a cobweb plot (Figure 4.4) where the function as well as the line $y = x$ is plotted. The last function value becomes the new x-value illustrated by the horizontal lines. Then the function is evaluated there illustrated by the vertical lines.

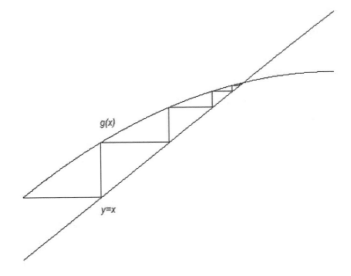

Figure 4.4 Cobweb plot for fixed point iteration

4.8   Mixed Methods

The *Bus and Dekker algorithm* is a hybrid of the secant and the bisection method. Starting from an initial interval $[a_0, b_0]$, where $f(a_0)$ and $f(b_0)$ have opposite signs, the algorithm constructs a sequence of sub-intervals, each containing the zero. The current approximation to the root is b_n called the *iterate*, and a_n is called its *contrapoint*. In each iteration two values are calculated

$$m = \frac{a_n + b_n}{2},$$

$$s = b_n - \frac{b_n - b_{n-1}}{f(b_n) - f(b_{n-1})} f(b_n),$$

where b_{-1} is taken to be a_0. The first value m is the midpoint of the interval, while the second value s is the approximation to the root given by the secant method. If s lies between b_n and m, it becomes the next iterate, that is $b_{n+1} = s$, otherwise the midpoint is chosen, $b_{n+1} = m$. If $f(a_n)$ and $f(b_{n+1})$ have opposite signs, the new contrapoint is $a_{n+1} = a_n$, otherwise $a_{n+1} = b_n$, since $f(b_n)$ and $f(b_{n+1})$ must have opposite signs in this case, since $f(a_n)$ and $f(b_n)$ had opposite signs in the previous iteration. Additionally, if the modulus of $f(a_{n+1})$ is less than the modulus of $f(b_{n+1})$, a_{n+1} is considered a better approximation to the root and it becomes the new iterate while b_{n+1} becomes the new contrapoint. Thus the iterate is always the better approximation.

This method performs generally well, but there are situations where every iteration employs the secant method and convergence is very slow, requiring far more iterations than the bisection method. In particular, $b_n - b_{n-1}$ might become arbitrarily small while the length of the interval given by a_n and b_n decreases very slowly. The following method tries to alleviate this problem.

Brent's method combines the bisection method, the secant method, and inverse quadratic interpolation. It is also known as the *Wijngaarden–Dekker–Brent method*. At every iteration, Brent's method decides which method out of these three is likely to do best, and proceeds by doing a step according to that method. This gives a robust and fast method.

A numerical tolerance ϵ is chosen. The method ensures that the bisection method is used, if consecutive iterates are too close together. More specifically, if the previous step performed the bisection method and $|b_n - b_{n-1}| \leq \epsilon$, then the bisection method will be performed again. Similarly, if the previous step performed interpolation (either linear interpolation for the secant method or inverse quadratic interpolation) and $|b_{n-1} - b_{n-2}| \leq \epsilon$, then the bisection method will be performed again. Thus b_n and b_{n-1} are allowed to become arbitrarily close at most two times in a row.

Additionally, the intersection s from interpolation (either linear or inverse quadratic) is only accepted as new iterate if $|s - b_n| < \frac{1}{2}|b_n - b_{n-1}|$, if the previous step used bisection, or if $|s - b_n| < \frac{1}{2}|b_{n-1} - b_{n-2}|$, if the previous step used interpolation (linear or inverse quadratic). These conditions enforce that consecutive interpolation steps halve the step size every two iterations until the step size becomes less than ϵ after at most $2\log_2(|b_{n-1} - b_{n-2}|/\epsilon)$ iterations which invokes a bisection.

Brent's algorithm uses linear interpolation, that is, the secant method, if any of $f(b_n), f(a_n)$, or $f(b_{n-1})$ coincide. If they are all distinct, inverse quadratic interpolation is used. However, the requirement for s to lie between m and b_n is changed: s has to lie between $(3a_n + b_n)/4$ and b_n.

Exercise 4.3. *Implement Brent's algorithm. It should terminate if either $f(b_n)$ or $f(s)$ is zero or if $|b_n - a_n|$ is small enough. Use the bisection rule if s is not between $(3a_n + b_n)/4$ and b_n for both linear and inverse quadratic interpolation or if any of Brent's conditions arises. Try your program on $f(x) = x^3 - x^2 - 4x + 4$, which has zeros at $-2, 1$, and 2. Start with the interval $[-4, 2.5]$, which contains all roots. List which method is used in each iteration.*

4.9 Revision Exercises

Exercise 4.4. *The secant method for finding the solution of $f(x) = 0$ is given by*

$$
\begin{aligned}
x^{(n+1)} &= \frac{f(x^{(n)})x^{(n-1)} - f(x^{(n-1)})x^{(n)}}{f(x^{(n)}) - f(x^{(n-1)})} \\
&= x^{(n)} - f(x^{(n)})\frac{x^{(n)} - x^{(n-1)}}{f(x^{(n)}) - f(x^{(n-1)})}.
\end{aligned}
$$

(a) By means of a sketch graph describe how the method works in a simple case and give an example where it might fail to converge.

(b) Let x^* be the root and let $e^{(n)}$ denote the error $e^{(n)} = x^{(n)} - x^*$ at the nth iteration. Express the error at the $(n+1)^{th}$ iteration in terms of the errors in the previous two iterations.

(c) Approximate $f(x^* + e^{(n-1)})/e^{(n-1)}$, $f(x^* + e^{(n)})/e^{(n)}$, and $f(x^* + e^{(n)}) - f(x^* + e^{(n-1)})$ using Taylor expansion. You may assume that $x^{(n)}$ and $x^{(n-1)}$ are close enough to the root such that the terms $O([e^{(n)}]^2)$ and $O([e^{(n-1)}]^2)$ can be neglected.

(d) Using the derived approximation and the expression derived for $e^{(n+1)}$, show that the error at $(n+1)^{th}$ iteration is approximately

$$e^{(n+1)} \approx e^{(n)}e^{(n-1)}\frac{f''(x^*)}{2f'(x^*)}.$$

(e) From $|e^{(n+1)}| = O(|e^{(n)}||e^{(n-1)}|)$ derive p such that $|e^{(n+1)}| = O(|e^{(n)}|^p)$.

(f) Derive the Newton method from the secant method.

(g) Let $f(x) = x^2$. Letting $x^{(1)} = \frac{1}{2}x^{(0)}$, for both the secant and the Newton method express $x^{(2)}$, $x^{(3)}$, and $x^{(4)}$ in terms of $x^{(0)}$.

Exercise 4.5. *Newton's method for finding the solution of $f(x) = 0$ is given by*

$$x^{(n+1)} = x^{(n)} - \frac{f(x^{(n)})}{f'(x^{(n)})},$$

where $x^{(n)}$ is the approximation to the root x^* in the nth iteration. The starting point $x^{(0)}$ is already close enough to the root.

(a) By means of a sketch graph describe how the method works in a simple case and give an example where it might fail to converge.

(b) Using the Taylor expansion of $f(x^*) = 0$ about $x^{(n)}$, relate the error in the next iteration to the error in the current iteration and show that the convergence of Newton's method is quadratic.

(c) Generalize Newton's method to higher dimensions.

(d) Let

$$\mathbf{f}(\mathbf{x}) = \mathbf{f}(x, y) = \begin{pmatrix} \frac{1}{2}x^2 + y \\ \frac{1}{2}y^2 + x \end{pmatrix}.$$

The roots lie at $(0,0)$ and $(-2,-2)$. Calculate the Jacobian of $\mathbf{f}$ and its inverse.

(e) Why does Newton's method fail near $(1,1)$ and $(-1,-1)$?

(f) Let $\mathbf{x}^{(0)} = (1,0)$. Calculate $\mathbf{x}^{(1)}, \mathbf{x}^{(2)}$ and $\mathbf{x}^{(3)}$, and their Euclidean norms.

(g) *The approximations converge to $(0,0)$. Show that the speed of convergence agrees with the theoretical quadratic speed of convergence.*

Exercise 4.6. *The Newton–Raphson iteration for solving $f(x) = 0$ is*

$$\hat{x} = x - \frac{f(x)}{f'(x)}.$$

(a) *By drawing a carefully labeled graph, explain the graphical interpretation of this formula.*

(b) *What is the order of convergence?*

(c) *Under which conditions can this order be achieved?*

(d) *Consider $f(x) = x^3 + x^2 - 2$. The following table shows successive iterations for each of the three starting values (i) $x = 1.5$, (ii) $x = 0.2$, (iii) $x = -0.5$. Note that, to the accuracy shown, each iteration finds the root at $x = 1$.*

n	(i)	(ii)	(iii)
0	1.50000×10^0	2.00000×10^{-1}	-5.00000×10^{-1}
1	1.12821×10^0	3.95384×10^0	-8.00000×10^0
2	1.01152×10^0	2.57730×10^0	-5.44318×10^0
3	1.00010×10^0	1.70966×10^0	-3.72976×10^0
4	1.00000×10^0	1.22393×10^0	-2.56345×10^0
5	1.00000×10^0	1.03212×10^0	-1.72202×10^0
6		1.00079×10^0	-9.62478×10^{-1}
7		1.00000×10^0	1.33836×10^0
8		1.00000×10^0	1.06651×10^0
9			1.00329×10^0
10			1.00000×10^0
11			1.00000×10^0

Sketch the graph of $f(x)$ and sketch the first iteration for cases (i) and (ii) to show why (i) converges faster than (ii).

(e) *In a separate (rough) sketch, show the first two iterations for case (iii).*

(f) *Now consider $f(x) = x^4 - 3x^2 - 2$. Calculate two Newton–Raphson iterations from the starting value $x = 1$. Comment on the prospects for convergence in this case.*

(g) *Give further examples where the method might fail to converge or converges very slowly.*

Exercise 4.7. *The following reaction occurs when water vapor is heated*

$$H_2O \rightleftharpoons H_2 + \frac{1}{2}O_2.$$

The fraction $x \in [0, 1]$ of H_2O that is consumed satisfies the equation

$$K = \frac{x}{1 - x} \sqrt{\frac{2p_t}{2 + x}}, \tag{4.8}$$

where K and p_t are given constants. The following figure illustrates this:

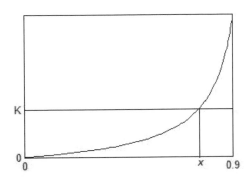

(a) Rephrase the problem of determining x as finding the root of a function $f(x)$ and state $f(x)$. Sketch a graph illustrating the rephrased problem.

(b) Describe the bisection method to find the root of a function. Comment on the robustness and speed of convergence of the method.

(c) Given an approximation $x^{(n)}$ to the root x^* of the function $f(x)$, give the formula how Newton's method calculates the next approximation $x^{(n+1)}$, explain what this means geometrically, and expand your sketch with an example of how Newton's method works.

(d) What is the order of convergence of Newton's method?

(e) What happens to the right-hand side of Equation (4.8) if x approaches 1 and what does this mean for Newton's method, if the starting point is chosen close to 1?

(f) The derivative of $f(x)$ at 0 is 1. What is the next approximation if 0 is chosen as the starting point? Depending on K, what problem might this cause?

(g) Give another example to demonstrate when Newton's method might fail to converge?

CHAPTER 5

Numerical Integration

In this part of the course we consider the numerical evaluation of an integral of the form

$$I = \int_a^b f(x)dx.$$

A *quadrature rule* approximates it by

$$I \approx Q_n(f) = \sum_{i=1}^n w_i f(x_i).$$

The points x_i, $i = 1, \ldots, n$ are called the *abscissae* chosen such that $a \le x_1 < \ldots < x_n \le b$. The coefficients w_i are called the *weights*. Quadrature rules are derived by integrating a polynomial interpolating the function values at the abscissae. Usually only positive weights are allowed since whether something is added or subtracted should be determined by the sign of the function at this point. This also avoids loss of significance.

5.1   Mid-Point and Trapezium Rule

The simplest rule arises from interpolating the function by a constant, which is a polynomial of degree zero, and it is reasonable to let this constant take the value of the function at the midpoint of the interval $[a, b]$. This is known as the *mid-point-rule* or *mid-ordinate rule*. It is illustrated in Figure 5.1. The formula is

$$I \approx (b - a)f(\frac{a + b}{2}).$$

The mid-point is the only abscissa and its weight is $b - a$.

Theorem 5.1. *The mid-point rule has the following error term*

$$\int_a^b f(x)dx = (b - a)f(\frac{a + b}{2}) + \frac{1}{24}f''(\xi)(b - a)^3,$$

where ξ is some point in the interval $[a, b]$

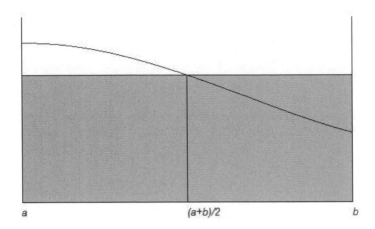

Figure 5.1 The mid-point rule

Proof. We use Taylor expansion of f around the midpoint $\frac{a+b}{2}$

$$f(x) = f(\frac{a+b}{2}) + (x - \frac{a+b}{2})f'(\frac{a+b}{2}) + \frac{1}{2}(x - \frac{a+b}{2})^2 f''(\xi),$$

where ξ lies in the interval $[a, b]$. Integrating this over the interval $[a, b]$ gives

$$\left[xf(\frac{a+b}{2}) + \frac{1}{2}(x - \frac{a+b}{2})^2 f'(\frac{a+b}{2}) + \frac{1}{6}(x - \frac{a+b}{2})^3 f''(\xi) \right]_a^b$$

$$= (b-a)f(\frac{a+b}{2}) + \frac{1}{24}f''(\xi)(b-a)^3,$$

since the term involving the first derivative becomes zero. □

Alternatively, we can integrate the linear polynomial fitted to the two endpoints of the interval as Figure 5.2 illustrates. This rule is the *trapezium rule* described by the formula

$$I \approx \frac{b-a}{2}[f(a) + f(b)].$$

Theorem 5.2. *The trapezium rule has the following error term*

$$\int_a^b f(x)dx = \frac{b-a}{2}[f(a) + f(b)] - \frac{1}{12}f''(\xi)(b-a)^3,$$

where ξ is some point in the interval $[a, b]$

Proof. We use Taylor expansion of $f(a)$ around the point x

$$f(a) = f(x) + (a - x)f'(x) + \frac{1}{2}(a - x)^2 f''(\xi),$$

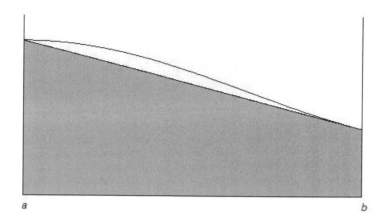

Figure 5.2 The trapezium rule

where ξ lies in the interval $[a, b]$. This can be rearranged as

$$f(x) = xf'(x) + f(a) - af'(x) - \frac{1}{2}(x - a)^2 f''(\xi).$$

Integrating this over the interval $[a, b]$ gives

$$
\begin{aligned}
\int_a^b f(x)dx &= \int_a^b xf'(x)dx + \left[xf(a) - af(x) - \frac{1}{6}(x - a)^3 f''(\xi) \right]_a^b \\
&= [xf(x)]_a^b - \int_a^b f(x)dx + (b - a)f(a) - a[f(b) - f(a)] \\
&\quad - \frac{1}{6}(b - a)^3 f''(\xi),
\end{aligned}
$$

where we used integration by parts for the first integral on the right-hand side. Solving for $\int_a^b f(x)$ this becomes

$$
\begin{aligned}
\int_a^b f(x)dx &= \frac{1}{2}[bf(b) - af(a) + bf(a) - af(b) - \frac{1}{6}(b - a)^3 f''(\xi)] \\
&= \frac{b - a}{2}[f(a) + f(b)] - \frac{1}{12}(b - a)^3 f''(\xi).
\end{aligned}
$$

$\square$

5.2   The Peano Kernel Theorem

So far we deduced the error term from first principles, but the Peano kernel theorem introduced in the chapter on interpolation proves very useful. In the case of a quadrature the functional $L(f)$ describing the approximation error is the difference between the value of the integral and the value given by the quadrature,

$$L(f) = \int_a^b f(x)dx - \sum_{i=1}^n w_i f(x_i).$$

Thus L maps the space of functions $C^{k+1}[a, b]$ to $\mathbb{R}$. L is a *linear functional*, i.e., $L(\alpha f + \beta g) = \alpha L(f) + \beta L(g)$ for all $\alpha, \beta \in \mathbb{R}$, since integration and weighted summation are linear operations themselves. We assume that the quadrature is constructed in such a way that it is correct for all polynomials of degree at most k, i.e., $L(p) = 0$ for all $p \in \mathbb{P}_k[x]$. Recall

Definition 5.1 (Peano kernel). *The Peano kernel K of L is the function defined by*

$$K(\theta) := L[(x - \theta)_+^k] \text{ for } \theta \in [a, b].$$

and

Theorem 5.3 (Peano kernel theorem). *Let L be a linear functional such that $L(p) = 0$ for all $p \in \mathbb{P}_k[x]$. Provided that the exchange of L with the integration is valid, then for $f \in C^{k+1}[a, b]$*

$$L(f) = \frac{1}{k!} \int_a^b K(\theta) f^{(k+1)}(\theta) d\theta.$$

Here the order of L and the integration can be swapped, since L consists of an integration and a weighted sum of function evaluations. The theorem has the following extension:

Theorem 5.4. *If K does not change sign in (a, b), then for $f \in C^{k+1}[a, b]$*

$$L(f) = \frac{1}{k!} \left[\int_a^b K(\theta) d\theta \right] f^{(k+1)}(\xi)$$

for some $\xi \in (a, b)$.

We derive the result about the error term for the trapezium rule again this time applying the Peano kernel theorem. By construction the trapezium rule is exact for all linear polynomials, thus $k = 1$. The kernel is given by

$$
\begin{aligned}
K(\theta) = L[(x - \theta)_+] &= \int_a^b (x - \theta)_+ dx - \frac{b-a}{2}[(a - \theta)_+ + (b - \theta)_+] \\
&= \int_\theta^b (x - \theta) dx - \frac{(b-a)(b-\theta)}{2} \\
&= \left[\frac{(x - \theta)^2}{2} \right]_\theta^b - \frac{(b-a)(b-\theta)}{2} \\
&= \frac{(b-\theta)^2}{2} - \frac{(b-a)(b-\theta)}{2} = \frac{(b-\theta)(a-\theta)}{2},
\end{aligned}
$$

where we used the fact that $(a - \theta)_+ = 0$, since $a \leq \theta$ for all $\theta \in [a, b]$ and $(b - \theta)_+ = b - \theta$, since $b \geq \theta$ for all $\theta \in [a, b]$.

The kernel $K(\theta)$ does not change sign for $\theta \in (a, b)$, since $b - \theta > 0$ and $a - \theta < 0$ for $\theta \in (a, b)$. The integral over $[a, b]$ of the kernel is given by

$$\int_a^b K(\theta)d\theta = \int_a^b \frac{(b - \theta)(a - \theta)}{2}d\theta = -\frac{1}{12}(b - a)^3,$$

which can be easily verified. Thus the error for the trapezium rule is

$$L(f) = \frac{1}{1!}\left[\int_a^b K(\theta)d\theta\right]f''(\xi) = -\frac{1}{12}(b - a)^3 f''(\xi)$$

for some $\xi \in [a, b]$.

5.3 Simpson's Rule

The error term can be reduced significantly, if a quadratic polynomial is fitted through the end-points and the mid-point. This is known as *Simpson's rule* and the formula is

$$I \approx \frac{b - a}{6}[f(a) + 4f(\frac{a + b}{2}) + f(b)].$$

The rule is illustrated in Figure 5.3.

Theorem 5.5. *Simpson's rule has the following error term*

$$\int_a^b f(x)dx = \frac{b - a}{6}[f(a) + 4f(\frac{a + b}{2}) + f(b)] - \frac{1}{90}f^{(4)}(\xi)(\frac{b - a}{2})^5,$$

where ξ is some point in the interval $[a, b]$

Proof. We apply Peano's kernel theorem to prove this result. Firstly, Simpson's rule is correct for all quadratic polynomials by construction. However, it is also correct for cubic polynomials, which can be proven by applying it to the monomial x^3. The value of the integral of x^3 over the interval $[a, b]$ is $(b^4 - a^4)/4$. Simpson's rule applied to x^3 gives

$$\frac{(b - a)}{6}\left(a^3 + 4(a + b)^3/2^3 + b^3\right)$$

$$= \frac{(b - a)}{12}\left(2a^3 + a^3 + 3a^2b + 3ab^2 + b^3 + 2b^3\right)$$

$$= \frac{(b - a)}{4}\left(a^3 + a^2b + ab^2 + b^3\right)$$

$$= \frac{1}{4}(a^3b + a^2b^2 + ab^3 + b^4 - a^4 - a^3b - a^2b^2 - ab^3)$$

$$= \frac{(b^4 - a^4)}{4}.$$

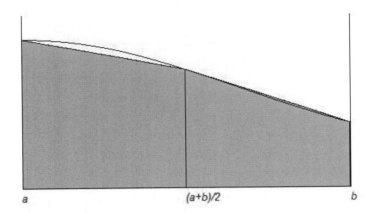

Figure 5.3 Simpson's rule

So Simpson's rule is correct for all cubic polynomials, since they form a linear space, and we have $k = 3$. However, it is not correct for polynomials of degree four. Let for example $a = 0$ and $b = 1$, then the integral of x^4 over the interval $[0, 1]$ is $1/5$ while the approximation by Simpson's rule is $1/6 * (0^4 + 4(1/2)^4 + 1^4) = 5/24$.

The kernel is given by

$$
\begin{aligned}
K(\theta) &= L[(x - \theta)^3_+] \\
&= \int_a^b (x - \theta)^3_+ dx - \tfrac{b-a}{6}[(a - \theta)^3_+ + 4(\tfrac{a+b}{2} - \theta)^3_+ + (b - \theta)^3_+] \\
&= \int_\theta^b (x - \theta)^3 dx - \tfrac{(b-a)}{6}[4(\tfrac{a+b}{2} - \theta)^3_+ + (b - \theta)^3] \\
&= \begin{cases} \tfrac{(b-\theta)^4}{4} - \tfrac{(b-a)}{6}[4(\tfrac{a+b}{2} - \theta)^3 + (b - \theta)^3] & \theta \in [a, \tfrac{a+b}{2}] \\ \tfrac{(b-\theta)^4}{4} - \tfrac{(b-a)}{6}(b - \theta)^3 & \theta \in [\tfrac{a+b}{2}, b]. \end{cases}
\end{aligned}
$$

For $\theta \in [\tfrac{a+b}{2}, b]$, the result can be simplified as

$$ K(\theta) = (b - \theta)^3 [(b - \theta)/4 - (b - a)/6] = (b - \theta)^3 [b/12 - \theta/4 + a/6]. $$

The first term $(b - \theta)^3$ is always positive, since $\theta < b$. The expression in the square brackets is decreasing and has a zero at the point $\tfrac{2}{3}a + \tfrac{1}{3}b = \tfrac{1}{2}(a + b) - \tfrac{1}{6}(b - a) < \tfrac{1}{2}(a + b)$ and thus is negative. Hence $K(\theta)$ is negative on $[\tfrac{a+b}{2}, b]$.

For $\theta \in [a, \tfrac{a+b}{2}]$ the additional term $2/3(b - a)[(a + b)/2 - \theta]^3$, which is positive in this interval, is subtracted. Thus, $K(\theta)$ is also negative here.

Hence the kernel does not change sign. We need to integrate $K(\theta)$ to obtain the error term

$$
\begin{aligned}
\int_a^b K(\theta)d\theta &= \int_a^b \frac{(b-\theta)^4}{4} - \frac{(b-a)}{6}(b-\theta)^3 d\theta - \int_a^{(a+b)/2} \frac{(b-a)}{6} 4\left(\frac{a+b}{2} - \theta\right)^3 d\theta \\
&= \left[-\frac{(b-\theta)^5}{20}\right]_a^b - \frac{b-a}{6}\left[-\frac{(b-\theta)^4}{4}\right]_a^b - \frac{b-a}{6}\left[-\left(\frac{a+b}{2} - \theta\right)^4\right]_a^{(a+b)/2} \\
&= \frac{(b-a)^5}{20} - \frac{(b-a)^5}{24} - \frac{b-a}{6}\left(\frac{b-a}{2}\right)^4 \\
&= \left(\frac{8}{5} - \frac{4}{3} - \frac{1}{3}\right)\left(\frac{b-a}{2}\right)^5 = -\frac{1}{15}\left(\frac{b-a}{2}\right)^5.
\end{aligned}
$$

Thus the error for Simpson's rule is

$$
L(f) = \frac{1}{3!}\left[\int_a^b K(\theta)d\theta\right] f^{(4)}(\xi) = -\frac{1}{90}\left(\frac{b-a}{2}\right)^5 f^{(4)}(\xi)
$$

for some $\xi \in [a,b]$. □

Exercise 5.1. *The mid-point rule $(b-a)f(\frac{1}{2}(a+b))$ is exact for polynomials of degree 1. Use Peano's kernel theorem to find a formula for $L(f)$. (Hint: This is similar to the trapezium rule, except that it is harder to prove that $K(\theta)$ does not change sign in $[a,b]$.)*

5.4 Newton–Cotes Rules

The idea of approximating f by its interpolating polynomial at equally spaced points and then integrating can be extended to n points. These rules are called *Newton–Cotes rules* or *Newton–Cotes formulae*. We distinguish two types of Newton–Cotes rules, the closed type, which uses the function values at the end points of the interval, and the open type, which does not use the function values at the end points of the interval.

The closed Newton–Cotes rules are

Trapezium	$(b-a)/2[f(a) + f(b)]$
Simpson's	$(b-a)/6[f(a) + 4f((a+b)/2) + f(b)]$
Simpson's 3/8	$(b-a)/8[f(a) + 3f((2a+b)/3) + 3f((a+2b)/3) + f(b)]$
Boole's	$(b-a)/90[7f(a) + 32f((3a+b)/4) + 12f((a+b)/2)$ $+32f((a+3b)/4) + 7f(b)]$

and their error terms are

Trapezium	$-(b-a)^3/12 f''(\xi)$
Simpson's	$-(b-a)^5/2880 f^{(4)}(\xi)$
Simpson's 3/8	$-(b-a)^5/6480 f^{(4)}(\xi)$
Boole's	$-(b-a)^7/1935360 f^{(6)}(\xi)$

The open Newton–Cotes rules are

Mid-point	$(b-a)f((a+b)/2)$
unnamed	$(b-a)/2[f((2a+b)/3)+f((a+2b)/3)]$
Milne's	$(b-a)/3[2f((3a+b)/4)-f((a+b)/2)+2f((a+3b)/4)]$

and their error terms are

Mid-point	$(b-a)^3/24 f''(\xi)$
unnamed	$(b-a)^3/36 f''(\xi)$
Milne's	$7(b-a)^5/23040 f^{(4)}(\xi)$

High-order Newton–Cotes Rules are rarely used, for two reasons. Firstly, for larger n some of the weights are negative, which leads to numerical instability. Secondly, methods based on high-order polynomials with equally spaced points have the same disadvantages as high-order polynomial interpolation, as we have seen with Runge's phenomenon.

5.5 Gaussian Quadrature

Since equally spaced abscissae cause problems in quadrature rules, the next step is to choose the abscissae $x_1, \ldots, x_n$ such that the formula is accurate for the highest possible degree of polynomial. These rules are known as *Gaussian quadrature* or *Gauss–Christoffel quadrature*.

To construct these quadrature rules, we need to revisit orthogonal polynomials. Let $\mathbb{P}_n[x]$ be the space of polynomials of degree at most n. We say that $p_n \in \mathbb{P}_n[x]$ is the nth orthogonal polynomial over (a, b) with respect to the weight function $w(x)$, where $w(x) > 0$ for all $x \in [a, b]$, if for all $p \in \mathbb{P}_{n-1}[x]$

$$\int_a^b p_n(x)p(x)w(x)dx = 0.$$

Theorem 5.6. *All roots of p_n are real, have multiplicity one, and lie in (a, b).*

Proof. Let $x_1, \ldots, x_m$ be the places where p_n changes sign in (a, b). These are roots of p_n, but p_n does not necessarily change sign at every root (e.g., if the root has even multiplicity where the curve just touches the x-axis, but does not cross it). For p_n to change sign the root must have odd multiplicity. There could be no places in (a, b) where p_n changes sign, in which case $m = 0$. We know that $m \leq n$, since p_n has at most n real roots.

The polynomial $(x - x_1)(x - x_2) \cdots (x - x_m)$ changes sign in the same way as p_n. Hence the product of the two does not change sign at all in (a, b). Now p_n is orthogonal to all polynomials of degree less than n. Thus,

$$\int_a^b (x - x_1)(x - x_2) \cdots (x - x_m) p_n(x) w(x) dx = 0,$$

unless $m = n$. However, the integrand is non-zero and never changes sign. Therefore we must have $n = m$, and $x_1, \ldots, x_n$ are the n distinct zeros of p_n in (a, b). $\square$

Gaussian quadrature based on orthogonal polynomials is developed as follows. Let $x_1, \ldots, x_n$ be the roots of the nth orthogonal polynomial p_n. Let L_i be the ith Lagrange interpolating polynomial for these points, i.e., L_i is the unique polynomial of degree $n - 1$ such that $L_i(x_i) = 1$ and $L_i(x_k) = 0$ for $k \neq i$. The weights for the quadrature rule are then calculated according to

$$w_i = \int_a^b L_i(x) w(x) dx$$

and we have

$$\int_a^b f(x) w(x) dx \approx \sum_{i=1}^n w_i f(x_i).$$

Theorem 5.7. *The above approximation is exact if f is a polynomial of degree at most $2n - 1$*

Proof. Suppose f is a polynomial of degree at most $2n - 1$. Then $f = p_n q + r$, where $q, r \in \mathbb{P}_{n-1}[x]$. Since p_n is orthogonal to all polynomials of degree at most $n - 1$, we have

$$\int_a^b f(x) w(x) dx = \int_a^b [p_n(x) q(x) + r(x)] w(x) dx = \int_a^b r(x) w(x) dx.$$

On the other hand, $r(x) = \sum_{i=1}^n r(x_i) L_i(x)$ and $f(x_i) = p_n(x_i) q(x_i) + r(x_i) = r(x_i)$, since p_n vanishes at $x_1, \ldots, x_n$. Therefore

$$\int_a^b r(x) w(x) dx = \sum_{i=1}^n f(x_i) \int_a^b L_i(x) w(x) dx = \sum_{i=1}^n w_i f(x_i)$$

and the assertion follows. $\square$

Exercise 5.2. *Calculate the weights w_0 and w_1 and the abscissae x_0 and x_1 such that the approximation*

$$\int_0^1 f(x)dx \approx w_0 f(x_0) + w_1 f(x_1)$$

is exact when f is a cubic polynomial. You may use the fact that x_0 and x_1 are the zeros of a quadratic polynomial which is orthogonal to all linear polynomials. Verify your calculation by testing the formula when $f(x) = 1, x, x^2$ and x^3.

The consequence of the theorem is that using equally spaced points, the resulting method is only necessarily exact for polynomials of degree at most $n - 1$. By picking the abscissae carefully, however, a method results which is exact for polynomials of degree up to $2n - 1$. For the price of storing the same number of points, one gets much more accuracy for the same number of function evaluations. However, if the values of the integrand are given as empirical data, where it was not possible to choose the abscissae, Gaussian quadrature is not appropriate.

The weights of Gaussian quadrature rules are positive. Consider $L_k^2(x)$, which is a polynomial of degree $2n - 2$, and thus the quadrature formula is exact for it and we have

$$0 < \int_a^b L_k^2(x)w(x)dx = \sum_{i=0}^n w_i L_k^2(x_i) = w_k.$$

Gaussian quadrature rules based on a weight function $w(x)$ work very well for functions that behave like a polynomial times the weight, something which occurs in physical problems. However, a change of variables may be necessary for this condition to hold.

The most common Gaussian quadrature formula is the case where $(a, b) = (-1, 1)$ and $w(x) \equiv 1$. The orthogonal polynomials are then called *Legendre polynomials*. To construct the quadrature rule, one must determine the roots of the Legendre polynomial of degree n and then calculate the associated weights.

As an example, let $n = 2$. The quadratic Legendre polynomial is $x^2 - 1/3$ with roots $\pm 1/\sqrt{3}$. The two interpolating Lagrange polynomials are $(\sqrt{3}x + 1)/2$ and $(-\sqrt{3}x+1)/2$ and both integrate to 1 over $[-1, 1]$. Thus the *two-point Gauss–Legendre rule* is given by

$$\int_{-1}^1 f(x)dx \approx f(-\frac{1}{\sqrt{3}}) + f(\frac{1}{\sqrt{3}})$$

and it is correct for all cubic polynomials.

For $n \leq 5$, the largest degree of polynomial for which the quadrature is correct, the abscissae and corresponding weights are given in the following table,

n	$2n-1$	abscissae	weights
2	3	$\pm 1/\sqrt{3}$	1
3	5	0	8/9
		$\pm\sqrt{3/5}$	5/9
4	7	$\pm\sqrt{(3-2\sqrt{6/5})/7}$	$(18+\sqrt{30})/36$
		$\pm\sqrt{(3+2\sqrt{6/5})/7}$	$(18-\sqrt{30})/36$
5	9	0	128/225
		$\pm\frac{1}{3}\sqrt{5-2\sqrt{10/7}}$	$(322+13\sqrt{70})/900$
		$\pm\frac{1}{3}\sqrt{5+2\sqrt{10/7}}$	$(322-13\sqrt{70})/900$

Notice that the abscissae are not uniformly distributed in the interval $[-1,1]$. They are symmetric about zero and cluster near the end points.

Exercise 5.3. *Implement the Gauss–Legendre quadrature for $n = 2,\ldots,5$, approximate $\int_{-1}^{1} x^j dx$ for $j = 1,\ldots,10$, and compare the results to the true solution. Interpret your results.*

Other choices of weight functions are listed in the following table

Name	Notation	Interval	Weight function
Legendre	P_n	$[-1,1]$	$w(x) \equiv 1$
Jacobi	$P_n^{(\alpha,\beta)}$	$(-1,1)$	$w(x) = (1-x)^\alpha (1+x)^\beta$
Chebyshev (first kind)	T_n	$(-1,1)$	$w(x) = (1-x^2)^{-1/2}$
Chebyshev (second kind)	U_n	$[-1,1]$	$w(x) = (1-x^2)^{1/2}$
Laguerre	L_n	$[0,\infty)$	$w(x) = e^{-x}$
Hermite	H_n	$(-\infty,\infty)$	$w(x) = e^{-x^2}$

where $\alpha,\beta > -1$.

Next we turn to the estimation of the error of Gaussian quadrature rules.

Theorem 5.8. *If $f \in C^{2n}[a,b]$ and the integral is approximated by a Gaussian quadrature rule, then*

$$\int_a^b f(x)w(x)dx - \sum_{i=1}^{n} w_i f(x_i) = \frac{f^{(2n)}(\xi)}{(2n)!} \int_a^b [\hat{p}_n(x)]^2 w(x)dx$$

for some $\xi \in (a,b)$, where $\hat{p}_n$ is the nth orthogonal polynomial with respect to $w(x)$, scaled such that the leading coefficient is 1.

Proof. Let q be the polynomial of degree at most $2n-1$, which satisfies the conditions

$$q(x_i) = f(x_i) \text{ and } q'(x_i) = f'(x_i), \qquad i = 1,\ldots,n.$$

This is possible, since these are $2n$ conditions and q has $2n$ degrees of freedom. Since the degree of q is at most $2n - 1$, the quadrature rule is exact

$$\int_a^b q(x)w(x)dx = \sum_{i=1}^n w_i q(x_i) = \sum_{i=1}^n w_i f(x_i).$$

Therefore we can rewrite the error term as

$$\int_a^b f(x)w(x)dx - \sum_{i=1}^n w_i f(x_i) = \int_a^b (f(x) - q(x))w(x)dx.$$

We know that $f(x) - q(x)$ vanishes at $x_1, \ldots, x_n$. Let x be any other fixed point in the interval, say between x_k and x_{k+1}. For $t \in [a, b]$ we define the function

$$\phi(t) := [f(t) - q(t)] \prod_{i=1}^n (x - x_i)^2 - [f(x) - q(x)] \prod_{i=1}^n (t - x_i)^2.$$

For $t = x_j$, $j = 1, \ldots, n$, the first term vanishes, since $f(x_j) = q(x_j)$, and by construction the product in the second term vanishes. We also have $\phi(x) = 0$, since then the two terms cancel. Hence ϕ has at least $n + 1$ distinct zeros in $[a, b]$. By *Rolle's theorem*, if a function with continuous derivative vanishes at two distinct points, then its derivative vanishes at an intermediate point. We deduce that ϕ' has at least one zero between x_j and x_{j+1} for $j = 1, \ldots, k - 1$, and $j = k+1, \ldots, n-1$. It also has two further zeros, one between x_k and x and one between x and x_{k+1}. These are n zeros. However, since $f'(x_i) = q'(x_i)$ for $i = 1, \ldots, n$ we have a further n zeros and thus altogether $2n$ zeros. Applying Rolle's theorem again, the second derivative of ϕ has at least $2n - 1$ zeros. Continuing in this manner, the $(2n)^{\text{th}}$ derivative has at least one zero, say at $\xi(x)$. There is a dependence on x since x was fixed.

Therefore

$$0 = \phi^{(2n)}(\xi(x)) = [f^{(2n)}(t) - q^{(2n)}(t)] \prod_{i=1}^n (x - x_i)^2 - [f(x) - q(x)](2n)!,$$

or after rearranging and using $(x - x_1)^2 \ldots (x - x_n)^2 = [\hat{p}_n(x)]^2$,

$$f(x) - q(x) = \frac{f^{(2n)}(\xi(x))}{(2n)!} [\hat{p}_n(x)]^2.$$

Next the function

$$\frac{f^{(2n)}(\xi(x))}{(2n)!} = \frac{f(x) - q(x)}{[\hat{p}_n(x)]^2}$$

is continuous on $[a, b]$, since the zeros of the denominator are also zeros of the numerator with the same multiplicity. We can apply the mean value theorem of integral calculus:

$$\int_a^b (f(x) - q(x))w(x)dx = \int_a^b \frac{f^{(2n)}(\xi(x))}{(2n)!}[\hat{p}_n(x)]^2 w(x)dx$$

$$= \frac{f^{(2n)}(\xi)}{(2n)!}\int_a^b [\hat{p}_n(x)]^2 w(x)dx$$

for some $\xi \in (a, b)$. □

There are two variations of Gaussian quadrature rules. The *Gauss–Lobatto rules*, also known as *Lobatto quadrature*, explicitly include the end points of the interval as abscissae, $x_1 = a$ and $x_n = b$, while the remaining $n - 2$ abscissae are chosen optimally. The quadrature is then accurate for polynomials up to degree $2n - 3$. For $w(x) \equiv 1$ and $[a, b] = [-1, 1]$, the remaining abscissae are the zeros of the derivative of the $(n - 1)^{\text{th}}$ Legendre polynomial $P_{n-1}(x)$. The Lobatto quadrature of $f(x)$ on $[-1, 1]$ is

$$\int_{-1}^1 f(x)dx \approx \frac{2}{n(n - 1)}[f(1) + f(-1)] + \sum_{i=2}^{n-1} w_i f(x_i),$$

where the weights are given by

$$w_i = \frac{2}{n(n - 1)[P_{n-1}(x_i)]^2}, \qquad i = 2, \ldots, n - 1.$$

Simpson's rule is the simplest Lobatto rule. The following table lists the Lobatto rules with their abscissae and weights until $n = 5$ and the degree of polynomial they are correct for.

n	$2n - 3$	abscissae	weights
3	3	0	4/3
		±1	1/3
4	5	±1/√5	5/6
		±1	1/6
5	7	0	32/45
		±√(3/7)	49/90
		±1	1/10

The second variation are the *Gauss–Radau rules* or *Radau quadratures*. Here one end point is included as abscissa. Therefore we distinguish left and right Radau rules. The remaining $n - 1$ abscissae are chosen optimally. The quadrature is accurate for polynomials of degree up to $2n - 2$. For $w(x) \equiv 1$ and $[a, b] = [-1, 1]$ and $x_1 = -1$, the remaining abscissae are the zeros of the polynomial given by

$$\frac{P_{n-1}(x) + P_n(x)}{1 + x}.$$

The Radau quadrature of $f(x)$ on $[-1, 1]$ is

$$\int_{-1}^{1} f(x)dx \approx \frac{2}{n^2}f(-1) + \sum_{i=2}^{n} w_i f(x_i),$$

where the other weights are given by

$$w_i = \frac{1 - x_i}{n^2[P_{n-1}(x_i)]^2}, \qquad i = 2, \ldots, n.$$

The following table lists the left Radau rules with their abscissae and weights until $n = 5$ and the degree of polynomial they are correct for. For $n = 4$ and 5 only approximations to the abscissae and weights are given.

n	$2n - 2$	abscissae	weights
2	2	-1	$1/2$
		$1/3$	$3/2$
3	4	-1	$2/9$
		$1/5(1 \pm \sqrt{6})$	$1/18(16 \mp \sqrt{6})$
4	6	-1	$1/8$
		-0.575319	0.657689
		0.181066	0.776387
		0.822824	0.440924
5	8	0	$2/25$
		-0.72048	0.446208
		-0.167181	0.623653
		0.446314	0.562712
		0.885792	0.287427

One drawback of Gaussian quadrature is the need to pre-compute the necessary abscissae and weights. Often the abscissae and weights are given in look-up tables for specific intervals. If one has a quadrature rule for the interval $[c, d]$, it can be adapted to the interval $[a, b]$ with a simple change of variables. Let $t(x)$ be the linear transformation taking $[c, d]$ to $[a, b]$ and $t^{-1}(y)$ its inverse,

$$y = t(x) = a + \frac{b - a}{d - c}(x - c),$$

$$x = t^{-1}(y) = c + \frac{d - c}{b - a}(y - a),$$

$$\frac{dy}{dx} = \frac{b - a}{d - c}.$$

The integration is then transformed:

$$\int_{a}^{b} f(y)w(y)dy = \int_{t^{-1}(a)}^{t^{-1}(b)} f(t(x))w(t(x))t'(x)dx$$

$$= \frac{b - a}{d - c}\int_{c}^{d} f(t(x))w(t(x))dx.$$

The integral is then approximated by

$$\int_a^b f(x)w(x)dx \approx \frac{b-a}{d-c}\sum_{i=1}^n w_i f(t(x_i)),$$

where x_i and w_i, $i = 0, \ldots, n$, are the abscissae and weights of a quadrature approximating integrals over $[c, d]$ with weight function $w(t(x))$. Thus the abscissae and weights of the quadrature over $[a, b]$ are

$$\hat{x}_i = t(x_i) \text{ and } \hat{w}_i = \frac{b-a}{d-c}w_i.$$

It is important to note that the change of variables alters the weight function. This does not play a role in the Gauss–Legendre quadrature, since there $w(x) \equiv 1$, but it does for the other Gaussian quadrature rules.

The change of variables technique is important for decomposing the interval of integration into smaller intervals, over each of which Gaussian quadrature rules can be applied. The strategy of splitting the integral into a set of panels is the subject of the next section.

5.6   Composite Rules

A *composite rule* is also known as a *repeated, compound, iterated, or extended rule*. It is constructed by splitting the integral into smaller integrals (usually of the same length) and applying (usually) the same quadrature rule in each sub-interval and summing the results. For example, consider the simple case of the midpoint rule. Figure 5.4 illustrates its composite rule.

Let there be N sub-intervals of equal length and let h denote the width of each sub-interval, i.e., $h = (a - b)/N$. The composite midpoint rule then has the formula

$$\int_a^b f(x)dx \approx h\sum_{i=1}^N f(a + (i - \frac{1}{2})h).$$

The error term is

$$\sum_{i=1}^N \frac{1}{24}f''(\xi_i)h^3 \le \frac{1}{24}\max_{\xi \in [a,b]}|f''(\xi)|Nh^3 = O(h^2),$$

where each ξ_i lies in the interval $[a + (i - 1)h, a + ih]$.

Figure 5.5 illustrates the composite trapezium rule, where the sub-intervals are generated in the same way. It has the formula

$$\int_a^b f(x)dx \approx \frac{h}{2}f(a) + h\sum_{i=1}^{N-1} f(a + ih) + \frac{h}{2}f(b),$$

because the function values at the interior abscissae are needed twice to approximate the integrals on the intervals on the left and right of them. Since the

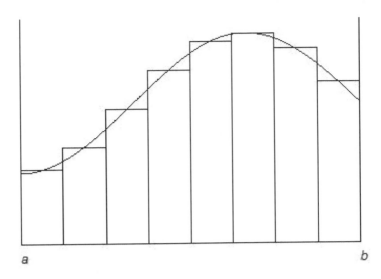

Figure 5.4 The composite midpoint rule

error in each sub-interval is of magnitude $O(h^3)$ and there are N sub-intervals, the overall error is $O(h^2)$.

The composite Simpson's rule has the formula

$$\int_a^b f(x)dx \approx \frac{h}{6}\left[f(a) + 2\sum_{i=1}^{N-1} f(a+ih) + 4\sum_{i=1}^{N} f(a+(i-\frac{1}{2})h) + f(b)\right].$$

On each sub-interval the error is $O(h^5)$, and since there are N sub-intervals, the overall error is $O(h^4)$.

Because evaluating an arbitrary function can be potentially expensive, the efficiency of quadrature rules is usually measured by the number of function evaluations required to achieve a desired accuracy. In composite rules it is therefore advantageous, if the endpoints are abscissae, since the function value at the end-point of one sub-interval will be used again in the next sub-interval. Therefore Lobatto rules play an important role in the construction of composite rules. In the following we put this on a more theoretical footing.

Definition 5.2 (Riemann integral). *For each $n \in \mathbb{N}$ let there be a set of numbers $a = \xi_0 < \xi_1 < \ldots < \xi_n = b$. A Riemann integral is defined by*

$$\int_a^b f(x)dx = \lim_{n\to\infty, \Delta\xi\to 0} \sum_{i=1}^{n} (\xi_i - \xi_{i-1})f(x_i),$$

where $x_i \in [\xi_{i-1}, \xi_i]$ and $\Delta\xi = \max_{1\leq i\leq n} |\xi_i - \xi_{i-1}|$. The sum on the right-hand side is called a Riemann sum.

Some simple quadrature rules are clearly Riemann sums. For example,

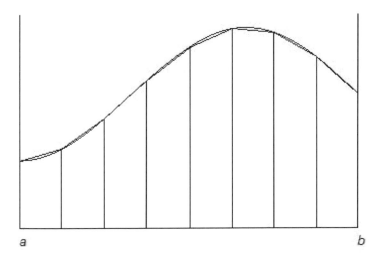

a   b

Figure 5.5 The composite trapezium rule

take the *composite rectangle rule*, which approximates the value on each sub-interval by the function value at the right end-point times the length of the interval

$$Q_N(f) = h \sum_{i=1}^{N} f(a + ih),$$

where $h = (b - a)/N$. This is illustrated in Figure 5.6.

Other quadrature rules of the form

$$Q_n(f) = \sum_{i=1}^{n} w_i f(x_i)$$

can be expressed as Riemann sums as long as we can set $\xi_0 = a$ and $\xi_n = b$ and there exist $\xi_1 < \xi_2 < \ldots < \xi_{n-1}$ such that $x_i \in [\xi_{i-1}, \xi_i]$ and $w_i = \xi_i - \xi_{i-1}$.

To show that a sequence of quadrature rules $Q_n(f)$, $n = 1, 2, \ldots$ converges to the integral of f, it is sufficient to prove that the sequence of quadrature rules is a sequence of Riemann sums where $\Delta \xi \to 0$ as $n \to \infty$.

Let Q_n be an n-point quadrature rule. We denote by $(M \times Q_n)$ the rule Q_n applied to M sub-intervals. If Q_n is an *open rule*, that is, it does not include the endpoints, then $(M \times Q_n)$ uses Mn points. However, if Q_n is a *closed rule*, that is, it includes both end-points, then $M \times Q_n$ uses only $(n-1)M + 1$ points, which is $M - 1$ less function evaluations.

Let the quadrature rule be given on $[c, d]$ by

$$Q_n(f) = \sum_{j=1}^{n} w_j f(x_j).$$

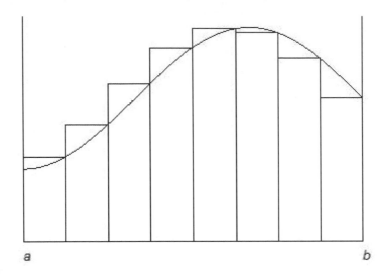

a b

Figure 5.6   The composite rectangle rule

Then the composite rule on $[a, b]$ is given by

$$(M \times Q_n)(f) = \frac{b-a}{M(d-c)} \sum_{i=1}^{M} \sum_{j=1}^{n} w_j f(x_{ij}), \tag{5.1}$$

where x_{ij} is the jth abscissa in the ith sub-interval calculated as $x_{ij} = t_i(x_j)$, where t_i is the transformation taking $[c, d]$ of length $d - c$ to $[a + (i - 1)(b - a)/M, a + i(b - a)/M]$ of length $(b - a)/M$.

Theorem 5.9. *Let Q_n be a quadrature rule that integrates constants exactly, i.e., $Q_n(1) = \int_c^d 1 dx = d-c$. If f is bounded on $[a, b]$ and is Riemann integrable then*

$$\lim_{M \to \infty} (M \times Q_n)(f) = \int_a^b f(x)dx.$$

Proof. Firstly note that

$$Q_n(1) = \sum_{j=1}^{n} w_j = d - c.$$

That is, the weights sum to $d - c$. Swapping the summations and taking everything independent of M out of the limit in (5.1) leads to

$$\lim_{M \to \infty}(M \times Q_n)(f) = \frac{1}{d-c} \sum_{j=1}^{n} w_j \lim_{M \to \infty} \left[\frac{b-a}{M} \sum_{i=1}^{M} f(x_{ij}) \right]$$

$$= \int_a^b f(x)dx,$$

since the expression in the square brackets is a Riemann sum with $\xi_i = a + i(b-a)/M$ for $i = 1, \ldots, M$. □

Definition 5.3. *The quadrature rule Q_n on $[a,b]$ is called a* simplex rule *if the error can be expressed as*

$$E_{Q_n}(f) = \int_a^b f(x)w(x)dx - Q_n(f) = C(b-a)^{k+1}f^{(k)}(\xi),$$

for $f \in C^k[a,b]$, $\xi \in (a,b)$, and some constant C.

We have already seen that the error can be expressed in such a form for all quadrature rules we have encountered so far.

Theorem 5.10. *Let Q_n be a simplex rule as defined above and let $E_{M \times Q_n}(f)$ denote the error of $(M \times Q_n)(f)$. Then*

$$\lim_{M \to \infty} M^k E_{M \times Q_n}(f) = C(b-a)^k \left[f^{(k-1)}(b) - f^{(k-1)}(a) \right].$$

That is, $(M \times Q_n)(f)$ converges to $\int_a^b f(x)w(x)dx$ like M^{-k} for sufficiently large M.

Proof. The error of the composite rule is the sum of the errors in each sub-interval. Thus

$$E_{M \times Q_n}(f) = \sum_{i=1}^M C \left(\frac{b-a}{M} \right)^{k+1} f^{(k)}(\xi_i),$$

where ξ_i lies inside the ith sub-interval. Multiplying by M^k and taking the limit gives

$$\lim_{M \to \infty} M^k E_{M \times Q_n}(f) = C(b-a)^k \lim_{M \to \infty} \left[\frac{b-a}{M} \sum_{i=1}^M f^{(k)}(\xi_i) \right].$$

The expression in the square brackets is a Riemann sum and therefore

$$\begin{aligned}
\lim_{M \to \infty} M^k E_{M \times Q_n}(f) &= C(b-a)^k \int_a^b f^{(k)}(x)dx \\
&= C(b-a)^k \left[f^{(k-1)}(b) - f^{(k-1)}(a) \right].
\end{aligned}$$

□

So far we have only looked at one-dimensional integration. The next section looks at several dimensions.

5.7 Multi-Dimensional Integration

Consider the problem of evaluating

$$I = \int_{l_1}^{u_1} \int_{l_2(x_1)}^{u_2(x_1)} \cdots \int_{l_d(x_1,\ldots,x_{d-1})}^{u_d(x_1,\ldots,x_{d-1})} f(x_1, x_2, \ldots, x_d) dx_d dx_{d-1} \ldots dx_1, \quad (5.2)$$

where there are d integrals and where the boundaries of each integral may depend on variables not used in that integral. In the kth dimension the interval of integration is $[l_k(x_1, \ldots, x_{k-1}), u_k(x_1, \ldots, x_{k-1})]$. Such problems often arise in practice, mostly for two or three dimensions, but sometimes for 10 or 20 dimensions. The problem becomes considerably more expensive with each extra dimension. Therefore different methods have been developed for different ranges of dimensions.

We first consider the transformation into standard regions with the hypercube as an example. Other standard regions are the hypersphere, the surface of the hypersphere, or a simplex, where a simplex is the generalization of the triangle or tetrahedron to higher dimensions. Different methods have been developed for different standard regions. Returning to the hypercube, it can be transformed to the region of the integral given in (5.2) by

$$x_i = \frac{1}{2} [u_i(x_1, \ldots, x_{i-1}) + l_i(x_1, \ldots, x_{i-1})]$$

$$+ y_i \frac{1}{2} [u_i(x_1, \ldots, x_{i-1}) - l_i(x_1, \ldots, x_{i-1})].$$

If $y_i = -1$, then $x_i = l_i(x_1, \ldots, x_{i-1})$, the lower limit of the integration. If $y_i = 1$, then $x_i = u_i(x_1, \ldots, x_{i-1})$, the upper limit of the integration. For $y_i = 0$ x_i is the midpoint of the interval. The derivative of this transformation is given by

$$\frac{dx_i}{dy_i} = \frac{1}{2} [u_i(x_1, \ldots, x_{i-1}) - l_i(x_1, \ldots, x_{i-1})].$$

Using the transformation, we write $f(x_1, \ldots, x_d) = g(y_1, \ldots, y_d)$ and the integral I becomes

$$\int_{-1}^{1} \cdots \int_{-1}^{1} \frac{1}{2^d} \prod_{i=1}^{d} [u_i(x_1, \ldots, x_{i-1}) - l_i(x_1, \ldots, x_{i-1})] g(y_1, \ldots, y_d) dy_1 \ldots dy_d.$$

We can now use a standard method. In the following we give an example of how such a method can be derived.

If $I_1 = [a, b]$ and $I_2 = [c, d]$ are intervals, then the product $I_1 \times I_2$ denotes the rectangle $a \leq x \leq b, c \leq y \leq d$. Let Q_1 and Q_2 denote quadrature rules over I_1 and I_2 respectively,

$$Q_1(f) = \sum_{i=1}^{n} w_{1i} f(x_{1i}) \text{ and } Q_2(g) = \sum_{j=1}^{m} w_{2j} g(x_{2j}),$$

where $x_{1i} \in [a, b]$, $i = 1, \ldots, n$, are the abscissae of Q_1 and $x_{2j} \in [c, d]$, $j = 1, \ldots, m$, are the abscissae of Q_2.

Definition 5.4. *The* product rule $Q_1 \times Q_2$ *to integrate a function* $F : I_1 \times I_2 \to \mathbb{R}$ *is defined by*

$$(Q_1 \times Q_2)(F) = \sum_{i=1}^{n} \sum_{j=1}^{m} w_{1i} w_{2j} F(x_{1i}, x_{2j}).$$

Exercise 5.4. *Let* Q_1 *integrate* f *exactly over the interval* I_1 *and let* Q_2 *integrate* g *exactly over the interval* I_2. *Prove that* $Q_1 \times Q_2$ *integrates* $f(x)g(y)$ *over* $I_1 \times I_2$ *exactly.*

A consequence of the above definition and exercise is that we can combine all the one-dimensional quadrature rules we encountered before to create rules in two dimensions. If Q_1 is correct for polynomials of degree at most k_1 and Q_2 is correct for polynomials of degree at most k_2, then the product rule is correct for any linear combination of the monomials $x^i y^j$, where $i = 0, \ldots, k_1$, and $j = 0, \ldots, k_2$. As an example let Q_1 and Q_2 both be Simpson's rule. The product rule is then given by

$$\begin{aligned} Q_1 \times Q_2 \ = \ \frac{(b-a)(d-c)}{36} \ &\Big[F(a, c) + 4F(\tfrac{a+b}{2}, c) + F(b, c) \\ &+ 4F(a, \tfrac{c+d}{2}) + 16F(\tfrac{a+b}{2}, \tfrac{c+d}{2}) + 4F(b, \tfrac{c+d}{2}) \\ & F(a, d) + 4F(\tfrac{a+b}{2}, d) + F(b, d) \Big], \end{aligned}$$

and it is correct for $1, x, y, x^2, xy, y^2, x^3, x^2 y, xy^2, y^3, x^3 y, x^2 y^2, xy^3, x^3 y^2, x^2 y^3$, and $x^3 y^3$.

Note that the product of two Gauss rules is not a Gauss rule. The optimal abscissae in more than one dimension have different positions. Gauss rules in more than one dimension exist, but are rarely used, since Gauss rules in higher dimensions do not always have positive weights and the abscissae might lie outside the region of integration, which causes a problem if the integrand is not defined there.

The idea of product rules generalize to d dimensions, but they become increasingly inefficient due to the *dimensional effect*. The error of most methods behaves like $n^{-k/d}$ as the number of dimensions increases, and where k is a constant depending on the method. Product rules are generally not used for more than three dimensions.

To improve accuracy in one dimension we normally half the sub-intervals. Thus n sub-intervals become $2n$ sub-intervals. If we have n hypercubes in d dimensions which subdivide the region and we half each of its sides, then we have $2^d n$ hypercubes. With each subdivision the number of function evaluations rises quickly, even if the abscissae lie on the boundaries and the function values there can be reused. Consequently for large d different methods are required. In a large number of dimensions analytical strategies such as changing the order of integration or even just reducing the number of dimensions by one can have a considerable effect. In high dimensional problems, high accuracy is often not required. Often merely the magnitude of the integral is sufficient. Note that high dimensional integration is a problem well-suited to parallel computing.

5.8   Monte Carlo Methods

In the following we will shortly introduce *Monte Carlo methods*. However, their probabilistic analysis is beyond this course and we will just state a few facts. Monte Carlo methods cannot be used to obtain high accuracy, but they are still very effective when dealing with many dimensions. Abscissae are generated pseudo-randomly. The function values at these abscissae are then averaged and multiplied by the volume of the region to estimate the integral.

The accuracy of Monte Carlo methods is generally very poor, even in one dimension. However, the dimensional effect, i.e., the loss of efficiency when extending to many dimensions, is less severe than in other methods.

As an example consider the one-dimensional case where we seek to calculate $I = \int_0^1 f(x)dx$. Let x_i, $i = 1, \ldots, n$ be a set of pseudo-random variables. Note that pseudo-random numbers are not really random, but are sets of numbers which are generated by an entirely deterministic causal process, which is easier than producing genuinely random numbers. The advantage is that the resulting numbers are reproducable, which is important for testing and fixing software. Pseudo randomness is defined as the set of numbers being non-distinguishable from the uniform distribution by any of several tests. It is an open question whether a test could be developed to distinguish a set of pseudo-random numbers from the uniform distribution.

The *strong law of large numbers* states that the probability of

$$\lim_{n \to \infty} \frac{1}{n} \sum_{i=1}^n f(x_i) = I$$

is one. That is, using truly random abscissae and equal weights, convergence is *almost sure*.

Statistical error estimates are available, but these depend on the variance of the function f. As an example for the efficiency of the Monte Carlo method, to obtain an error less than 0.01 with 99% certainty in the estimation of I, we need to average 6.6×10^4 function values. To gain an extra decimal place with the same certainty requires 6.6×10^6 function evaluations.

The advantage of the Monte Carlo method is that the error behaves like $n^{-1/2}$ and not like $n^{-k/d}$, that is, it is independent of the number of dimensions. However, there is still a dimensional effect, since higher dimensional functions have larger variances.

As a final algorithm the *Korobov–Conroy* method needs to be mentioned. Here the abscissae are not chosen pseudo-randomly, but are in some sense optimal. They are derived from results in number theory (See for example [15] H. Niederreiter, *Random Number Generation and Quasi-Monte Carlo Methods*).

5.9   Revision Exercises

Exercise 5.5. *Consider the numerical evaluation of an integral of the form*

$$I = \int_a^b f(x)w(x)dx.$$

(a) *Define Gaussian quadrature and state how the abscissae are obtained. Give a formula for the weights. If f is a polynomial, what is the maximum degree of f for which the Gaussian quadrature rule is correct?*

(b) *In the following let the interval be $[a, b] = [-2, 2]$ and $w(x) = 4 - x^2$. Thus we want to approximate the integral*

$$\int_{-2}^2 (4 - x^2) f(x)dx.$$

Let the number of abscissae be 2. Calculate the abscissae.

(c) *Calculate the weights.*

(d) *To approximate the integral*

$$\int_{-1}^1 (1 - x^2) f(x)dx$$

by a Gaussian quadrature the orthogonal polynomials are the Jacobi polynomials for $\alpha = 1$ and $\beta = 1$. For $n = 2$ the abscissae are $x_1 = -1/\sqrt{5}$ and $x_2 = 1/\sqrt{5}$. The weights are $w_1 = w_2 = 2/3$. The interval of integration is changed from $[-1, 1]$ to $[-2, 2]$. What are the new abscissae and weights? Explain why the weights are different from the weights derived in the previous part.

Exercise 5.6. *Let $f \in \mathbb{C}^{k+1}[a, b]$, that is, f is $k + 1$ times continuously differentiable.*

(a) *Expand $f(x)$ in a Taylor series using the integral form of the remainder.*

(b) *The integral of f over $[a, b]$ is approximated numerically in such a way that the approximation is correct whenever f is a polynomial of degree k or less. Let $L(f)$ denote the approximation error. Show that $L(f)$ can be calculated as*

$$L(f) = \frac{1}{k!} \int_a^b K(\theta) f^{(k+1)}(\theta)d\theta.$$

In the process specify $K(\theta)$ and state which conditions $L(f)$ has to satisfy.

(c) *If K does not change sign in (a, b), how can the expression for $L(f)$ be further simplified?*

(d) *In the following we let $a = 0$ and $b = 2$ and let*

$$L(f) = \int_0^2 f(x)dx - \frac{1}{3}[f(0) + 4f(1) + f(2)].$$

Find the highest degree of polynomials for which this approximation is correct.

(e) *Calculate $K(\theta)$ for $\theta \in [0, 2]$.*

(f) *Given that $K(\theta)$ is negative for $\theta \in [0, 2]$, obtain c such that*

$$L(f) = cf^{(4)}(\xi)$$

for some $\xi \in (0, 2)$.

(g) *The above is the Simpson rule of numerical integration on the interval $[0, 2]$. State two other rules of numerical integration on the interval $[0, 2]$.*

Exercise 5.7. *(a) Describe what is meant by a composite rule of integration.*

(b) *Give two examples of composite rules and their formulae.*

(c) *Let a quadrature rule be given on $[c, d]$ by*

$$Q_n(f) = \sum_{j=1}^{n} w_j f(x_j) \approx \int_c^d f(x)dx.$$

We denote by $(M \times Q_n)$ the composite rule Q_n applied to M subintervals of $[a, b]$. Give the formula for $(M \times Q_n)$.

(d) *Describe the difference between open and closed quadrature rules and how this affects the composite rule.*

(e) *Show that if Q_n is a quadrature rule that integrates constants exactly, i.e., $Q_n(1) = \int_c^d 1dx = d - c$, and if f is bounded on $[a, b]$ and is Riemann integrable, then*

$$\lim_{M \to \infty} (M \times Q_n)(f) = \int_a^b f(x)dx.$$

(f) *Let $[c, d] = [-1, 1]$. Give the constant, linear, and quadratic monic polynomials which are orthogonal with respect to the inner product given by*

$$\langle f, g \rangle = \int_{-1}^1 f(x)g(x)dx$$

and check that they are orthogonal to each other.

(g) *Give the abscissae of the two-point Gauss–Legendre rule on the interval* $[-1, 1]$.

(h) *The weights of the two-point Gauss–Legendre rule are 1 for both abscissae. State the two-point Gauss–Legendre rule and give the formula for the composite rule on $[a, b]$ employing the two-point Gauss–Legendre rule.*

Exercise 5.8. *The integral*

$$\int_{-1}^{1} (1 - x^2) f(x) dx$$

is approximated by a Gaussian quadrature rule of the form

$$\sum_{i=1}^{n} w_i f(x_i),$$

which is exact for all $f(x)$ that are polynomials of degree less than or equal to $2n - 1$.

(a) *Explain how the weights w_i are calculated, writing down explicit expressions in terms of integrals.*

(b) *Explain why it is necessary that the x_i are the zeros of a (monic) polynomial p_n of degree n that satisfies $\int_{-1}^{1}(1 - x^2)p_n(x)q(x)dx = 0$ for any polynomial $q(x)$ of degree less than n.*

(c) *The first such polynomials are $p_0 = 1, p_1 = x, p_2 = x^2 - \frac{1}{5}, p_3 = x^3 - \frac{3}{7}x$. Show that the Gaussian quadrature formulae for $n = 2, 3$ are*

$$n = 2: \quad \frac{2}{3}\left[f(-\frac{1}{\sqrt{5}}) + f(\frac{1}{\sqrt{5}}) \right],$$

$$n = 3: \quad \frac{14}{45}\left[f(-\sqrt{\frac{3}{7}}) + f(\sqrt{\frac{3}{7}}) \right] + \frac{32}{45} f(0).$$

(d) *Verify the result for $n = 3$ by considering $f(x) = 1, x^2, x^4$.*

Exercise 5.9. *The integral*

$$\int_{0}^{2} f(x) dx$$

shall be approximated by a two point Gaussian quadrature formula.

(a) *Find the monic quadratic polynomial $g(x)$ which is orthogonal to all linear polynomials with respect to the scalar product*

$$\langle f, g \rangle = \int_{0}^{2} f(x)g(x)dx,$$

where $f(x)$ denotes an arbitrary linear polynomial.

(b) *Calculate the zeros of the polynomial found in (a) and explain how they are used to construct a Gaussian quadrature rule.*

(c) *Describe how the weights are calculated for a Gaussian quadrature rule and calculate the weights to approximate $\int_0^2 f(x)dx$.*

(d) *For which polynomials is the constructed quadrature rule correct?*

(e) *State the functional $L(f)$ acting on f describing the error when the integral $\int_0^2 f(x)dx$ is approximated by the quadrature rule.*

(f) *Define the Peano kernel and state the Peano kernel theorem.*

(g) *Calculate the Peano kernel for the functional $L(f)$ in (e).*

(h) *The Peano kernel does not change sign in $[0, 2]$ (not required to be proven). Derive an expression for $L(f)$ of the form constant times a derivative of f. (Hint: $(a + b)^4 = a^4 + 4a^3b + 6a^2b^2 + 4ab^3 + b^4$.)*

ODEs

We wish to approximate the exact solution of the *ordinary differential equation (ODE)*

$$\frac{\partial \mathbf{y}}{\partial t} = \mathbf{y}' = \mathbf{f}(t, \mathbf{y}), \qquad t \geq 0, \tag{6.1}$$

where $\mathbf{y} \in \mathbb{R}^N$ and the function $\mathbf{f} : \mathbb{R} \times \mathbb{R}^N \to \mathbb{R}^N$ is sufficiently well behaved. The equation is accompanied by the initial condition $\mathbf{y}(0) = \mathbf{y}_0$.

The following definition is central in the analysis of ODEs.

Definition 6.1 (Lipschitz continuity). $\mathbf{f}$ *is Lipschitz continuous, if there exists a bound* $\lambda \geq 0$ *such that*

$$\|\mathbf{f}(t, \mathbf{v}) - \mathbf{f}(t, \mathbf{w})\| \leq \lambda \|\mathbf{v} - \mathbf{w}\|, \qquad t \in [0, t^*], \qquad \mathbf{v}, \mathbf{w} \in \mathbb{R}^N.$$

Lipschitz continuity means that the slopes of all secant lines to the function between possible points $\mathbf{v}$ and $\mathbf{w}$ are bounded above by a positive constant. Thus a Lipschitz continuous function is limited in how much and how fast it can change. In the theory of differential equations, Lipschitz continuity is the central condition of the *Picard–Lindelöf theorem*, which guarantees the existence and uniqueness of a solution to an initial value problem.

For our analysis of numerical solutions we henceforth assume that $\mathbf{f}$ is analytic and we are always able to expand locally into a Taylor series.

We want to calculate $\mathbf{y}_{n+1} \approx \mathbf{y}(t_{n+1})$, $n = 0, 1, \ldots$, from $\mathbf{y}_0, \mathbf{y}_1, \ldots, \mathbf{y}_n$, where $t_n = nh$ and the *time step* $h > 0$ is small.

6.1 One-Step Methods

In a *one-step method* $\mathbf{y}_{n+1}$ is only allowed to depend on t_n, $\mathbf{y}_n$, h and the ODE given by Equation (6.1). The slope of $\mathbf{y}$ at $t = 0$ is given by $\mathbf{y}'(0) = \mathbf{f}(0, \mathbf{y}(0)) = \mathbf{f}(0, \mathbf{y}_0)$. The most obvious approach is to truncate the Taylor series expansion $\mathbf{y}(h) = \mathbf{y}(0) + h\mathbf{y}'(0) + \frac{1}{2}h^2\mathbf{y}''(0) + \cdots$ before the h^2 term. Thus we set $\mathbf{y}_1 = \mathbf{y}_0 + h\mathbf{f}(0, \mathbf{y}_0)$. Following the same principle, we advance from h to $2h$ by letting $\mathbf{y}_2 = \mathbf{y}_1 + h\mathbf{f}(t_1, \mathbf{y}_1)$. However, note that the second

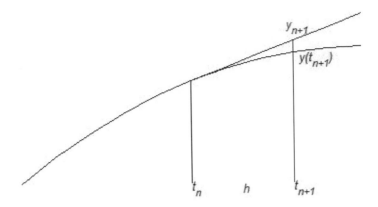

Figure 6.1 The forward Euler method

term is an approximation to $\mathbf{y}'(h)$, since $\mathbf{y}_1$ is an approximation to $\mathbf{y}(h)$. Carrying on, we obtain the *Euler method*

$$\mathbf{y}_{n+1} = \mathbf{y}_n + h\mathbf{f}(t_n, \mathbf{y}_n), \qquad n = 0, 1, \dots. \qquad (6.2)$$

Figure 6.1 illustrates the method.

Exercise 6.1. *Let* $h = \frac{1}{M}$, *where* M *is a positive integer. The following ODEs are given*

$$y' = -\frac{y}{1+t} \qquad \text{and} \qquad y' = \frac{2y}{1+t}, \qquad 0 \le t \le 1,$$

with starting conditions $y_0 = y(0) = 1$ *in both cases. Forward Euler is used to calculate the estimates* y_n, $n = 1, \dots M$. *By using induction and by canceling as many terms as possible in the resultant products, deduce simple explicit expressions for* y_n, $n = 1, \dots M$, *which should be free from summations and products. By considering the limit for* $h \to 0$, *deduce the exact solutions of the equations. Verify that the error* $|y_n - y(t_n)|$ *is at most* $O(h)$.

Definition 6.2 (Convergence). *Let* $t^* > 0$ *be given. A method which produces for every* $h > 0$ *the solution sequence* $\mathbf{y}_n = \mathbf{y}_n(h)$, $n = 0, 1, \dots, \lfloor t^*/h \rfloor$ *converges, if, as* $h \to 0$ *and* $n_k(h)h \xrightarrow{k \to \infty} t$, *it is true that* $\mathbf{y}_{n_k} \to \mathbf{y}(t)$, *the exact solution of Equation (6.1), uniformly for* $t \in [0, t^*]$.

Theorem 6.1. *Suppose that* $\mathbf{f}$ *is Lipschitz continuous as in definition 6.1. Then the Euler method given by (6.2) converges.*

Proof. Let $\mathbf{e}_n = \mathbf{y}_n - \mathbf{y}(t_n)$ be error at step n, where $0 \le n \le \lfloor t^*/h \rfloor$. Thus

$$\mathbf{e}_{n+1} = \mathbf{y}_{n+1} - \mathbf{y}(t_{n+1}) = [\mathbf{y}_n + h\mathbf{f}(t_n, \mathbf{y}_n)] - [\mathbf{y}(t_n) + h\mathbf{y}'(t_n) + O(h^2)].$$

The $O(h^2)$ term can be bounded uniformly for all $[0, t^*]$ by ch^2 for some $c > 0$. Using (6.1) and the triangle inequality, we can deduce

$$
\begin{aligned}
\|\mathbf{e}_{n+1}\| &\le \|\mathbf{y}_n - \mathbf{y}(t_n)\| + h\|\mathbf{f}(t_n, \mathbf{y}_n) - \mathbf{f}(t_n, \mathbf{y}(t_n))\| + ch^2 \\
&\le \|\mathbf{y}_n - \mathbf{y}(t_n)\| + h\lambda\|\mathbf{y}_n - \mathbf{y}(t_n)\| + ch^2 = (1 + h\lambda)\|\mathbf{e}_n\| + ch^2,
\end{aligned}
$$

where we used the Lipschitz condition in the second inequality.

By induction, and bearing in mind that $\mathbf{e}_0 = 0$, we have

$$\|\mathbf{e}_{n+1}\| \le ch^2 \sum_{j=0}^{n} (1 + h\lambda)^j = ch^2 \frac{(1 + h\lambda)^{n+1} - 1}{(1 + h\lambda) - 1} \le \frac{ch}{\lambda}(1 + h\lambda)^{n+1}.$$

Looking at the expansion of $e^{h\lambda}$, we see that $1 + h\lambda \le e^{h\lambda}$, since $h\lambda > 0$. Thus

$$\frac{ch}{\lambda}(1 + h\lambda)^{n+1} \le \frac{ch}{\lambda}e^{(n+1)h\lambda} \le \frac{ce^{t^*\lambda}}{\lambda}h = Ch,$$

where we use the fact that $(n + 1)h \le t^*$. Thus $\|\mathbf{e}_n\|$ converges uniformly to zero and the theorem is true. $\qquad\square$

Exercise 6.2. *Assuming that* $\mathbf{f}$ *is Lipschitz continuous and possesses a bounded third derivative in* $[0, t^*]$, *use the same method of analysis to show that the trapezoidal rule*

$$\mathbf{y}_{n+1} = \mathbf{y}_n + \frac{1}{2}h[\mathbf{f}(t_n, \mathbf{y}_n) + \mathbf{f}(t_{n+1}, \mathbf{y}_{n+1})]$$

converges and that $\|\mathbf{y}_n - \mathbf{y}(t_n)\| \le ch^2$ *for some* $c > 0$ *and all* n *such that* $0 \le nh \le t^*$.

6.2   Multistep Methods, Order, and Consistency

Generally we define a multistep method in terms of a map ϕ_h such that the next approximation is given by

$$\mathbf{y}_{n+1} = \phi_h(t_n, \mathbf{y}_0, \mathbf{y}_1, \ldots \mathbf{y}_{n+1}). \tag{6.3}$$

If $\mathbf{y}_{n+1}$ appears on the right-hand side, then the method is called *implicit*, otherwise *explicit*.

Definition 6.3 (Order, local truncation/discretization error). *The* order *of a numerical method given by (6.3) to obtain solutions for (6.1) is the largest integer* $p \ge 0$ *such that*

$$\delta_{n+1,h} = \mathbf{y}(t_{n+1}) - \phi_h(t_n, \mathbf{y}(t_0), \mathbf{y}(t_1), \ldots \mathbf{y}(t_{n+1})) = O(h^{p+1})$$

for all $h > 0$, $n \ge 0$ *and all sufficiently smooth functions* $\mathbf{f}$. *The expression* $\delta_{n+1,h}$ *is called the* local truncation error *or* local discretization error.

Note that the arguments in ϕ_h are the true function values of $\mathbf{y}$. Thus the local truncation error is the difference between the true solution and the method applied to the true solution. Hence it only gives an indication of the error if all previous steps have been exact.

Definition 6.4 (Consistency). *The numerical method given by (6.3) to obtain solutions for (6.1) is called* consistent *if*

$$\lim_{h \to 0} \frac{\delta_{n+1,h}}{h} = 0.$$

Thinking back to the definition of the O-notation, consistency is equivalent to saying that the order is at least one. For convergence, $p \geq 1$ is necessary.

For Euler's method we have $\phi_h(t_n, \mathbf{y}_n) = \mathbf{y}_n + h\mathbf{f}(t_n, \mathbf{y}_n)$. Using again Taylor series expansion,

$$
\begin{aligned}
\mathbf{y}(t_{n+1}) - [\mathbf{y}(t_n) + h\mathbf{f}(t_n, \mathbf{y}(t_n))] &= [\mathbf{y}(t_n) + h\mathbf{y}'(t_n) + \tfrac{1}{2}h^2\mathbf{y}''(t_n) + \cdots] \\
&\quad -[\mathbf{y}(t_n) + h\mathbf{y}'(t_n)] \\
&= O(h^2),
\end{aligned}
$$

and we deduce that Euler's method is of order 1.

Definition 6.5 (Theta methods). *These are methods of the form*

$$\mathbf{y}_{n+1} = \mathbf{y}_n + h[\theta\mathbf{f}(t_n, \mathbf{y}_n) + (1 - \theta)\mathbf{f}(t_{n+1}, \mathbf{y}_{n+1})], \qquad n = 0, 1, \ldots,$$

where $\theta \in [0, 1]$ is a parameter.

1. For $\theta = 1$, we recover the Euler method.

2. The choice $\theta = 0$, is known as *backward Euler*

$$\mathbf{y}_{n+1} = \mathbf{y}_n + h\mathbf{f}(t_{n+1}, \mathbf{y}_{n+1}).$$

3. For $\theta = \frac{1}{2}$ we have the *trapezoidal rule*

$$\mathbf{y}_{n+1} = \mathbf{y}_n + \frac{1}{2}h[\mathbf{f}(t_n, \mathbf{y}_n) + \mathbf{f}(t_{n+1}, \mathbf{y}_{n+1})].$$

The order of the theta method can be determined as follows

$$\mathbf{y}(t_{n+1}) - \mathbf{y}(t_n) - h(\theta\mathbf{y}'(t_n) + (1 - \theta)\mathbf{y}'(t_{n+1})) =$$

$$(\mathbf{y}(t_n) + h\mathbf{y}'(t_n) + \frac{1}{2}h^2\mathbf{y}''(t_n) + \frac{1}{6}h^3\mathbf{y}'''(t_n)) + O(h^4) -$$

$$-\mathbf{y}(t_n) - h\theta\mathbf{y}'(t_n) - (1 - \theta)h(\mathbf{y}'(t_n) + h\mathbf{y}''(t_n) + \frac{1}{2}h^2\mathbf{y}'''(t_n)) =$$

$$(\theta - \frac{1}{2})h^2\mathbf{y}''(t_n) + (\frac{1}{2}\theta - \frac{1}{3})h^3\mathbf{y}'''(t_n) + O(h^4).$$

Therefore all theta methods are of order 1, except that the trapezoidal rule ($\theta = 1/2$) is of order 2.

If $\theta < 1$, then the theta method is *implicit*. That means each time step requires the solution of N (generally non-linear) algebraic equations to find the unknown vector $\mathbf{y}_{n+1}$. This can be done by iteration and generally the first estimate $\mathbf{y}_{n+1}^{[0]}$ for $\mathbf{y}_{n+1}$ is set to $\mathbf{y}_n$, assuming that the function does not change rapidly between time steps.

To obtain further estimates for $\mathbf{y}_{n+1}$ one can use *direct iteration*;

$$\mathbf{y}_{n+1}^{[j+1]} = \phi_h(t_n, \mathbf{y}_0, \mathbf{y}_1, \ldots \mathbf{y}_n, \mathbf{y}_{n+1}^{[j]}).$$

Other methods arise by viewing the problem of finding $\mathbf{y}_{n+1}$ as finding the zero of the function $F : \mathbb{R}^N \to \mathbb{R}^N$ defined by

$$F(\mathbf{y}) = \mathbf{y} - \phi_h(t_n, \mathbf{y}_0, \mathbf{y}_1, \ldots \mathbf{y}_n, \mathbf{y}).$$

This is the subject of the chapter on non-linear systems. Assume we already have an estimate $\mathbf{y}^{[j]}$ for the zero. Let

$$F(\mathbf{y}) = \begin{pmatrix} F_1(\mathbf{y}) \\ \vdots \\ F_N(\mathbf{y}) \end{pmatrix}.$$

and let $\mathbf{h} = (h_1, \ldots h_N)^T$ be a small perturbation vector. The multidimensional Taylor expansion of each function component F_i, $i = 1, \ldots, N$ is

$$F_i(\mathbf{y}^{[j]} + \mathbf{h}) = F_i(\mathbf{y}^{[j]}) + \sum_{k=1}^{N} \frac{\partial F_i(\mathbf{y}^{[j]})}{\partial y_k} h_k + O(\|\mathbf{h}\|^2).$$

The Jacobian matrix $JF(\mathbf{y}^{[j]})$ has the entries $\frac{\partial F_i(\mathbf{y}^{[j]})}{\partial y_k}$ and thus we can write in matrix notation

$$F(\mathbf{y}^{[j]} + \mathbf{h}) = F(\mathbf{y}^{[j]}) + JF(\mathbf{y}^{[j]})\mathbf{h} + O(\|\mathbf{h}\|^2).$$

Neglecting the $O(\|\mathbf{h}\|^2)$, we equate this to zero (since we are looking for a better approximation of the zero) and solve for $\mathbf{h}$. We let the new estimate be

$$\mathbf{y}^{[j+1]} = \mathbf{y}^{[j]} + \mathbf{h} = \mathbf{y}^{[j]} - [JF(\mathbf{y}^{[j]})]^{-1} F(\mathbf{y}^{[j]}).$$

This method is called the *Newton–Raphson method*.

Of course the inverse of the Jacobian is not computed explicitly, instead the equation

$$[JF(\mathbf{y}^{[j]})]\mathbf{h} = -F(\mathbf{y}^{[j]})$$

is solved for $\mathbf{h}$.

The method can be simplified further by using the same Jacobian $JF(\mathbf{y}^{[0]})$ in the computation of the new estimate $\mathbf{y}^{[j+1]}$, which is then called *modified Newton–Raphson*.

Exercise 6.3. *Implement the backward Euler method in MATLAB or a different programming language of your choice.*

6.3   Order Conditions

We now consider multistep methods which can be expressed as linear combinations of approximations to the function values and first derivatives. Assuming that $\mathbf{y}_n, \mathbf{y}_{n+1}, \ldots, \mathbf{y}_{n+s-1}$ are available, where $s \geq 1$, we say that

$$\sum_{l=0}^{s} \rho_l \mathbf{y}_{n+l} = h \sum_{l=0}^{s} \sigma_l \mathbf{f}(t_{n+l}, \mathbf{y}_{n+l}), \qquad n = 0, 1, \ldots \qquad (6.4)$$

is an s-step method. Here $\rho_s = 1$. If $\sigma_s = 0$, the method is *explicit*, otherwise *implicit*. If $s \geq 2$ we need to obtain additional *starting values* $\mathbf{y}_1, \ldots, \mathbf{y}_{s-1}$ by a different time-stepping method.

We define the following complex polynomials

$$\rho(w) = \sum_{l=0}^{s} \rho_l w^l, \qquad\qquad \sigma(w) = \sum_{l=0}^{s} \sigma_l w^l.$$

Theorem 6.2. *The s-step method (6.4) is of order $p \geq 1$ if and only if*

$$\rho(e^z) - z\sigma(e^z) = O(z^{p+1}), \qquad z \to 0.$$

Proof. Define the operator D_t to be differentiation with respect to t. Substituting the exact solution and expanding into Taylor series about t_n, we obtain

$$\sum_{l=0}^{s} \rho_l \mathbf{y}(t_{n+l}) - h \sum_{l=0}^{s} \sigma_l \mathbf{y}'(t_{n+l}) =$$

$$\sum_{l=0}^{s} \rho_l \sum_{k=0}^{\infty} \frac{(lh)^k}{k!} \mathbf{y}^{(k)}(t_n) - h \sum_{l=0}^{s} \sigma_l \sum_{k=0}^{\infty} \frac{(lh)^k}{k!} \mathbf{y}^{(k+1)}(t_n) =$$

$$\sum_{l=0}^{s} \rho_l \sum_{k=0}^{\infty} \frac{(lhD_t)^k}{k!} \mathbf{y}(t_n) - h \sum_{l=0}^{s} \sigma_l \sum_{k=0}^{\infty} \frac{(lhD_t)^k}{k!} D_t \mathbf{y}(t_n) =$$

$$\sum_{l=0}^{s} \rho_l e^{lhD_t} \mathbf{y}(t_n) - hD_t \sum_{l=0}^{s} \sigma_l e^{lhD_t} \mathbf{y}(t_n) =$$

$$[\rho(e^{hD_t}) - hD_t \sigma(e^{hD_t})]\mathbf{y}(t_n).$$

Sorting the terms with regards to derivatives we have

$$\sum_{l=0}^{s} \rho_l \mathbf{y}(t_{n+l}) - h \sum_{l=0}^{s} \sigma_l \mathbf{y}'(t_{n+l}) =$$

$$\left(\sum_{l=0}^{s} \rho_l \right) \mathbf{y}(t_n) + \sum_{k=1}^{\infty} \frac{1}{k!} \left(\sum_{l=0}^{s} l^k \rho_l - k \sum_{l=0}^{s} l^{k-1} \sigma_l \right) h^k \mathbf{y}^{(k)}(t_n).$$

Thus, to achieve $O(h^{p+1})$ regardless of the choice of $\mathbf{y}$, it is necessary and sufficient that

$$\sum_{l=0}^{s} \rho_l = 0, \qquad \sum_{l=0}^{s} l^k \rho_l = k \sum_{l=0}^{s} l^{k-1} \sigma_l, \qquad k = 1, 2, \ldots, p. \qquad (6.5)$$

This is equivalent to saying that $\rho(e^z) - z\sigma(e^z) = O(z^{p+1})$. $\qquad\square$

Definition 6.6 (Order conditions). *The formulae given in Equation (6.5) are the* order conditions *to achieve order p.*

We illustrate this with the 2-step *Adams–Bashforth method* defined by

$$\mathbf{y}_{n+2} - \mathbf{y}_{n+1} = h\left[\frac{3}{2}\mathbf{f}(t_{n+1}, \mathbf{y}_{n+1}) - \frac{1}{2}\mathbf{f}(t_n, \mathbf{y}_n)\right].$$

Here we have $\rho(w) = w^2 - w$ and $\sigma(w) = \frac{3}{2}w - \frac{1}{2}$, and therefore

$$
\begin{aligned}
\rho(e^z) - z\sigma(e^z) &= [1 + 2z + 2z^2 + \frac{4}{3}z^3] - [1 + z + \frac{1}{2}z^2 + \frac{1}{6}z^3] \\
&\quad - \frac{3}{2}z[1 + z + \frac{1}{2}z^2] + \frac{1}{2}z + O(z^4) \\
&= \frac{5}{12}z^3 + O(z^4).
\end{aligned}
$$

Hence the method is of order 2.

Exercise 6.4. *Calculate the coefficients of the multistep method*

$$\mathbf{y}_{n+3} + \rho_2\mathbf{y}_{n+2} + \rho_1\mathbf{y}_{n+1} + \rho_0\mathbf{y}_n = \sigma_2 f(t_{n+2}, \mathbf{y}_{n+2})$$

such that it is of order 3.

Let us consider the 2-step method

$$\mathbf{y}_{n+2} - 3\mathbf{y}_{n+1} + 2\mathbf{y}_n = \frac{1}{12}h\left[13\mathbf{f}(t_{n+2}, \mathbf{y}_{n+2}) - 20\mathbf{f}(t_{n+1}, \mathbf{y}_{n+1}) - 5\mathbf{f}(t_n, \mathbf{y}_n)\right]. \tag{6.6}$$

Exercise 6.5. *Show that the above method is at least of order 2 just like the 2-step Adams–Bashforth method.*

However, we consider the trivial ODE given by $y' = 0$ and $y(0) = 1$, which has the exact solution $y(t) \equiv 1$. The right-hand side in (6.6) vanishes and thus a single step reads $y_{n+2} - 3y_{n+1} + 2y_n = 0$. This linear recurrence relation can be solved by considering the roots of the quadratic $x^2 - 3x + 2$, which are $x_1 = 1$ and $x_2 = 2$. A general solution is given by $y_n = c_1 x_1^n + c_2 x_2^n = c_1 + c_2 2^n$, $n = 0, 1, \ldots$, where c_1 and c_2 are constants determined by $y_0 = 1$ and the choice of y_1. If $c_2 \neq 0$, then for $h \to 0$, $nh \to t$ and thus $n \to \infty$, we have $|y_n| \to \infty$ and we are moving away from the exact solution. A choice of $y_1 = 1$ yields $c_2 = 0$, but even then this method poses problems because of the presence of round-off errors if the right-hand side is nonzero.

Thus method (6.6) does not converge. The following theorem provides a theoretical tool to allow us to check for convergence.

Theorem 6.3 (The Dahlquist equivalence theorem). *The multistep method (6.4) is convergent if and only if it is of order $p \geq 1$ and the polynomial ρ obeys the root condition, which means all its zeros lie within $|w| \leq 1$ and all zeros of unit modulus are simple zeros. In this case the method is called zero-stable.*

Proof. The proof of this result is long and technical. Details can be found in [10] W. Gautschi, *Numerical Analysis* or [11] P. Henrici, *Discrete Variable Methods in Ordinary Differential Equations.* □

Exercise 6.6. *Show that the multistep method given by*

$$\sum_{j=0}^{3} \rho_j \mathbf{y}_{n+j} = h \sum_{j=0}^{2} \sigma_j \mathbf{f}(t_{n+j}, \mathbf{y}_{n+j})$$

is fourth order only if the conditions $\rho_0 + \rho_2 = 8$ and $\rho_1 = -9$ are satisfied. Hence deduce that this method cannot be both fourth order and satisfy the root condition.

Theorem 6.4 (The first Dahlquist barrier). *Convergence implies that the order p can be at most $s + 2$ for even s and $s + 1$ for odd s.*

Proof. Again the proof is technical and beyond the scope of this course. See again [11] P. Henrici, *Discrete Variable Methods in Ordinary Differential Equations.* □

6.4   Stiffness and A-Stability

Consider the linear system $\mathbf{y}' = A\mathbf{y}$ for a general $N \times N$ constant matrix A. By defining

$$e^{tA} = \sum_{k=0}^{\infty} \frac{1}{k!} t^k A^k$$

the exact solution of the ODE can be represented explicitly as $\mathbf{y}(t) = e^{tA}\mathbf{y}_0$.

We solve the ODE with the forward Euler method. Then

$$\mathbf{y}_{n+1} = (I + hA)\mathbf{y}_n,$$

and therefore

$$\mathbf{y}_n = (I + hA)^n \mathbf{y}_0.$$

Let the eigenvalues of A be $\lambda_1, \ldots, \lambda_N$ with corresponding linear independent eigenvectors $\mathbf{v}_1, \ldots, \mathbf{v}_N$. Further let D be the diagonal matrix with the eigenvalues being the entries on the diagonal and $V = [\mathbf{v}_1, \ldots, \mathbf{v}_N]$, whence $A = VDV^{-1}$.

We assume further that $\text{Re}\lambda_l < 0, l = 1, \ldots, N$. In this case $\lim_{t \to \infty} \mathbf{y}(t) =$

0, since $e^{tA} = Ve^{tD}V^{-1}$ and

$$
e^{tD} = \begin{bmatrix}
e^{t\lambda_1} & 0 & \cdots & 0 \\
0 & e^{t\lambda_2} & \ddots & \vdots \\
\vdots & \ddots & \ddots & 0 \\
0 & \cdots & 0 & e^{t\lambda_N}
\end{bmatrix}.
$$

On the other hand, however, the forward Euler method $\mathbf{y}_n = V(I + hD)^n V^{-1}\mathbf{y}_0$ and therefore $\lim_{n\to\infty}\mathbf{y}_n = 0$ for all initial values of $\mathbf{y}_0$ if and only if $|1 + h\lambda_l| < 1$, for $l = 1,\ldots,N$. We illustrate this with a concrete example

$$
\begin{bmatrix}
-\frac{1}{10} & 1 \\
0 & -100
\end{bmatrix}.
$$

The exact solution is a linear combination of $e^{-1/10t}$ and e^{-100t}: the first decays gently, whereas the second becomes practically zero almost at once. Thus we require $|1 - \frac{1}{10}h| < 1$ and $|1 - 100h| < 1$, hence $h < \frac{1}{50}$. This restriction on h has *nothing* to do with local accuracy. Its purpose is solely to prevent an unbounded growth in the numerical solution.

Definition 6.7 (Stiffness). *We say that the ODE $\mathbf{y}' = \mathbf{f}(t,\mathbf{y})$ is stiff if (for some methods) we need to depress h to maintain stability well beyond requirements of accuracy.*

The opposite of stiff is called *non-stiff*.

An important example of stiff systems is when an equation is linear and $\mathrm{Re}\lambda_l < 0$, $l = 1,\ldots,N$ and the quotient $\max_{l=1,\ldots,N} |\lambda_l|/ \min_{l=1,\ldots,N} |\lambda_l|$ is large. A ratio of 10^{20} is not unusual in real-life problems. Non-linear stiff equations occur throughout applications where we have two (or more) different timescales in the ODE, i.e., if different parts of the system change with different speeds. A typical example are equations of *chemical kinetics* where each timescale is determined by the speed of reaction between two compounds. Such speeds can differ by many orders of magnitude.

Exercise 6.7. *The stiff differential equation*

$$
y'(t) = -10^6(y - t^{-1}) - t^{-2}, \qquad t \geq 1, \qquad y(1) = 1,
$$

has the analytical solution $y(t) = t^{-1}$, $t \geq 1$. Let it be solved numerically by forward Euler $y_{n+1} = y_n + h_n f(t_n, y_n)$ and by backward Euler $y_{n+1} = y_n + h_n f(t_{n+1}, y_{n+1})$, where $h_n = t_{n+1} - t_n$ is allowed to depend on n and to be different for the two methods. Suppose that at a point $t_n \geq 1$ an accuracy of $|y_n - y(t_n)| \leq 10^{-6}$ is achieved and that we want to achieve the same accuracy in the next step, i.e., $|y_{n+1} - y(t_{n+1})| \leq 10^{-6}$. Show that forward Euler can fail if $h_n = 2 \times 10^{-6}$, but that backward Euler always achieves the desired accuracy if $h_n \leq t_n t_{n+1}^2$. (Hint: Find relations between $y_{n+1} - y(t_{n+1})$ and $y_n - y(t_n)$.)

Definition 6.8 (A-stability). *Suppose that a numerical method is applied to the test equation $y' = \lambda y$ with initial condition $y(0) = 1$ and produces the solution sequence $\{y_n\}_{n\in\mathbb{Z}^+}$ for constant h. We call the set*

$$D = \{h\lambda \in \mathbb{C} : \lim_{n\to\infty} y_n = 0\}$$

the linear stability domain *of the method. The set of $\lambda \in \mathbb{C}$ for which $y(t) \overset{t\to\infty}{\to} 0$ is the* exact stability domain *and is the left half-plane $\mathbb{C}^- = \{z \in \mathbb{C} : Rez < 0\}$. We say that the method is* A-stable *if $\mathbb{C}^- \subseteq D$.*

Note that A-stability does not mean that any step size can be chosen. We need to choose h small enough to achieve the desired accuracy, but we do not need to make it smaller to prevent instability.

We have already seen that for the forward Euler method $y_n \to 0$ if and only if $|1 + h\lambda| < 1$. Therefore the stability domain is $D = \{z \in \mathbb{C} : |1 + z| < 1\}$.

Solving $y' = \lambda y$ with the trapezoidal rule $y_{n+1} = y_n + \frac{1}{2}h\lambda(y_n + y_{n+1})$ gives

$$y_{n+1} = \frac{1 + \frac{1}{2}h\lambda}{1 - \frac{1}{2}h\lambda}y_n = \left(\frac{1 + \frac{1}{2}h\lambda}{1 - \frac{1}{2}h\lambda}\right)^{n+1} y_0.$$

Therefore the trapezoidal rule has the stability domain

$$D = \left\{z \in \mathbb{C} : \left|\frac{1 + \frac{1}{2}z}{1 - \frac{1}{2}z}\right| < 1\right\} = \mathbb{C}^-$$

and the method is A-stable.

Similarly it can be proven that the stability domain of the backward Euler method is $D = \{z \in \mathbb{C} : |1 - z| > 1\}$ and hence the method is also A-stable.

The stability domains of the Theta methods (to which forward, backward Euler, and the trapezoidal rule belong) are shown in Figure 6.2 for the values $\theta = 1, 0.9, 0.8, 0.7, 0.6, 0.5, 0.4, 0.3, 0.2, 0.1$. As θ decreases from 1, the dark stability region increases until the circle opens up and becomes the imaginary axis for $\theta = \frac{1}{2}$, the trapezoidal rule. Then the white instability region continues to shrink until it is the circle given by the backward Euler method ($\theta = 0$).

Exercise 6.8. *Find stability domain D for the explicit mid-point rule $\mathbf{y}_{n+2} = \mathbf{y}_n + 2h\mathbf{f}(t_{n+1}, \mathbf{y}_{n+1})$.*

The requirement of a method to be A-stable, however, limits the achievable order.

Theorem 6.5. *An s-step method given by (6.4) is A-stable if and only if the zeros of the polynomial given by*

$$\tau(w) = \sum_{l=0}^{s}(\rho_l - h\lambda\sigma_l)w^l \tag{6.7}$$

lie within the unit circle for all $h\lambda \in \mathbb{C}^-$ and the roots on the unit circle are simple roots (these are roots where the function vanishes, but not its derivative).

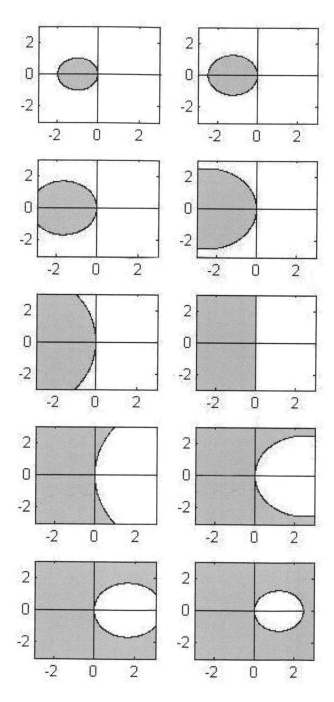

Figure 6.2 Stability domains for $\theta = 1, 0.9, 0.8, 0.7, 0.6, 0.5, 0.4, 0.3, 0.2,$ and 0.1

Proof. When the s-step method given by (6.4) is applied to the test equation $y' = \lambda y$, $y(0) = 1$, it reads

$$\sum_{l=0}^{s}(\rho_l - h\lambda\sigma_l)y_{n+l} = 0.$$

This recurrence relation has the characteristic polynomial given by (6.7). Let its zeros be $w_1(h\lambda), \ldots, w_{N(h\lambda)}(h\lambda)$ with multiplicities $\mu_1(h\lambda), \ldots, \mu_{N(h\lambda)}(h\lambda)$, respectively, where the multiplicities sum to the order of the polynomial τ. If the root of a function has multiplicity k, then it and its first $k-1$ derivatives vanish there. The solutions of the recurrence relation are given by

$$y_n = \sum_{j=1}^{N(h\lambda)} \sum_{i=0}^{\mu_j(h\lambda)-1} n^i w_j(h\lambda)^n \alpha_{ij}(h\lambda),$$

where $\alpha_{ij}(h\lambda)$ are independent of n but depend on the starting values $y_0, \ldots, y_{s-1}$. Hence the linear stability domain is the set of all $h\lambda \in \mathbb{C}$ such that all the zeros of (6.7) satisfy $|w_j(h\lambda)| \leq 1$ and if $|w_j(h\lambda)| = 1$, then $\mu_j(h\lambda) = 1$. □

The theorem implies that $h\lambda \in \mathbb{C}$ is in the stability region if the roots of the polynomial $\rho(w) - h\lambda\sigma(w)$ lie within the unit circle. It follows that if $h\lambda$ is on the boundary of the stability region, then $\rho(w) - h\lambda\sigma(w)$ must have at least one root with magnitude exactly equal to 1. Let this root be $e^{i\alpha}$ for some value α in the interval $[0, 2\pi]$. Since $e^{i\alpha}$ is a root we have

$$\rho(e^{i\alpha}) - h\lambda\sigma(e^{i\alpha}) = 0$$

and hence

$$h\lambda = \frac{\rho(e^{i\alpha})}{\sigma(e^{i\alpha})}.$$

Since every point $h\lambda$ on the boundary of the stability domain has to be of this form, we can determine the parametrized curve

$$z(\alpha) = \frac{\rho(e^{i\alpha})}{\sigma(e^{i\alpha})}$$

for $0 \leq \alpha \leq 2\pi$ which are all points that are potentially on the boundary of the stability domain. For simple methods this yields the stability domain directly after one determines on which side of the boundary the stability domain lies. This is known as the *boundary locus method*.

We illustrate this with the Theta methods, which are given by

$$y_{n+1} - y_n = h[(1-\theta)f(t_{n+1}, y_{n+1}) + \theta f(t_n, y_n)].$$

Thus $\rho(w) = w - 1$ and $\sigma(w) = (1-\theta)w + \theta$ and the parametrized curve is

$$z(\alpha) = \frac{\rho(e^{i\alpha})}{\sigma(e^{i\alpha})} = \frac{e^{i\alpha} - 1}{(1-\theta)e^{i\alpha} + \theta}.$$

For various values of θ, these curves were used to generate Figure 6.2.

Theorem 6.6 (the second Dahlquist barrier). *A-stability implies that the order p has to be less than or equal to 2. Moreover, the second order A-stable method with the least truncation error is the trapezoidal rule.*

So no multistep method of $p \geq 3$ may be A-stable, but there are methods which are satisfactory for most stiff equations. The point is that in many stiff linear systems in real world applications the eigenvalues are not just in $\mathbb{C}^-$, but also well away from $i\mathbb{R}$, that is, the imaginary axis. Therefore relaxed stability concepts are sufficient. Requiring stability only across a wedge in $\mathbb{C}^-$ of angle α results in *A(α)-stability*. A-stability is equivalent to A(90°)-stability. A(α)-stability is sufficient for most purposes. High-order A(α)-stable methods exist for $\alpha < 90°$. For $\alpha \to 90°$ the coefficients of high-order A(α)-stable methods begin to diverge. If λ in the test equation is purely imaginary, then the solution is a linear combination of $\sin(\lambda t)$ and $\cos(\lambda t)$ and it oscillates a lot for large λ. Therefore numerical pointwise solutions are useless anyway.

6.5 Adams Methods

Here we consider a technique to generate multistep methods which are convergent and of high order. In the proof of Theorem (6.2) we have seen that a necessary condition to achieve an order $p \geq 1$ is

$$\sum_{l=0}^{s} \rho_l = \rho(1) = 0.$$

The technique chooses an arbitrary s-degree polynomial ρ that obeys the root condition and has a simple root at 1 in order to achieve convergence. To achieve the maximum order, we let σ be the polynomial of degree s for implicit methods or the polynomial of degree $s - 1$ for explicit methods which arises from the truncation of the Taylor series expansion of $\dfrac{\rho(w)}{\log w}$ about the point $w = 1$. Thus, for example, for an implicit method,

$$\sigma(w) = \frac{\rho(w)}{\log w} + O(|w-1|^{s+1}) \qquad \Rightarrow \qquad \rho(e^z) - z\sigma(e^z) = O(z^{s+2}) \quad (6.8)$$

and the order is at least $s + 1$.

The *Adams methods* correspond to the choice

$$\rho(w) = w^{s-1}(w - 1).$$

For $\sigma_s = 0$ we obtain the *explicit Adams–Bashforth methods* of order s. Otherwise we obtain the *implicit Adams–Moulton methods* of order $s + 1$.

For example, letting $s = 2$ and $v = w - 1$ (equivalent $w = v + 1$) we expand

$$\begin{aligned}
\frac{w(w-1)}{\log w} &= \frac{v + v^2}{\log(1+v)} = \frac{v + v^2}{v - \frac{1}{2}v^2 + \frac{1}{3}v^3 - \cdots} = \frac{1+v}{1 - \frac{1}{2}v + \frac{1}{3}v^2 - \cdots} \\
&= (1+v)[1 + (\frac{1}{2}v - \frac{1}{3}v^2) + (\frac{1}{2}v - \frac{1}{3}v^2)^2 + O(v^3)]
\end{aligned}$$

$$= 1 + \frac{3}{2}v + \frac{5}{12}v^2 + O(v^3)$$

$$= 1 + \frac{3}{2}(w - 1) + \frac{5}{12}(w - 1)^2 + O(|w - 1|^3)$$

$$= -\frac{1}{12} + \frac{2}{3}w + \frac{5}{12}w^2 + O(|w - 1|^3).$$

Therefore the 2-step, 3rd order Adams–Moulton method is

$$\mathbf{y}_{n+2} - \mathbf{y}_{n+1} = h\left[-\frac{1}{12}\mathbf{f}(t_n, \mathbf{y}_n) + \frac{2}{3}\mathbf{f}(t_{n+1}, \mathbf{y}_{n+1}) + \frac{5}{12}\mathbf{f}(t_{n+2}, \mathbf{y}_{n+2})\right].$$

Exercise 6.9. *Calculate the actual values of the coefficients of the 3-step Adams–Bashforth method.*

Exercise 6.10 (Recurrence relation for Adams–Bashforth). *Let ρ_s and σ_s denote the polynomials generating the s-step Adams–Bashforth method. Prove that*

$$\sigma_s(w) = w\sigma_{s-1}(w) + \alpha_{s-1}(w - 1)^{s-1},$$

where $\alpha_s \neq 0$, $s = 1, 2, \ldots$, is a constant such that $\rho_s(z) - \log z\sigma_s(w) = \alpha_s(w - 1)^{s+1} + O(|w - 1|^{s+2})$ for w close to 1.

The Adams–Bashforth methods are as follows:

1. 1-step $\rho(w) = w - 1$, $\sigma(w) \equiv 1$, which is the forward Euler method.

2. 2-step $\rho(w) = w^2 - w$, $\sigma(w) = \frac{1}{2}(3w - 1)$.

3. 3-step $\rho(w) = w^3 - w^2$, $\sigma(w) = \frac{1}{12}(23w^2 - 16w + 5)$.

The Adams–Moulton methods are

1. 1-step $\rho(w) = w - 1$, $\sigma(w) = \frac{1}{2}(w + 1)$, which is the trapezoidal rule.

2. 2-step $\rho(w) = w^2 - w$, $\sigma(w) = \frac{1}{12}(5w^2 + 8w - 1)$.

3. 3-step $\rho(w) = w^3 - w^2$, $\sigma(w) = \frac{1}{24}(9w^3 + 19w^2 - 5w + 1)$.

Figure 6.3 depicts the stability domain for the 1-,2-, and 3-step Adams–Bashforth methods, while Figure 6.4 depicts the stability domain for the 1-,2-, and 3-step Adams–Moulton methods.

Another way to derive the Adams–Bashforth methods is by transforming the initial value problem given by

$$y' = f(t, y), \qquad y(t_0) = y_0,$$

into its integral form

$$y(t) = y_0 + \int_{t_0}^{t} f(\tau, y(\tau))d\tau$$

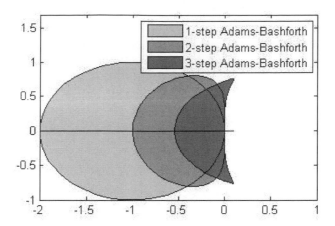

Figure 6.3 The stability domains for various Adams–Bashforth methods

and partitioning the interval equally into $t_0 < t_1 < \cdots < t_n < \cdots$ with step size h. Having already approximated $y_n, \ldots, y_{n+s-1}$, we use polynomial interpolation to find the polynomial p of degree $s - 1$ such that $p(t_{n+i}) = f(t_{n+i}, y_{n+i})$ for $i = 0, \ldots, s - 1$. Locally p is a good approximation to the right-hand side of $y' = f(t, y)$ that is to be solved, so we consider $y' = p(t)$ instead. This can be solved explicitly by

$$y_{n+s} = y_{n+s-1} + \int_{t_{n+s-1}}^{t_{n+s}} p(\tau)d\tau.$$

The coefficients of the Adams–Bashforth method can be calculated by substituting the Lagrange formula for p given by

$$
\begin{aligned}
p(t) &= \sum_{j=0}^{s-1} f(t_{n+j}, y_{n+j}) \prod_{\substack{i=0 \\ i \neq j}}^{s-1} \frac{t - t_{n+i}}{t_{n+j} - t_{n+i}} = \sum_{j=0}^{s-1} f(t_{n+j}, y_{n+j}) \prod_{\substack{i=0 \\ i \neq j}}^{s-1} \frac{t - t_{n+i}}{jh - ih} \\
&= \sum_{j=0}^{s-1} \frac{(-1)^{s-j-1} f(t_{n+j}, y_{n+j})}{j(j-1)\ldots(j-(j-1))(j-(j+1))\ldots(j-(s-1))h^{s-1}} \prod_{\substack{i=0 \\ i \neq j}}^{s-1} (t - t_{n+i}) \\
&= \sum_{j=0}^{s-1} \frac{(-1)^{s-j-1} f(t_{n+j}, y_{n+j})}{j!(s-j-1)!h^{s-1}} \prod_{\substack{i=0 \\ i \neq j}}^{s-1} (t - t_{n+i}).
\end{aligned}
$$

The one-step Adams–Bashforth method uses a zero-degree, i.e., constant polynomial interpolating the single value $f(t_n, y_n)$.

The Adams–Moulton methods arise in a similar way; however, the interpolating polynomial is of degree s and not only uses the points $t_n, \ldots t_{n+s-1}$, but also the point t_{n+s}. In this framework the backward Euler method is also often

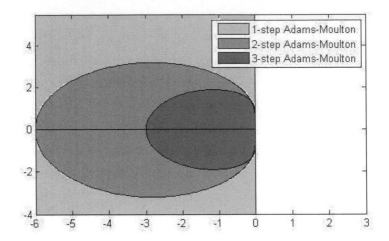

Figure 6.4 The stability domains for various Adams–Moulton methods

regarded as an Adams–Moulton method, since the interpolating polynomial is the constant interpolating the single value $f(t_{n+1}, y_{n+1})$.

6.6   Backward Differentiation Formulae

The *backward differentiation formulae (BDF)* are a family of implicit multistep methods. Here all but one coefficient on the right-hand side of (6.4) are zero, i.e., $\sigma(w) = \sigma_s w^s$ for some $s \neq 0$, $\sigma_s \neq 0$. In other words,

$$\sum_{l=0}^{s} \rho_l \mathbf{y}_{n+l} = h\sigma_s \mathbf{f}(t_{n+s}, \mathbf{y}_{n+s}), \qquad n = 0, 1, \ldots,$$

BDF are especially used for the solution of stiff differential equations.

To derive the explicit form of the s-step BDF we again employ the technique introduced in (6.8), this time solving for $\rho(w)$, since $\sigma(w)$ is given. Thus $\rho(w) = \sigma_s w^s \log w + O(|w-1|^{s+1})$. Dividing by w^s and setting $v = 1/w$, this becomes

$$\sum_{l=0}^{s} \rho_l v^{s-l} = -\sigma_s \log v + O(|v-1|^{s+1}).$$

The simple change form w to v in the O-term is possible since we are considering w close to 1 or written mathematically $w = O(1)$ and

$$O(|w-1|^{s+1}) = O(|w|^{s+1}|1 - \frac{1}{w}|^{s+1}) = O(|w|^{s+1})O(|1-v|^{s+1}) = O(|v-1|^{s+1}).$$

Now $\log v = \log(1 + (v - 1)) = \sum_{l=1}^{\infty} (-1)^{l-1}(v-1)^l/l$. Consequently we want to choose the coefficients $\rho_1, \ldots, \rho_s$ such that

$$\sum_{l=0}^{s} \rho_l v^{s-l} = \sigma_s \sum_{l=1}^{s} \frac{(-1)^l}{l} (v-1)^l O(|v-1|^{s+1}).$$

Restoring $w = v^{-1}$ and multiplying by w^s yields

$$\sum_{l=0}^{s} \rho_l w^l = \sigma_s w^s \sum_{l=1}^{\infty} \frac{(-1)^l}{l} w^{-l}(1-w)^l = \sigma_s \sum_{l=1}^{\infty} \frac{(-1)^l}{l} w^{s-l}(1-w)^l. \quad (6.9)$$

We expand

$$(1-w)^l = 1 - \binom{l}{1} w + \binom{l}{2} w^2 - \ldots + (-1)^l \binom{l}{l} w^l$$

and then pick σ_s so that $\rho_s = 1$ by collecting the powers of w^s on the right-hand side. This gives

$$\sigma_s = \left(\sum_{l=1}^{s} \frac{1}{l} \right)^{-1}. \quad (6.10)$$

As an example we let $s = 2$. Substitution in (6.10) gives $\sigma_2 = \frac{2}{3}$ and using (6.9) we have $\rho(w) = w^2 - \frac{4}{3}w + \frac{1}{3}$. Hence the 2-step BDF is

$$\mathbf{y}_{n+2} - \frac{4}{3}\mathbf{y}_{n+1} + \frac{1}{3}\mathbf{y}_n = \frac{2}{3}h\mathbf{f}(t_{n+2}, \mathbf{y}_{n+2}).$$

The BDF are as follows

- 1-step $\rho(w) = w - 1$, $\sigma(w) = w$, which is the backward Euler method (again!).

- 2-step $\rho(w) = w^2 - \frac{4}{3}w + \frac{1}{3}$, $\sigma(w) = \frac{2}{3}w^2$.

- 3-step $\rho(w) = w^3 - \frac{18}{11}w^2 + \frac{9}{11}w - \frac{2}{11}$, $\sigma(w) = \frac{6}{11}w^3$.

- 4-step $\rho(w) = w^4 - \frac{48}{25}w^3 + \frac{36}{25}w^2 - \frac{16}{25}w + \frac{3}{25}$, $\sigma(w) = \frac{12}{25}w^4$.

- 5-step $\rho(w) = w^5 - \frac{300}{137}w^4 + \frac{300}{137}w^3 - \frac{200}{137}w^2 + \frac{75}{137}w - \frac{12}{137}$, $\sigma(w) = \frac{60}{137}w^5$.

- 6-step $\rho(w) = w^6 - \frac{360}{147}w^5 + \frac{450}{147}w^4 - \frac{400}{147}w^3 + \frac{225}{147}w^2 - \frac{72}{147}w + \frac{10}{147}$, $\sigma(w) = \frac{60}{147}w^6$.

It can be proven that BDF are convergent if and only if $s \leq 6$. For higher values of s they must not be used. Figure (6.5) shows the stability domain for various BDF methods. As the number of steps increases the stability domain shrinks but it remains unbounded for $s \leq 6$. In particular, the 3-step BDF is A($86°2'$)-stable.

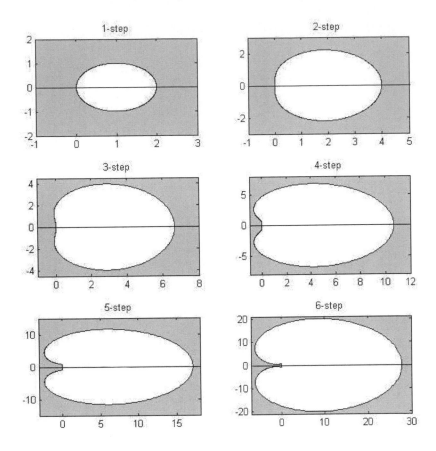

Figure 6.5 The stability domains of various BDF methods in grey. The instability regions are in white.

6.7 The Milne and Zadunaisky Device

The step size h is not some fixed quantity: it is a parameter of the method which can vary from step to step. The basic input of a well-designed computer package for ODEs is not the step size but the *error tolerance*, as required by the user. With the choice of $h > 0$ we can keep a local estimate of the error beneath the required tolerance in the solution interval. We don't just need a *time-stepping algorithm*, but also mechanisms for error control and for adjusting the step size.

Suppose we wish to estimate in each step the error of the trapezoidal rule (TR)

$$\mathbf{y}_{n+1} = \mathbf{y}_n + \frac{1}{2}h[\mathbf{f}(t_n, \mathbf{y}_n) + \mathbf{f}(t_{n+1}, \mathbf{y}_{n+1})].$$

Substituting the true solution, we deduce that

$$\mathbf{y}(t_{n+1}) - \{\mathbf{y}(t_n) + \frac{1}{2}h[\mathbf{y}'(t_n) + \mathbf{y}'(t_{n+1})]\} = -\frac{1}{12}h^3\mathbf{y}'''(t_n) + O(h^4)$$

and the order is 2. The *error constant* of TR is $c_{TR} = -\frac{1}{12}$. To estimate the error in a single step we assume that $\mathbf{y}_n = \mathbf{y}(t_n)$ and this yields

$$\mathbf{y}(t_{n+1}) - \mathbf{y}_{n+1} = c_{TR}h^3\mathbf{y}'''(t_n) + O(h^4). \tag{6.11}$$

Each multi-step method has its own error constant. For example, the 2nd-order 2-step Adams–Bashforth method (AB2)

$$\mathbf{y}_{n+1} - \mathbf{y}_n = \frac{1}{2}h[3\mathbf{f}(t_n, \mathbf{y}_n) - \mathbf{f}(t_{n-1}, \mathbf{y}_{n-1})],$$

has the error constant $c_{AB2} = \frac{5}{12}$.

The idea behind the *Milne device* is to use two multistep methods of the same order, one explicit and the second implicit, to estimate the local error of the implicit method. For example, *locally*,

$$\mathbf{y}_{n+1}^{AB2} \approx \mathbf{y}(t_{n+1}) - c_{AB2}h^3\mathbf{y}'''(t_n) = \mathbf{y}(t_{n+1}) - \frac{5}{12}h^3\mathbf{y}'''(t_n),$$

$$\mathbf{y}_{n+1}^{TR} \approx \mathbf{y}(t_{n+1}) - c_{TR}h^3\mathbf{y}'''(t_n) = \mathbf{y}(t_{n+1}) + \frac{1}{12}h^3\mathbf{y}'''(t_n).$$

Subtracting, we obtain the estimate

$$h^3\mathbf{y}'''(t_n) \approx -2(\mathbf{y}_{n+1}^{AB2} - \mathbf{y}_{n+1}^{TR}),$$

which we can insert back into (6.11) to give

$$\mathbf{y}(t_{n+1}) - \mathbf{y}_{n+1} \approx \frac{1}{6}(\mathbf{y}_{n+1}^{AB2} - \mathbf{y}_{n+1}^{TR})$$

and we use the right-hand side as an estimate of the local error.

The trapezoidal rule is the superior method, since it is A-stable and its global behaviour is hence better. The Adams–Bashforth method is solely employed to estimate the local error. Since it is explicit this adds little overhead.

In this example, the trapezoidal method was the *corrector*, while the Adams–Bashforth method was the *predictor*. The Adams–Bashforth and Adams–Moulton methods of the same order are often employed as predictor–corrector pairs. For example, for the order 3 we have the predictor

$$\mathbf{y}_{n+2} = \mathbf{y}_{n+1} + h[\frac{5}{12}\mathbf{f}(t_{n-1}, \mathbf{y}_{n-1}) - \frac{4}{3}\mathbf{f}(t_n, \mathbf{y}_n) + \frac{23}{12}\mathbf{f}(t_{n+1}, \mathbf{y}_{n+1})],$$

and the corrector

$$\mathbf{y}_{n+2} = \mathbf{y}_{n+1} + h[-\frac{1}{12}\mathbf{f}(t_n, \mathbf{y}_n) + \frac{2}{3}\mathbf{f}(t_{n+1}, \mathbf{y}_{n+1}) + \frac{5}{12}\mathbf{f}(t_{n+2}, \mathbf{y}_{n+2})].$$

The predictor is employed not just to estimate the error of the corrector, but also to provide an initial guess in the solution of the implicit corrector equations. Typically, for nonstiff equations, we iterate correction equations at most twice, while stiff equations require iteration to convergence, otherwise the typically superior stability features of the corrector are lost.

Exercise 6.11. *Consider the predictor–corrector pair given by*

$$\mathbf{y}_{n+3}^P = -\tfrac{1}{2}\mathbf{y}_n + 3\mathbf{y}_{n+1} - \tfrac{3}{2}\mathbf{y}_{n+2} + 3h\mathbf{f}(t_{n+2}, \mathbf{y}_{n+2}),$$

$$\mathbf{y}_{n+3}^C = \tfrac{1}{11}[2\mathbf{y}_n - 9\mathbf{y}_{n+1} + 18\mathbf{y}_{n+2} + 6h\mathbf{f}(t_{n+3}, \mathbf{y}_{n+3})].$$

The predictor is as in Exercise 6.4. The corrector is the three-step backward differentiation formula. Show that both methods are third order, and that the estimate of the error of the corrector formula by Milne's device has the value $\tfrac{6}{17}|\mathbf{y}_{n+3}^P - \mathbf{y}_{n+3}^C|$.

Let $\epsilon > 0$ be a user-specified *tolerance*, i.e., this is the maximal error allowed. After completion of a single step with a local error estimate $\mathbf{e}$, there are three outcomes.

1. $\tfrac{1}{10}\epsilon \le \|\mathbf{e}\| \le \epsilon$: Accept the step size and continue to calculate the next estimate with the same step size.

2. $\|\mathbf{e}\| \le \tfrac{1}{10}\epsilon$: Accept the step and increase the step length.

3. $\|\mathbf{e}\| > \epsilon$: Reject the step and return to t_n and use a smaller step size.

The quantity $\mathbf{e}/h$ is the estimate for the global error in an interval of unit length. This is usually required not to exceed ϵ, since good implementations of numerical ODEs should monitor the accumulation of *global error*. This is called *error estimation per unit step*.

Adjusting the step size can be done with polynomial interpolation to obtain estimates of the function values required, although this means that we need to store past values well in excess of what is necessary for simple implementation of both multistep methods.

Exercise 6.12. *Let* $\mathbf{p}$ *be the cubic polynomial that is defined by* $\mathbf{p}(t_j) = \mathbf{y}_j, j = n, n+1, n+2,$ *and by* $\mathbf{p}'(t_{n+2}) = \mathbf{f}(t_{n+2}, \mathbf{y}_{n+2})$. *Show that the predictor formula of the previous exercise is* $\mathbf{y}_{n+3}^P = \mathbf{p}(t_{n+2}+h)$. *Further, show that the corrector formula is equivalent to the equation*

$$\mathbf{y}_{n+3}^C = \mathbf{p}(t_{n+2}) + \frac{5}{11}h\mathbf{p}'(t_{n+2}) - \frac{1}{22}h^2\mathbf{p}''(t_{n+2})$$
$$- \frac{7}{66}h^3\mathbf{p}'''(t_{n+2}) + \frac{6}{11}h\mathbf{f}(t_{n+2}+h, \mathbf{y}_{n+3}).$$

The point is that $\mathbf{p}$ *can be derived from available data, and then the above forms of the predictor and corrector can be applied for any choice of* $h = t_{n+3} - t_{n+2}$.

The *Zadunaisky device* is another way to obtain an error estimate. Suppose we have calculated the (not necessarily equidistant) solution values $\mathbf{y}_n, \mathbf{y}_{n-1}, \ldots, \mathbf{y}_{n-p}$ with an arbitrary numerical method of order p. We form an interpolating p^{th} degree (vector valued) polynomial $\mathbf{p}$ such that $\mathbf{p}(t_{n-i}) = \mathbf{y}_{n-i}$, $i = 0, 1, \ldots, p$, and consider the differential equation

$$\mathbf{z}' = \mathbf{f}(t, \mathbf{z}) + \mathbf{p}'(t) - \mathbf{f}(t, \mathbf{p}), \qquad \mathbf{z}(t_n) = \mathbf{y}_n. \tag{6.12}$$

There are two important observations with regard to this differential equation. Firstly, it is a small perturbation on the original ODE, since the term $\mathbf{p}'(t) - \mathbf{f}(t, \mathbf{p})$ is usually small since locally $\mathbf{p}(t) - \mathbf{y}(t) = O(h^{p+1})$ and $\mathbf{y}'(t) = \mathbf{f}(t, \mathbf{y}(t))$. Secondly, the exact solution of (6.12) is obviously $\mathbf{z}(t) = \mathbf{p}(t)$. Now we calculate $\mathbf{z}_{n+1}$ using exactly the same method and implementation details. We then evaluate the error in $\mathbf{z}_{n+1}$, namely $\mathbf{z}_{n+1} - \mathbf{p}(t_{n+1})$, and use it as an estimate of the error in $\mathbf{y}_{n+1}$. The error estimate can then be used to assess the step and adjust the step size if necessary.

6.8 Rational Methods

In the following we consider methods which use higher derivatives. Repeatedly differentiating (6.1) we obtain formulae of the form $\mathbf{y}^{(k)} = \mathbf{f}_k(\mathbf{y})$ where

$$\mathbf{f}_0(\mathbf{y}) = \mathbf{y}, \qquad \mathbf{f}_1(\mathbf{y}) = \mathbf{f}(\mathbf{y}), \qquad \mathbf{f}_2(\mathbf{y}) = \frac{\partial \mathbf{f}(\mathbf{y})}{\partial \mathbf{y}} \mathbf{f}(\mathbf{y}), \qquad \ldots$$

This then motivates the *Taylor method*

$$\mathbf{y}_{n+1} = \sum_{k=0}^{p} \frac{1}{k!} h^k \mathbf{f}_k(\mathbf{y}_n), \qquad n \in \mathbb{Z}^+. \tag{6.13}$$

For example,

$$p = 1: \quad \mathbf{y}_{n+1} = \mathbf{y}_n + h\mathbf{f}(\mathbf{y}_n) \qquad\qquad \text{forward Euler,}$$
$$p = 2: \quad \mathbf{y}_{n+1} = \mathbf{y}_n + h\mathbf{f}(\mathbf{y}_n) + \frac{1}{2} h^2 \frac{\partial \mathbf{f}(\mathbf{y}_n)}{\partial \mathbf{y}} \mathbf{f}(\mathbf{y}_n).$$

Theorem 6.7. *The Taylor method given by (6.13) has error $O(h^{p+1})$.*

Proof. The proof is easily done by induction. Firstly we have $\mathbf{y}_0 = \mathbf{y}(0) = \mathbf{y}(0) + O(h^{p+1})$. Now assume $\mathbf{y}_n = \mathbf{y}(t_n) + O(h^{p+1}) = \mathbf{y}(nh) + O(h^{p+1})$. It follows that $\mathbf{f}_k(\mathbf{y}_n) = \mathbf{f}_k(\mathbf{y}(nh)) + O(h^{p+1})$, since $\mathbf{f}$ is analytic. Hence $\mathbf{y}_{n+1} = \mathbf{y}((n+1)h) + O(h^{p+1}) = \mathbf{y}(t_{n+1}) + O(h^{p+1})$. $\qquad\square$

Recalling the differentiation operator $D_t \mathbf{y}(t) = \mathbf{y}'(t)$, we see that

$$D_t^k \mathbf{y} = \mathbf{f}_k(\mathbf{y}).$$

Let $R(z) = \sum_{k=0}^{\infty} r_k z^k$ be an analytic function such that $R(z) = e^z + O(z^{p+1})$, i.e., $r_k = \frac{1}{k!}$ for $k = 0, 1, \ldots, p$. Then the formal method defined by

$$\mathbf{y}_{n+1} = R(hD_t)\mathbf{y}_n = \sum_{k=0}^{\infty} r_k h^k \mathbf{f}_k(\mathbf{y}_n), \qquad n \in \mathbb{Z}^+,$$

is of order p. Indeed, the Taylor method is one such method by dint of letting $R(z)$ be the pth section of the Taylor expansion of e^z.

We can let R be a rational function of the form

$$R(z) = \frac{\sum_{k=0}^{M} p_k z^k}{\sum_{k=0}^{N} q_k z^k},$$

which corresponds to the numerical method

$$\sum_{k=0}^{N} q_k h^k \mathbf{f}_k(\mathbf{y}_{n+1}) = \sum_{k=0}^{M} p_k h^k \mathbf{f}_k(\mathbf{y}_n), \tag{6.14}$$

where $q_0 = 1$. For $N \geq 1$ this is an algebraic system of equations which has to be solved by a suitable method; for example, Newton–Raphson.

Definition 6.9 (Padé approximants). *Given a function f which is analytic at the origin, the $[M/N]$ Padé approximant is the quotient of an Mth degree polynomial over an Nth degree polynomial which matches the Taylor expansion of f to the highest possible order.*

For $f(z) = e^z$ we have the Padé approximant $R_{M/N} = P_{M/N}/Q_{M/N}$, where

$$P_{M/N} = \sum_{k=0}^{M} \binom{M}{k} \frac{(M+N-k)!}{(M+N)!} z^k,$$

$$Q_{M/N} = \sum_{k=0}^{N} \binom{N}{k} \frac{(M+N-k)!}{(M+N)!} (-z)^k = P_{N/M}(-z).$$

Lemma 6.1. $R_{M/N}(z) = e^z + O(z^{M+N+1})$ *and no other rational function of this form can do better.*

Proof. Omitted. For further information on Padé approximants please see [1] G. A. Baker Jr. and P. Graves–Morris, *Padé Approximants.* □

It follows directly from the lemma above that the *Padé method*

$$\sum_{k=0}^{N} (-1)^k \binom{N}{k} \frac{(M+N-k)!}{(M+N)!} h^k \mathbf{f}_k(\mathbf{y}_{n+1}) = \sum_{k=0}^{M} \binom{M}{k} \frac{(M+N-k)!}{(M+N)!} h^k \mathbf{f}_k(\mathbf{y}_n)$$

is of order $M + N$.

For example, if $M = 1, N = 0$ we have $P_{1/0} = 1 + z$ and $Q_{1/0} = 1$ and this yields the forward Euler method

$$\mathbf{y}_{n+1} = \mathbf{y}_n + h\mathbf{f}(\mathbf{y}_n).$$

On the other hand, if $M = 0, N = 1$ we have $P_{0/1} = 1$ and $Q_{0/1} = 1 - z$ and this yields the backward Euler method

$$\mathbf{y}_{n+1} = \mathbf{y}_n + h\mathbf{f}(\mathbf{y}_{n+1}).$$

For $M = N = 1$, the polynomials are $P_{1/1} = 1 + \frac{1}{2}z$ and $Q_{1/1} = 1 - \frac{1}{2}z$ and we have the trapezoidal rule

$$\mathbf{y}_{n+1} = \mathbf{y}_n + \frac{1}{2}h(\mathbf{f}(\mathbf{y}_n) + \mathbf{f}(\mathbf{y}_{n+1})).$$

On the other hand, $M = 0, N = 2$ gives $P_{0/2} = 1$ and $Q_{0/2} = 1 - z + \frac{1}{2}z^2$ with the method

$$\mathbf{y}_{n+1} - h\mathbf{f}(\mathbf{y}_{n+1}) + \frac{1}{2}h^2 \frac{\partial \mathbf{f}(\mathbf{y}_{n+1})}{\partial \mathbf{y}} \mathbf{f}(\mathbf{y}_{n+1}) = \mathbf{y}_n,$$

which could be described as backward Taylor method.

Considering the test equation $y' = \lambda y$, $y(0) = 1$, then the method given by (6.14) calculates the approximations according to

$$\sum_{k=0}^{N} q_k h^k \lambda^k y_{n+1} = \sum_{k=0}^{M} p_k h^k \lambda^k y_n.$$

Hence $y_{n+1} = R(h\lambda)y_n = (R(h\lambda))^{n+1}y_0$. Thus the stability domain of this method is

$$\{z \in \mathbb{C} : |R(z)| < 1\}.$$

Theorem 6.8. *The rational method given by (6.14) is a A-stable if and only if firstly all the poles of R reside in the right half-plane $\mathbb{C}^+ := \{z \in \mathbb{C} : Rez > 0\}$ and secondly $|R(iy)| \leq 1$ for all $y \in \mathbb{R}$.*

Proof. This is proven by the maximum modulus principle. Since R is analytic in $\mathbb{C}^-$ and all poles are in $\mathbb{C}^+$, R takes its maximum modulus in $\mathbb{C}^-$ on the boundary of $\mathbb{C}^-$, which is the line iy, $y \in \mathbb{R}$. □

Moreover, according to a theorem of Wanner, Hairer, and Norsett, the Padé method is A-Stable if and only if $M \leq N \leq M + 2$.

6.9 Runge–Kutta Methods

In the section on Adams methods we have seen how they can be derived by approximating the derivative by an interpolating polynomial which is then integrated. This was inspired by replacing the initial value problem given by

$$\mathbf{y}' = \mathbf{f}(t, \mathbf{y}), \qquad \mathbf{y}(t_0) = \mathbf{y}_0,$$

into its integral form

$$y(t_{n+1}) = y(t_n) + \int_{t_n}^{t_{n+1}} f(\tau, y(\tau))d\tau. \tag{6.15}$$

The integral can be approximated by a *quadrature formula*. A quadrature formula is a linear combination of function values in a given interval which approximates the integral of the function over the interval. Quadrature formulae are often chosen so that they are exact for polynomials up to a certain degree ν.

More precisely, we approximate (6.15) by

$$\mathbf{y}_{n+1} = \mathbf{y}_n + h \sum_{l=1}^{\nu} b_l \mathbf{f}(t_n + c_l h, \mathbf{y}(t_n + c_l h)), \tag{6.16}$$

except that, of course, the vectors $\mathbf{y}(t_n + c_l h)$ are unknown! *Runge–Kutta methods* are a means of implementing (6.16) by replacing unknown values of $\mathbf{y}$ by suitable linear combinations. The general form of a ν-*stage explicit Runge–Kutta method (RK)* is

$$
\begin{aligned}
\mathbf{k}_1 &= \mathbf{f}(t_n, \mathbf{y}_n), \\
\mathbf{k}_2 &= \mathbf{f}(t_n + c_2 h, \mathbf{y}_n + h c_2 \mathbf{k}_1), \\
\mathbf{k}_3 &= \mathbf{f}(t_n + c_3 h, \mathbf{y}_n + h(a_{3,1}\mathbf{k}_1 + a_{3,2}\mathbf{k}_2)), \quad a_{3,1} + a_{3,2} = c_3, \\
&\vdots \\
\mathbf{k}_\nu &= \mathbf{f}\left(t_n + c_\nu h, \mathbf{y}_n + h \sum_{j=1}^{\nu-1} a_{\nu,j}\mathbf{k}_j\right), \qquad \sum_{j=1}^{\nu-1} a_{\nu,j} = c_\nu, \\
\mathbf{y}_{n+1} &= \mathbf{y}_n + h \sum_{l=1}^{\nu} b_l \mathbf{k}_l.
\end{aligned}
$$

$\mathbf{b} = (b_1, \ldots, b_\nu)^T$ are called the *Runge–Kutta weights* and satisfy the condition $\sum_{l=1}^{\nu} b_l = 1$. $\mathbf{c} = (c_1, \ldots, c_\nu)^T$ are called the *Runge–Kutta nodes*. The method is called *consistent* if

$$\sum_{j=1}^{\nu-1} a_{i,j} = c_i, \qquad i = 1, \ldots, \nu.$$

To specify a particular method, one needs to provide the integer ν (the number of stages), and the coefficients $a_{i,j}$ (for $1 \le i, j \le \nu$), b_i (for $i = 1, 2, \ldots, \nu$), and c_i (for $i = 1, 2, \ldots, \nu$). These data are usually arranged in a mnemonic device, known as a *Butcher tableau* shown in Figure 6.6.

The simplest Runge–Kutta method is the forward Euler method, which we can rewrite as

$$\mathbf{y}_{n+1} = \mathbf{y}_n + h\mathbf{k}_1, \qquad \mathbf{k}_1 = \mathbf{f}(t_n, \mathbf{y}_n),$$

$$
\begin{array}{c|ccccc}
0 & & & & & \\
c_2 & a_{2,1} & & & & \\
c_3 & a_{3,1} & a_{3,2} & & & \\
\vdots & \vdots & \vdots & \ddots & & \\
c_\nu & a_{\nu,1} & a_{\nu,2} & \cdots & a_{\nu,\nu-1} & \\
\hline
& b_1 & b_2 & \cdots & b_{\nu-1} & b_\nu
\end{array}
\qquad \Leftrightarrow \qquad
\begin{array}{c|c}
\mathbf{c} & A \\
\hline
& \mathbf{b}^T
\end{array}
$$

Figure 6.6 Butcher tableau

where $\mathbf{k}_1$ is an increment based on an estimate of $\mathbf{y}'$ at the point $(t_n, \mathbf{y}_n)$. This is the only consistent explicit Runge–Kutta method of one stage. The corresponding tableau is:

$$
\begin{array}{c|c}
0 & \\
\hline
& 1
\end{array}
$$

A 2-stage Runge–Kutta method can be derived by re-estimating $\mathbf{y}'$ based at the point $(t_{n+1}, \mathbf{y}_{n+1})$. This gives an increment, say $\mathbf{k}_2$, where

$$
\mathbf{k}_2 = \mathbf{f}(t_n + h, \mathbf{y}_n + h\mathbf{k}_1).
$$

The Runge–Kutta method uses the average of these two increments, i.e.,

$$
\mathbf{y}_{n+1} = \mathbf{y}_n + \frac{1}{2}h(\mathbf{k}_1 + \mathbf{k}_2).
$$

The corresponding tableau is

$$
\begin{array}{c|cc}
0 & & \\
1 & 1 & \\
\hline
& \frac{1}{2} & \frac{1}{2}
\end{array}
$$

This method is known as *Heun's method*.

Another 2-stage method is provided by the midpoint method specified by

$$
\mathbf{y}_{n+1} = \mathbf{y}_n + h\mathbf{f}(t_n + \frac{1}{2}h, \mathbf{y}_n + \frac{1}{2}h\mathbf{f}(t_n, \mathbf{y}_n)).
$$

with tableau

$$
\begin{array}{c|cc}
0 & & \\
\frac{1}{2} & \frac{1}{2} & \\
\hline
& 0 & 1
\end{array}
$$

Both of these methods belong to the family of explicit methods given by

$$
\mathbf{y}_{n+1} = \mathbf{y}_n + h\left[(1 - \frac{1}{2\alpha})\mathbf{f}(t_n, \mathbf{y}_n) + \frac{1}{2\alpha}\mathbf{f}(t_n + \alpha h, \mathbf{y}_n + \alpha h\mathbf{f}(t_n, \mathbf{y}_n))\right].
$$

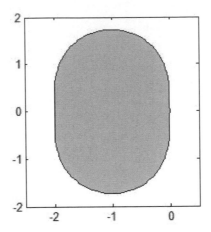

Figure 6.7 Stability domain of the methods given by (6.17)

They have the tableau

$$
\begin{array}{c|cc}
0 & & \\
\alpha & \alpha & \\
\hline
& (1 - \frac{1}{2\alpha}) & \frac{1}{2\alpha}
\end{array}
\tag{6.17}
$$

The choice $\alpha = \frac{1}{2}$ recovers the mid-point rule and $\alpha = 1$ is Heun's rule. All these methods have the same stability domain which is shown in Figure 6.7.

The choice of the *RK coefficients* $a_{l,j}$ is motivated at the first instance by order considerations. Thus we again have to perform Taylor expansions. As an example we derive a 2-stage Runge–Kutta method. We have $\mathbf{k}_1 = \mathbf{f}(t_n, \mathbf{y}_n)$ and to examine $\mathbf{k}_2$ we Taylor-expand about $(t_n, \mathbf{y}_n)$,

$$
\begin{aligned}
\mathbf{k}_2 &= \mathbf{f}(t_n + c_2 h, \mathbf{y}_n + h c_2 \mathbf{f}(t_n, \mathbf{y}_n)) \\
&= \mathbf{f}(t_n, \mathbf{y}_n) + h c_2 \left[\frac{\partial \mathbf{f}(t_n, \mathbf{y}_n)}{\partial t} + \frac{\partial \mathbf{f}(t_n, \mathbf{y}_n)}{\partial \mathbf{y}} \mathbf{f}(t_n, \mathbf{y}_n) \right] + O(h^2).
\end{aligned}
$$

On the other hand we have

$$
\mathbf{y}' = \mathbf{f}(t, \mathbf{y}) \qquad \Rightarrow \mathbf{y}'' = \frac{\partial \mathbf{f}(t, \mathbf{y})}{\partial t} + \frac{\partial \mathbf{f}(t, \mathbf{y})}{\partial \mathbf{y}} \mathbf{f}(t, \mathbf{y}).
$$

Therefore, substituting the exact solution $\mathbf{y}_n = \mathbf{y}(t_n)$, we obtain $\mathbf{k}_1 = \mathbf{y}'(t_n)$ and $\mathbf{k}_2 = \mathbf{y}'(t_n) + h c_2 \mathbf{y}''(t_n) + O(h^2)$. Consequently, the *local error* is

$$
\begin{aligned}
\mathbf{y}(t_{n+1}) - \mathbf{y}_{n+1} &= [\mathbf{y}(t_n) + h\mathbf{y}'(t_n) + \tfrac{1}{2}h^2 \mathbf{y}''(t_n) + O(h^3)] \\
&\quad - [\mathbf{y}(t_n) + h(b_1 + b_2)\mathbf{y}'(t_n) + h^2 b_2 c_2 \mathbf{y}''(t_n) + O(h^3)].
\end{aligned}
$$

We deduce that the Runge–Kutta method is of order 2 if $b_1 + b_2 = 1$ and $b_2 c_2 = \frac{1}{2}$. The methods given by (6.17) all fulfill this criterion.

Exercise 6.13. *Show that the truncation error of methods given by (6.17) is minimal for $\alpha = \frac{2}{3}$. Also show that no such method has order 3 or above.*

Different categories of Runge–Kutta methods are abbreviated as follows

1. *Explicit RK (ERK):* A is strictly lower triangular;

2. *Diagonally implicit RK (DIRK):* A lower triangular;

3. *Singly diagonally implicit RK (SDIRK):* A lower triangular, $a_{i,i} \equiv$ const $\neq 0$;

4. *Implicit RK (IRK):* Otherwise.

The original Runge–Kutta method is fourth-order and is given by the tableau

$$
\begin{array}{c|cccc}
0 \\
\frac{1}{2} & \frac{1}{2} \\
\frac{1}{2} & 0 & \frac{1}{2} \\
1 & 0 & 0 & 1 \\
\hline
& \frac{1}{6} & \frac{1}{3} & \frac{1}{3} & \frac{1}{6}
\end{array}
\tag{6.18}
$$

Its stability domain is shown in Figure 6.8.

Two 3-stage explicit methods of order 3 are named after Kutta and Nystrom, respectively.

$$
\text{Kutta} \quad
\begin{array}{c|ccc}
0 \\
\frac{1}{2} & \frac{1}{2} \\
1 & -1 & 2 \\
\hline
& \frac{1}{6} & \frac{2}{3} & \frac{1}{6}
\end{array}
\qquad
\text{Nystrom} \quad
\begin{array}{c|ccc}
0 \\
\frac{2}{3} & \frac{2}{3} \\
\frac{2}{3} & 0 & \frac{2}{3} \\
\hline
& \frac{1}{4} & \frac{3}{8} & \frac{3}{8}
\end{array}
$$

The one in the tableau of Kutta's method shows that this method explicitly employs an estimate of $\mathbf{f}$ at $t_n + h$. Both methods have the same stability domain shown in Figure 6.9.

An error control device specific to Runge–Kutta methods is *embedding* with *adaptive step size*. We embed a method in a larger method. For example, let

$$
A = \begin{bmatrix} \tilde{A} & \mathbf{0} \\ \mathbf{a}^T & a \end{bmatrix}, \qquad \mathbf{c} = \begin{bmatrix} \tilde{\mathbf{c}} \\ c \end{bmatrix},
$$

such that the method given by

$$
\begin{array}{c|c}
\mathbf{c} & A \\
\hline
& \mathbf{b}^T
\end{array}
$$

is of higher order than

$$
\begin{array}{c|c}
\tilde{\mathbf{c}} & \tilde{A} \\
\hline
& \tilde{\mathbf{b}}^T
\end{array}.
$$

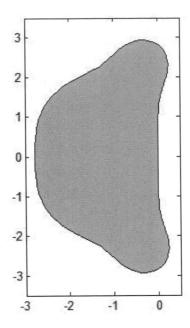

Figure 6.8 The stability domain of the original Runge–Kutta method given by (6.18)

Comparison of the two yields an estimate of the local error. We use the method with the smaller truncation error to estimate the error in the other method. More specifically, $\|\mathbf{y}_{n+1} - \mathbf{y}(t_{n+1})\| \approx \|\mathbf{y}_{n+1} - \tilde{\mathbf{y}}_{n+1}\|$. This is then used to adjust the step size to achieve a desired accuracy. The error estimate is used to improve the solution. Often the matrix $\tilde{A}$ and vector $\tilde{\mathbf{c}}$ are actually not extended. Both methods use the same matrix of coefficients and nodes. However, the weights $\tilde{\mathbf{b}}$ differ. The methods are described with an *extended Butcher tableau*, which is the Butcher tableau of the higher-order method with another row added for the weights of the lower-order method.

$$
\begin{array}{c|cccc}
c_1 & a_{1,1} & a_{1,2} & \cdots & a_{1,\nu} \\
c_2 & a_{2,1} & a_{2,2} & \cdots & a_{2,\nu} \\
\vdots & \vdots & \vdots & \ddots & \vdots \\
c_\nu & a_{\nu,1} & a_{\nu,2} & \cdots & a_{\nu,\nu} \\
\hline
& b_1 & b_2 & \cdots & b_\nu \\
& \tilde{b}_1 & \tilde{b}_2 & \cdots & \tilde{b}_\nu
\end{array}
$$

The simplest adaptive Runge–Kutta method involves combining the Heun method, which is order 2, with the forward Euler method, which is order 1.

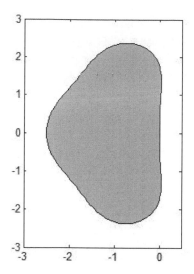

Figure 6.9 Stability domain of Kutta's and Nystrom's method

The result is the *Heun–Euler method* and its extended Butcher tableau is

$$
\begin{array}{c|cc}
0 & & \\
1 & 1 & \\
\hline
& \frac{1}{2} & \frac{1}{2} \\
& 1 & 0
\end{array}
$$

The zero in the bottom line of the tableau shows that the forward Euler method does not use the estimate k_2.

The *Bogacki–Shampine method* has two methods of orders 3 and 2. Its extended Butcher tableau is shown in Figure 6.10

$$
\begin{array}{c|cccc}
0 & & & & \\
\frac{1}{2} & \frac{1}{2} & & & \\
\frac{3}{4} & 0 & \frac{3}{4} & & \\
1 & \frac{2}{9} & \frac{1}{3} & \frac{4}{9} & \\
\hline
& \frac{2}{9} & \frac{1}{3} & \frac{4}{9} & 0 \\
& \frac{7}{24} & \frac{1}{4} & \frac{1}{3} & \frac{1}{8}
\end{array}
$$

Figure 6.10 Tableau of the Bogacki–Shampine method

There are several things to note about this method: firstly, the zero in the vector of weights of the higher-order method. Thus the higher-order method actually does not employ $\mathbf{k}_4$. Thus the lower-order method employs more estimates. Secondly, the calculation of $\mathbf{k}_4$ uses the same weights as the calculation of $\mathbf{y}_{n+1}$. Thus $\mathbf{k}_4$ actually equals $\mathbf{f}(t_n + h, \mathbf{y}_{n+1})$. This is known as the *First Same As Last (FSAL)* property, since $\mathbf{k}_4$ in this step will be $\mathbf{k}_1$ in the next step. So this method uses three function evaluations of $\mathbf{f}$ per step. This method is implemented in the `ode23` function in MATLAB.

The *Runge–Kutta–Fehlberg method (or Fehlberg method)* combines two methods of order 4 and 5. Figure 6.11 shows the tableau. This method is often implemented as `RKF45` in collections of numerical methods. The coefficients are chosen so that the error in the fourth-order method is minimized.

$$
\begin{array}{c|cccccc}
0 \\
\frac{1}{4} & \frac{1}{4} \\
\frac{3}{8} & \frac{3}{32} & \frac{9}{32} \\
\frac{12}{13} & \frac{1932}{2197} & -\frac{7200}{2197} & \frac{7296}{2197} \\
1 & \frac{439}{216} & -8 & \frac{3680}{513} & -\frac{845}{4104} \\
\frac{1}{2} & -\frac{8}{27} & 2 & -\frac{3544}{2565} & \frac{1859}{4104} & -\frac{11}{40} \\
\hline
& \frac{16}{135} & 0 & \frac{6656}{12825} & \frac{28561}{56430} & -\frac{9}{50} & \frac{2}{55} \\
& \frac{25}{216} & 0 & \frac{1408}{2565} & \frac{2197}{4104} & \frac{1}{5} & 0
\end{array}
$$

Figure 6.11 Tableau of the Fehlberg method

The *Cash–Karp method* takes the concept of error control and adaptive step size to a whole new level by not only embedding one lower method, methods of order $1, 2, 3,$ and 4 are embedded in a fifth-order method. The tableau is as displayed in Figure 6.12. Note that the order-one method is the forward Euler method.

The Cash–Karp method was motivated to deal with the situation when certain derivatives of the solution are very large for part of the region. These are, for example, regions where the solution has a sharp front or some derivative of the solution is discontinuous in the limit. In these circumstance the step size has to be adjusted. By computing solutions at several different orders, it is possible to detect sharp fronts or discontinuities before all the function evaluations defining the full Runge–Kutta step have been computed. We can then either accept a lower-order solution or abort the step (and try again with a smaller step-size), depending on which course of action seems appropriate. J. Cash provides the code for this algorithm in `Fortran,` `C`, and `MATLAB` on his homepage.

The *Dormand–Prince method* is similar to the Fehlberg method. It also embeds a fourth-order method in a fifth-order method. The coefficients are, however, chosen so that the error of the fifth-order solution is minimized and

$$
\begin{array}{c|cccccc}
0 \\
\frac{1}{5} & \frac{1}{5} \\
\frac{3}{10} & \frac{3}{40} & \frac{9}{40} \\
\frac{3}{5} & \frac{3}{10} & -\frac{9}{10} & \frac{6}{5} \\
1 & -\frac{11}{54} & \frac{5}{2} & -\frac{70}{27} & \frac{35}{27} \\
\frac{7}{8} & \frac{1631}{55296} & \frac{175}{512} & \frac{575}{13824} & \frac{44275}{110592} & \frac{253}{4096} \\
\hline
& \frac{37}{378} & 0 & \frac{250}{621} & \frac{125}{594} & 0 & \frac{512}{1771} & \text{Order 5} \\
& \frac{2825}{27648} & 0 & \frac{18575}{48384} & \frac{13525}{55296} & \frac{277}{14336} & \frac{1}{4} & \text{Order 4} \\
& \frac{19}{54} & 0 & -\frac{10}{27} & \frac{55}{54} & 0 & 0 & \text{Order 3} \\
& -\frac{3}{2} & \frac{5}{2} & 0 & 0 & 0 & 0 & \text{Order 2} \\
& 1 & 0 & 0 & 0 & 0 & 0 & \text{Order 1}
\end{array}
$$

Figure 6.12 Tableau of the Cash–Karp method

the difference between the solutions is used to estimate the error in the fourth-order method. The Dormand–Prince method has seven stages, but it uses only six function evaluations per step, because it has the FSAL property. The tableau is given in Figure 6.13. This method is currently the default in MATLAB's ode45 solver.

$$
\begin{array}{c|ccccccc}
0 \\
\frac{1}{5} & \frac{1}{5} \\
\frac{3}{10} & \frac{3}{40} & \frac{9}{40} \\
\frac{4}{5} & \frac{44}{45} & -\frac{56}{15} & \frac{32}{9} \\
\frac{8}{9} & \frac{19372}{6561} & -\frac{25360}{2187} & \frac{644482}{6561} & -\frac{212}{729} \\
1 & \frac{9017}{3168} & -\frac{355}{33} & \frac{46732}{5247} & \frac{49}{176} & -\frac{5103}{18656} \\
1 & \frac{35}{384} & 0 & \frac{500}{1113} & \frac{125}{192} & -\frac{2187}{6784} & \frac{11}{84} \\
\hline
& \frac{35}{384} & 0 & \frac{500}{1113} & \frac{125}{192} & -\frac{2187}{6784} & \frac{11}{84} & 0 \\
& \frac{5179}{57600} & 0 & \frac{7571}{16695} & \frac{393}{640} & -\frac{92097}{339200} & \frac{187}{2100} & \frac{1}{40}
\end{array}
$$

Figure 6.13 Tableau of the Dormand–Prince method

Runge–Kutta methods are a field of ongoing research. New high-order methods are invented. For recent research see for example the work by J.H. Verner whose work concerns the derivation of Runge–Kutta pairs.

Next we will turn our attention to implicit Runge–Kutta methods. A general ν-stage Runge–Kutta method is

$$\mathbf{k}_l = \mathbf{f}\left(t_n + c_l h, \mathbf{y}_n + h\sum_{j=1}^{\nu} a_{l,j}\mathbf{k}_j\right), \quad \sum_{j=1}^{\nu} a_{l,j} = c_l \tag{6.19}$$

$$\mathbf{y}_{n+1} = \mathbf{y}_n + h\sum_{l=1}^{\nu} b_l \mathbf{k}_l.$$

Obviously, $a_{l,j} = 0$ for all $l < j$ yields the standard *explicit RK*. Otherwise, an RK method is said to be *implicit*.

One way to derive implicit Runge–Kutta methods are *collocation methods*. However, not all Rung–Kutta methods are collocation methods. The idea is to choose a finite-dimensional space of candidate solutions (usually, polynomials up to a certain degree) and a number of points in the domain (called *collocation points*), and to select that solution which satisfies the given equation at the collocation points. More precisely, we want to find a ν-degree polynomial $\mathbf{p}$ such that $\mathbf{p}(t_n) = \mathbf{y}_n$ and

$$\mathbf{p}'(t_n + c_l h) = \mathbf{f}(t_n + c_l h, \mathbf{p}(t_n + c_l h)), \qquad l = 1, \ldots, \nu,$$

where $c_l, l = 1, \ldots, \nu$ are the collocation points. This gives $\nu + 1$ conditions, which matches the $\nu + 1$ parameters needed to specify a polynomial of degree ν. The new estimate $\mathbf{y}_{n+1}$ is defined to be $\mathbf{p}(t_{n+1})$.

As an example pick the two collocation points $c_1 = 0$ and $c_2 = 1$ at the beginning and the end of the interval $[t_n, t_{n+1}]$. The collocation conditions are

$$\mathbf{p}(t_n) = \mathbf{y}_n,$$
$$\mathbf{p}'(t_n) = \mathbf{f}(t_n, \mathbf{p}(t_n)),$$
$$\mathbf{p}'(t_n + h) = \mathbf{f}(t_n + h, \mathbf{p}(t_n + h)).$$

For these three conditions we need a polynomial of degree 2, which we write in the form

$$\mathbf{p}(t) = \alpha(t - t_n)^2 + \beta(t - t_n) + \gamma.$$

The collocation conditions can be solved to give the coefficients

$$\alpha = \frac{1}{2h}[\mathbf{f}(t_n + h, \mathbf{p}(t_n + h)) - \mathbf{f}(t_n, \mathbf{p}(t_n))],$$
$$\beta = \mathbf{f}(t_n, \mathbf{p}(t_n)),$$
$$\gamma = \mathbf{y}_n.$$

Putting these coefficients back into the definition of $\mathbf{p}$ and evaluating it at $t = t_{n+1}$ gives the method

$$\mathbf{y}_{n+1} = \mathbf{p}(t_n + h) = \mathbf{y}_n + \frac{1}{2}h(\mathbf{f}(t_n + h, \mathbf{p}(t_n + h)) + \mathbf{f}(t_n, \mathbf{p}(t_n)))$$

$$= \mathbf{y}_n + \frac{1}{2}h(\mathbf{f}(t_n + h, \mathbf{y}_{n+1}) + \mathbf{f}(t_n, \mathbf{y}_n)),$$

and we have recovered Heun's method.

Theorem 6.9 (Guillou and Soulé). *Let the collocation points $c_l, l = 1, \ldots, \nu$ be distinct. Let $w(t)$ be the ν-degree polynomial defined by $w(t) := \prod_{l=1}^{\nu}(t - c_l)$ and let $w_l(t)$ the $(\nu - 1)$ degree polynomial defined by $w_l(t) := \frac{w(t)}{t - c_l}$, $l = 1, \ldots, \nu$. The collocation method is then identical to the ν-stage Runge–Kutta method with coefficients*

$$a_{k,l} = \frac{1}{w_l(c_l)} \int_0^{c_k} w_l(\tau)d\tau, \qquad b_l = \frac{1}{w_l(c_l)} \int_0^1 w_l(\tau)d\tau, \qquad k, l = 1, \ldots, \nu.$$

Proof. The polynomial $\mathbf{p}'$ is the $(\nu - 1)^{\text{th}}$ degree Lagrange interpolation polynomial. Define the lth *Lagrange cardinal polynomial* of degree $\nu - 1$ as

$$L_l(t) = \frac{w_l(t)}{w_l(c_l)}.$$

Thus $L_l(c_l) = 1$ and $L_l(c_i) = 0$ for all $i \neq l$. We then have

$$
\begin{aligned}
\mathbf{p}'(t) &= \sum_{l=1}^{\nu} L_l((t - t_n)/h)\mathbf{f}(t_n + c_l h, \mathbf{p}(t_n + c_l h)) \\
&= \sum_{l=1}^{\nu} \frac{w_l((t - t_n)/h)}{w_l(c_l)}\mathbf{f}(t_n + c_l h, \mathbf{p}(t_n + c_l h))
\end{aligned}
$$

and by integration we obtain

$$
\begin{aligned}
\mathbf{p}(t) &= \mathbf{p}(t_n) + \int_{t_n}^{t} \mathbf{p}'(\tilde{\tau})d\tilde{\tau} \\
&= \mathbf{p}(t_n) + \sum_{l=1}^{\nu} \mathbf{f}(t_n + c_l h, \mathbf{p}(t_n + c_l h)) \int_{t_n}^{t} \frac{w_l((\tilde{\tau} - t_n)/h)}{w_l(c_l)}d\tilde{\tau} \\
&= \mathbf{y}_n + h\sum_{l=1}^{\nu} \mathbf{f}(t_n + c_l h, \mathbf{p}(t_n + c_l h)) \int_0^{(t - t_n)/h} \frac{w_l(\tau)}{w_l(c_l)}d\tau.
\end{aligned}
$$

Now we have

$$
\begin{aligned}
\mathbf{p}(t_n + c_k h) &= \mathbf{y}_n + h\sum_{l=1}^{\nu} \mathbf{f}(t_n + c_l h, \mathbf{p}(t_n + c_l h)) \frac{1}{w_l(c_l)} \int_0^{c_k} w_l(\tau)d\tau \\
&= \mathbf{y}_n + h\sum_{l=1}^{\nu} a_{k,l}\mathbf{f}(t_n + c_l h, \mathbf{p}(t_n + c_l h)).
\end{aligned}
$$

This and defining $\mathbf{k}_l = \mathbf{f}(t_n + c_l h, \mathbf{p}(t_n + c_l h))$ gives the intermediate stages of the Runge–Kutta method. Additionally we have

$$\mathbf{y}_{n+1} = \mathbf{p}(t_n + h) = \mathbf{y}_n + h\sum_{l=1}^{\nu} \mathbf{k}_l \frac{1}{w_l(c_l)} \int_0^1 w_l(\tau)d\tau = \mathbf{p}(t_n) + h\sum_{l=1}^{\nu} b_l \mathbf{k}_l.$$

This and the definition of the Runge–Kutta method proves the theorem. □

Collocation methods have the advantage that we obtain a continuous approximation to the solution $\mathbf{y}(t)$ in each of the intervals $[t_n, t_{n+1}]$. Different choices of collocation points lead to different Runge–Kutta methods. For this we have to look more closely at the concept of numerical quadrature. In terms of order the optimal choice is to let the collocation points be the roots of a *shifted Legendre polynomial*. The Legendre polynomials p_s, $s = 0, 1, 2, \ldots$, are defined on the interval $[-1, 1]$ by

$$p_s(x) = \frac{1}{2^s s!} \frac{d^s}{dx^s} [(x^2 - 1)^s].$$

Thus p_s is an s-degree polynomial. Shifting the Legendre polynomials from $[-1, 1]$ to $[0, 1]$ by $x \mapsto 2x - 1$ leads to

$$\tilde{p}_s(x) = p_s(2x - 1) = \frac{1}{s!} \frac{d^s}{dx^s} [x^s (x - 1)^s]$$

and using its roots as collocation points gives the *Gauss–Legendre Runge–Kutta methods* which are of order $2s$, if the roots of p_s are employed. For $s = 1, 2$, and 3, the collocation points in this case are

- $c_1 = \frac{1}{2}$, order 2 with tableau $\begin{array}{c|c} \frac{1}{2} & \frac{1}{2} \\ \hline & 1 \end{array}$, which is the implicit midpoint rule,

- $c_1 = \frac{1}{2} - \frac{\sqrt{3}}{6}, c_2 = \frac{1}{2} + \frac{\sqrt{3}}{6}$ order 4 with tableau

$$\begin{array}{c|cc} \frac{1}{2} - \frac{\sqrt{3}}{6} & \frac{1}{4} & \frac{1}{4} - \frac{\sqrt{3}}{6} \\ \frac{1}{2} + \frac{\sqrt{3}}{6} & \frac{1}{4} + \frac{\sqrt{3}}{6} & \frac{1}{4} \\ \hline & \frac{1}{2} & \frac{1}{2} \end{array}$$

- $c_1 = \frac{1}{2} - \frac{\sqrt{15}}{10}, c_2 = \frac{1}{2}, c_3 = \frac{1}{2} + \frac{\sqrt{15}}{10}$ order 6 with tableau

$$\begin{array}{c|ccc} \frac{1}{2} - \frac{\sqrt{15}}{10} & \frac{5}{36} & \frac{2}{9} - \frac{\sqrt{15}}{15} & \frac{5}{36} - \frac{\sqrt{15}}{30} \\ \frac{1}{2} & \frac{5}{36} + \frac{\sqrt{15}}{24} & \frac{2}{9} & \frac{5}{36} - \frac{\sqrt{15}}{24} \\ \frac{1}{2} + \frac{\sqrt{15}}{10} & \frac{5}{36} + \frac{\sqrt{15}}{30} & \frac{2}{9} + \frac{\sqrt{15}}{15} & \frac{5}{36} \\ \hline & \frac{5}{18} & \frac{4}{9} & \frac{5}{18} \end{array}$$

Further methods were developed by considering the roots of *Lobatto and Radau polynomials*. Butcher introduced a classification of type I, II, or III which still survives in the naming conventions.

I $\quad \dfrac{d^{s-1}}{dx^{s-1}} [x^s (x - 1)^{s-1}]$ left Radau

II $\quad \dfrac{d^{s-1}}{dx^{s-1}} [x^{s-1} (x - 1)^s]$ right Radau

III $\dfrac{d^{s-2}}{dx^{s-2}}[x^{s-1}(x-1)^{s-1}]$ Lobatto

The methods are of order $2s-1$ in the Radau case and of order $2s-2$ in the Lobatto case. Note that the Radau I methods have 0 as one of their collocation points, while the Radau II method has 1 as one of its collocation points. This means that for Radau I the first row in the tableau always consists entirely of zeros while for Radau II the last row is identical to the vector of weights. The 2-stage methods are given in Figure 6.14. The letters correspond to certain conditions imposed on A which are however beyond this course (for further information see [3] J. C. Butcher, *The numerical analysis of ordinary differential equations: Runge–Kutta and general linear methods.*

Radau I	0	0	0
	$\frac{2}{3}$	$\frac{1}{3}$	$\frac{1}{3}$
		$\frac{1}{4}$	$\frac{3}{4}$

Lobatto III(A)	0	0	0
	1	$\frac{1}{2}$	$\frac{1}{2}$
		$\frac{1}{2}$	$\frac{1}{2}$

Radau IA	0	$\frac{1}{4}$	$-\frac{1}{4}$
	$\frac{2}{3}$	$\frac{1}{4}$	$\frac{5}{12}$
		$\frac{1}{4}$	$\frac{3}{4}$

Lobatto IIIB	0	$\frac{1}{2}$	0
	1	$\frac{1}{2}$	0
		$\frac{1}{2}$	$\frac{1}{2}$

Radau II(A)	$\frac{1}{3}$	$\frac{5}{12}$	$-\frac{1}{12}$
	1	$\frac{3}{4}$	$\frac{1}{4}$
		$\frac{3}{4}$	$\frac{1}{4}$

Lobatto IIIC	0	$\frac{1}{2}$	$-\frac{1}{2}$
	1	$\frac{1}{2}$	$\frac{1}{2}$
		$\frac{1}{2}$	$\frac{1}{2}$

Figure 6.14 The 2-stage Radau and Lobatto methods

For the three stages we have the Radau methods as specified in Figure 6.15 and the Lobatto methods as in Figure 6.16.

Next we examine the stability domain of Runge–Kutta methods by considering the test equation $y' = \lambda y = f(t, y)$. Firstly, we get for the internal stages the relations

$$\begin{aligned} k_i &= f(t + c_i h, y_n + h(a_{i,1}k_{11} + a_{i,2}k_2 + \ldots a_{i,\nu}k_\nu)) \\ &= \lambda\left[y_n + h(a_{i,1}k_{11} + a_{i,2}k_2 + \ldots a_{i,\nu}k_\nu)\right]. \end{aligned}$$

Defining $\mathbf{k} = (k_1, \ldots k_\nu)^T$ as the vector of stages and the $\nu \times \nu$ matrix $A = (a_{i,j})$, and denoting $(1, \ldots, 1)^T$ as $\mathbf{1}$, this can be rewritten in matrix from as

$$\mathbf{k} = \lambda\left(y_n \mathbf{1} + hA\mathbf{k}\right).$$

Radau I	0	0	0	0
	$\frac{6-\sqrt{6}}{10}$	$\frac{9+\sqrt{6}}{75}$	$\frac{24+\sqrt{6}}{120}$	$\frac{168-73*\sqrt{6}}{600}$
	$\frac{6+\sqrt{6}}{10}$	$\frac{9-\sqrt{6}}{75}$	$\frac{168+73*\sqrt{6}}{600}$	$\frac{24-\sqrt{6}}{120}$
		$\frac{1}{9}$	$\frac{16+\sqrt{6}}{36}$	$\frac{16-\sqrt{6}}{36}$

Radau IA	0	$\frac{1}{9}$	$\frac{-1-\sqrt{6}}{18}$	$\frac{-1+\sqrt{6}}{18}$
	$\frac{6-\sqrt{6}}{10}$	$\frac{1}{9}$	$\frac{88+7\sqrt{6}}{360}$	$\frac{88-43\sqrt{6}}{360}$
	$\frac{6+\sqrt{6}}{10}$	$\frac{1}{9}$	$\frac{88+43*\sqrt{6}}{360}$	$\frac{88-7\sqrt{6}}{360}$
		$\frac{1}{9}$	$\frac{16+\sqrt{6}}{36}$	$\frac{16-\sqrt{6}}{36}$

Radau II(A)	$\frac{4-\sqrt{6}}{10}$	$\frac{88-7\sqrt{6}}{360}$	$\frac{296-169\sqrt{6}}{1800}$	$\frac{-2+3\sqrt{6}}{225}$
	$\frac{4+\sqrt{6}}{10}$	$\frac{296+169\sqrt{6}}{1800}$	$\frac{88+7\sqrt{6}}{360}$	$\frac{-2-3\sqrt{6}}{225}$
	1	$\frac{16-\sqrt{6}}{36}$	$\frac{16+\sqrt{6}}{36}$	$\frac{1}{9}$
		$\frac{16-\sqrt{6}}{36}$	$\frac{16+\sqrt{6}}{36}$	$\frac{1}{9}$

Figure 6.15 The 3-stage Radau methods

Solving for **k**,

$$\mathbf{k} = \lambda y_n \left(I - h\lambda A\right)^{-1} \mathbf{1}.$$

Further, we have

$$y_{n+1} = y_n + h\sum_{l=1}^{\nu} b_l k_l = y_n + h\mathbf{b}^T \mathbf{k},$$

if we define $\mathbf{b} = (b_1, \ldots, b_\nu)^T$. Using the above equation for **k**, we obtain

$$y_{n+1} = y_n \left(1 + h\lambda \mathbf{b}^T \left(I - h\lambda A\right)^{-1} \mathbf{1}\right).$$

Since $\sum_{l=1}^{\nu} b_l = \mathbf{b}^T \mathbf{1} = 1$, this can be rewritten as

$$
\begin{aligned}
y_{n+1} &= y_n \mathbf{b}^T \left[I + h\lambda \left(I - h\lambda A\right)^{-1}\right] \mathbf{1} \\
&= y_n \mathbf{b}^T \left(I - h\lambda A\right)^{-1} \left[I - h\lambda(A - I)\right] \mathbf{1} \\
&= y_n \frac{1}{\det(I - h\lambda A)} \mathbf{b}^T \mathrm{adj}(I - h\lambda A)\left[I - h\lambda(A - I)\right] \mathbf{1}.
\end{aligned}
$$

Lobatto III(A)	$\begin{array}{c\|ccc} 0 & 0 & 0 & 0 \\ \frac{1}{2} & \frac{5}{24} & \frac{1}{3} & -\frac{1}{24} \\ 1 & \frac{1}{6} & \frac{2}{3} & \frac{1}{6} \\ \hline & \frac{1}{6} & \frac{2}{3} & \frac{1}{6} \end{array}$
Lobatto IIIB	$\begin{array}{c\|ccc} 0 & \frac{1}{6} & -\frac{1}{6} & 0 \\ \frac{1}{2} & \frac{1}{6} & \frac{1}{3} & 0 \\ 1 & \frac{1}{6} & \frac{5}{6} & 0 \\ \hline & \frac{1}{6} & \frac{2}{3} & \frac{1}{6} \end{array}$
Lobatto IIIC	$\begin{array}{c\|ccc} 0 & \frac{1}{6} & -\frac{1}{3} & \frac{1}{6} \\ \frac{1}{2} & \frac{1}{6} & \frac{5}{12} & -\frac{1}{12} \\ 1 & \frac{1}{6} & \frac{2}{3} & \frac{1}{6} \\ \hline & \frac{1}{6} & \frac{2}{3} & \frac{1}{6} \end{array}$

Figure 6.16 The 3-stage Lobatto methods

Now $\det(I - h\lambda A)$ is a polynomial in $h\lambda$ of degree ν, each entry in the adjoint matrix of $I - h\lambda A$ is a polynomial in $h\lambda$ of degree $\nu - 1$, and each entry in $I - h\lambda(A - I)$ is a polynomial in $h\lambda$ of degree 1. So we can deduce that $y_{n+1} = R(h\lambda)y_n$, where $R = P/Q$ is a rational function with P and Q being polynomials of degree ν. R is called the *stability function* of the Runge–Kutta method. For explicit methods $Q \equiv 1$, since the matrix A is strictly lower triangular and thus the inverse of $I - h\lambda A$ can be easily calculated without involving the determinant and the adjoint matrix.

By induction we deduce $y_n = [R(h\lambda)]^n y_0$ and we deduce that the stability domain is given by
$$D = \{z \in \mathbb{C} : |R(z)| < 1\}.$$
For A-stability it is therefore necessary that $|R(z)| < 1$ for every $z \in \mathbb{C}^-$. For explicit methods we have $Q \equiv 1$ and the stability domain of an explicit method is a bounded set. We have seen the stability regions for various Runge–Kutta methods in Figures 6.7, 6.8, 6.9, and 6.17.

To illustrate we consider the 3rd-order method given by

$$
\begin{aligned}
\mathbf{k}_1 &= \mathbf{f}(t_n, \mathbf{y}_n + \frac{1}{4}h(\mathbf{k}_1 - \mathbf{k}_2)), \\
\mathbf{k}_2 &= \mathbf{f}(t_n + \frac{2}{3}h, \mathbf{y}_n + \frac{1}{12}h(3\mathbf{k}_1 + 5\mathbf{k}_2)), \\
\mathbf{y}_{n+1} &= \mathbf{y}_n + \frac{1}{4}h(\mathbf{k}_1 + 3\mathbf{k}_2).
\end{aligned}
\tag{6.20}
$$

Applying it to $y' = \lambda y$, we have

$$
\begin{aligned}
hk_1 &= h\lambda(y_n + \frac{1}{4}hk_1 - \frac{1}{4}hk_2), \\
hk_2 &= h\lambda(y_n + \frac{1}{4}hk_1 + \frac{5}{12}hk_2).
\end{aligned}
$$

This is a linear system which has the solution

$$
\begin{aligned}
\begin{bmatrix} hk_1 \\ hk_2 \end{bmatrix} &= \begin{bmatrix} 1 - \frac{1}{4}h\lambda & \frac{1}{4}h\lambda \\ -\frac{1}{4}h\lambda & 1 - \frac{5}{12}h\lambda \end{bmatrix}^{-1} \begin{bmatrix} h\lambda y_n \\ h\lambda y_n \end{bmatrix} \\
&= \frac{h\lambda y_n}{1 - \frac{2}{3}h\lambda + \frac{1}{6}(h\lambda)^2} \begin{bmatrix} 1 - \frac{2}{3}h\lambda \\ 1 \end{bmatrix},
\end{aligned}
$$

therefore

$$
y_{n+1} = y_n + \frac{1}{4}hk_1 + \frac{3}{4}hk_2 = \frac{1 + \frac{1}{3}h\lambda}{1 - \frac{2}{3}h\lambda + \frac{1}{6}(h\lambda)^2} y_n.
$$

Hence the stability function is given by

$$
R(z) = \frac{1 + \frac{1}{3}z}{1 - \frac{2}{3}z + \frac{1}{6}z^2}.
$$

Figure 6.17 illustrates the stability region given by this stability function.

We prove A-stability by the following technique. According to the *maximum modulus principle*, if g is analytic in the closed complex domain V, then $|g|$ attains its maximum on the boundary ∂V. We let $g = R$. This is a rational function, hence its only singularities are the poles $2 \pm i\sqrt{2}$, which are the roots of the denominator and g is analytic in $V = \mathrm{cl}\mathbb{C}^- = \{z \in \mathbb{C} : \mathrm{Re}z \leq 0\}$. Therefore it attains its maximum on $\partial V = i\mathbb{R}$ and the following statements are equivalent

$$
\text{A-stability} \quad \Leftrightarrow \quad |R(z)| < 1, \quad z \in \mathbb{C}^- \quad \Leftrightarrow \quad |R(it)| \leq 1, \quad t \in \mathbb{R}.
$$

In this example we have

$$
|R(it)| \leq 1 \quad \Leftrightarrow \quad |1 - \frac{2}{3}it - \frac{1}{6}t^2|^2 - |1 + \frac{1}{3}it|^2 \geq 0.
$$

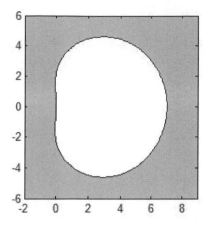

Figure 6.17 Stability domain of the method given in (6.20). The instability region is white.

But $|1 - \frac{2}{3}it - \frac{1}{6}t^2|^2 - |1 + \frac{1}{3}it|^2 = (1 - \frac{1}{6}t^2)^2 + (\frac{2}{3}t)^2 - 1 - (\frac{1}{3}t)^2 = \frac{1}{36}t^4 \geq 0$ and it follows that the method is A-stable.

As another example we consider the 2-stage *Gauss–Legendre method* given by

$$\mathbf{k}_1 = \mathbf{f}(t_n + (\frac{1}{2} - \frac{\sqrt{3}}{6})h, \mathbf{y}_n + \frac{1}{4}h\mathbf{k}_1 + (\frac{1}{4} - \frac{\sqrt{3}}{6})h\mathbf{k}_2),$$

$$\mathbf{k}_2 = \mathbf{f}(t_n + (\frac{1}{2} + \frac{\sqrt{3}}{6})h, \mathbf{y}_n + (\frac{1}{4} + \frac{\sqrt{3}}{6})h\mathbf{k}_1 + \frac{1}{4}h\mathbf{k}_2),$$

$$\mathbf{y}_{n+1} = \mathbf{y}_n + \frac{1}{2}h(\mathbf{k}_1 + \mathbf{k}_2).$$

It is possible to prove that it is of order 4. [You can do this for $y' = f(y)$ by expansion, but it becomes messy for $\mathbf{y}' = \mathbf{f}(t, \mathbf{y})$.] It can be easily verified that for $y' = \lambda y$ we have $y_n = [R(h\lambda)]^n y_0$, where $R(z) = (1 + \frac{1}{2}z + \frac{1}{12}z^2)/(1 - \frac{1}{2}z + \frac{1}{12}z^2)$. The poles of R are at $3 \pm i\sqrt{3}$, and thus lie in $\mathbb{C}^+$. In addition, $|R(it)| \equiv 1$, $t \in \mathbb{R}$, because the denominator is the complex conjugate of the numerator. We can again use the maximum modulus principle to argue that $D \supseteq \mathbb{C}^-$ and the 2-stage Gauss–Legendre method is A-stable, indeed $D = \mathbb{C}^-$.

Unlike other multistep methods, **implicit** high-order Runge–Kutta methods may be A-stable, which makes them particularly suitable for stiff problems.

In the following we will consider the order of Runge–Kutta methods. As an example we consider the 2-stage method given by

$$
\begin{aligned}
\mathbf{k}_1 &= \mathbf{f}(t_n, \mathbf{y}_n + \tfrac{1}{4}h(\mathbf{k}_1 - \mathbf{k}_2)), \\
\mathbf{k}_2 &= \mathbf{f}(t_n + \tfrac{2}{3}h, \mathbf{y}_n + \tfrac{1}{12}h(3\mathbf{k}_1 + 5\mathbf{k}_2)), \\
\mathbf{y}_{n+1} &= \mathbf{y}_n + \tfrac{1}{4}h(\mathbf{k}_1 + 3\mathbf{k}_2).
\end{aligned}
$$

In order to analyze the order of this method, we restrict our attention to scalar, autonomous equations of the form $y' = f(y)$. (This procedure simplifies the process but might lead to loss of generality for methods of order ≥ 5.) For brevity, we use the convention that all functions are evaluated at $y = y_n$, e.g., $f_y = df(y_n)/dy$. Thus,

$$
\begin{aligned}
k_1 &= f + \tfrac{1}{4}h(k_1 - k_2)f_y + \tfrac{1}{32}h^2(k_1 - k_2)^2 f_{yy} + O(h^3), \\
k_2 &= f + \tfrac{1}{12}h(3k_1 + 5k_2)f_y + \tfrac{1}{288}h^2(3k_1 + 5k_2)^2 f_{yy} + O(h^3).
\end{aligned}
$$

We see that both k_1 and k_2 equal $f + O(h)$ and substituting this in the above equations yields $k_1 = f + O(h^2)$ (since $(k_1 - k_2)$ in the second term) and $k_2 = f + \tfrac{2}{3}hf_yf + O(h^2)$. Substituting this result again, we obtain

$$
\begin{aligned}
k_1 &= f - \tfrac{1}{6}h^2 f_y^2 f + O(h^3), \\
k_2 &= f + \tfrac{2}{3}hf_yf + h^2(\tfrac{5}{18}f_y^2 f + \tfrac{2}{9}f_{yy}f^2) + O(h^3).
\end{aligned}
$$

Inserting these results into the definition of y_{n+1} gives

$$
y_{n+1} = y + hf + \tfrac{1}{2}h^2 f_y f + \tfrac{1}{6}h^3(f_y^2 f + f_{yy}f^2) + O(h^4).
$$

But $y' = f \Rightarrow y'' = f_y f \Rightarrow y''' = f_y^2 f + f_{yy}f^2$ and we deduce from Taylor's theorem that the method is at least of order 3. (By applying it to the equation $y' = \lambda y$, it is easy to verify that it is not of order 4.)

Exercise 6.14. *The following four-stage Runge–Kutta method has order four,*

$$
\begin{aligned}
\mathbf{k}_1 &= \mathbf{f}(t_n, \mathbf{y}_n), \\
\mathbf{k}_2 &= \mathbf{f}(t_n + \tfrac{1}{3}h, \mathbf{y}_n + \tfrac{1}{3}h\mathbf{k}_1), \\
\mathbf{k}_3 &= \mathbf{f}(t_n + \tfrac{2}{3}h, \mathbf{y}_n - \tfrac{1}{3}h\mathbf{k}_1 + h\mathbf{k}_2), \\
\mathbf{k}_4 &= \mathbf{f}(t_n + h, \mathbf{y}_n + h\mathbf{k}_1 - h\mathbf{k}_2 + h\mathbf{k}_3), \\
\mathbf{y}_{n+1} &= \mathbf{y}_n + \tfrac{1}{8}h(\mathbf{k}_1 + 3\mathbf{k}_2 + 3\mathbf{k}_3 + \mathbf{k}_4).
\end{aligned}
$$

By applying it to the equation $y' = y$, show that the order is at most four.

Then, for scalar functions, prove that the order is at least four in the easy case when f is independent of y, and that the order is at least three in the relatively easy case when f is independent of t. (Thus you are not expected to do Taylor expansions when f depends on both y and t.)

Any Runge–Kutta method can be analyzed using Taylor series expansions. The following lemma is helpful.

Lemma 6.2. *Every non-autonomous system dependent on t can be transferred into an autonomous system independent of t. If $\sum_{i=1}^{\nu} b_i = 1$ and $c_i = \sum_{j=1}^{\nu} a_{ij}$, and if the Runge–Kutta method gives a unique approximation, then this equivalence of autonomous and non-autonomous systems is preserved by this method.*

Proof. Set

$$\begin{aligned}
\tau &:= t, \; \tau_0 := t_0 \\
\mathbf{z} &:= (t, \mathbf{y})^T \\
\mathbf{g}(\mathbf{z}) &:= (1, \mathbf{f}(t, \mathbf{y}))^T = (1, \mathbf{f}(\mathbf{z}))^T \\
\mathbf{z}(\tau_0) &= (\tau_0, \mathbf{y}(\tau_0))^T = (t_0, \mathbf{y}_0)^T = \mathbf{z}_0.
\end{aligned}$$

Note that $\dfrac{dt}{d\tau} = 1$.

$$\frac{d\mathbf{z}}{d\tau} = \begin{pmatrix} \frac{dt}{d\tau} \\ \frac{d\mathbf{y}}{d\tau} \end{pmatrix} = \begin{pmatrix} 1 \\ \frac{d\mathbf{y}}{dt} \end{pmatrix} = \begin{pmatrix} 1 \\ \mathbf{f}(\mathbf{y}, t) \end{pmatrix} = \mathbf{g}(\mathbf{z}).$$

A solution to this system gives a solution for (6.1) by removing the first component.

Suppose $\mathbf{z}_n = (t_n, \mathbf{y}_n)^T$. Now let $\mathbf{l}_j = (1, \mathbf{k}_j)$.

$$\begin{aligned}
\mathbf{l}_i &= \mathbf{g}\left(\mathbf{z}_n + h\sum_{j=1}^{\nu} a_{i,j}\mathbf{l}_j\right) \\
&= \mathbf{g}\left(t_n + h\sum_{j=1}^{\nu} a_{i,j}, \mathbf{y}_n + h\sum_{j=1}^{\nu} a_{i,j}\mathbf{k}_j\right) \\
&= \mathbf{g}\left(t_n + hc_i, \mathbf{y}_n + h\sum_{j=1}^{\nu} a_{i,j}\mathbf{k}_j\right) \\
&= \left(1, \mathbf{f}\left(t_n + hc_i, \mathbf{y}_n + h\sum_{j=1}^{\nu} a_{i,j}\mathbf{k}_j\right)\right)^T = \begin{pmatrix} 1 \\ \mathbf{k}_i \end{pmatrix}
\end{aligned}$$

since we assumed that the Runge–Kutta method gives a unique solution.

Additionally we have

$$
\begin{aligned}
\mathbf{z}_{n+1} &= \mathbf{z}_n + h\sum_{i=1}^{\nu} b_i \mathbf{l}_i = (t_n, \mathbf{y}_n)^T + h\sum_{i=1}^{\nu} b_i (1, \mathbf{k}_i)^T \\
&= (t_n + h\sum_{i=1}^{\nu} b_i, \mathbf{y}_n + h\sum_{i=1}^{\nu} b_i \mathbf{k}_i)^T \\
&= (t_{n+1}, \mathbf{y}_n + h\sum_{i=1}^{\nu} b_i \mathbf{k}_i)^T,
\end{aligned}
$$

since $\sum_{i=1}^{\nu} b_i = 1$ and $t_{n+1} = t_n + h$. □

For simplification we restrict ourselves to one-dimensional, autonomous systems in the following analysis. The Taylor expansion will produce terms of the form $y' = f$, $y'' = f_y f$, $y''' = f_{yy} f^2 + f_y (f_y f)$ and $y^{(4)} = f_{yyy} f^3 + 4f_{yy} f_y f^2 + f_y^3 f$. The $f, f_y f, f_y^2 f, f_{yy} f^2$, etc., are called *elementary differentials*. Every derivative of y is a linear combination with positive integer coefficients of elementary differentials.

A convenient way to represent elementary differentials is by rooted trees. For example, f is represented by

$$
T_0 = \boxed{f}
$$

while $f_y f$ is represented by

$$
T_1 = \boxed{f_y}
$$

For the third derivative, we have

$f_{yy} f^2$	$f_y (f_y f)$
$T_2 =$	$T_3 =$

With the fourth derivative of $\mathbf{y}$, it becomes more interesting, since the elementary differential $f_{yy}f_y f^2$ arises in the differentiation of the first term as well as the second term of $\mathbf{y}'''$. This corresponds to two different trees and the distinction is important since in several variables these correspond to differentiation matrices which do not commute. In Figure 6.18 we see the two different trees for the same elementary differential.

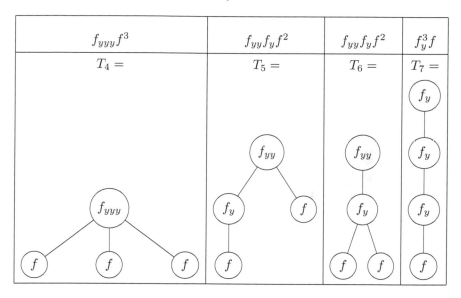

Figure 6.18 Fourth order elementary differentials

The correspondence to Runge–Kutta methods is created by labeling the vertices in the tree with indices, say j and k. Every edge from j to k is associated with the sum $\sum_{j,k=1}^{\nu} a_{j,k}$. If the root is labeled by say i, it is associated with the sum $\sum_{i=1}^{\nu} b_i$. For every tree T this gives an expression $\Phi(T)$. The identity $\sum_{j,k=1}^{\nu} a_{j,k} = c_j$ can be used to simplify $\Phi(T)$.

For the trees we have considered so far, we have

$$\Phi(T_0) = \sum_{i=1}^{\nu} b_i$$

$$\Phi(T_1) = \sum_{i,j=1}^{\nu} b_i a_{i,j} = \sum_{i=1}^{\nu} b_i c_i$$

$$\Phi(T_2) = \sum_{i,j,k=1}^{\nu} b_i a_{i,j} a_{i,k} = \sum_{i=1}^{\nu} b_i c_i^2$$

$$\Phi(T_3) = \sum_{i,j,k=1}^{\nu} b_i a_{i,j} a_{j,k} = \sum_{i,j=1}^{\nu} b_i a_{i,j} c_j$$

$$\Phi(T_4) = \sum_{i,j,k,l=1}^{\nu} b_i a_{i,j} a_{i,k} a_{i,l} = \sum_{i=1}^{\nu} b_i c_i^3$$

$$\Phi(T_5) = \sum_{i,j,k,l=1}^{\nu} b_i a_{i,j} a_{j,k} a_{i,l} = \sum_{i,j=1}^{\nu} b_i c_i a_{i,j} c_j$$

$$\Phi(T_6) = \sum_{i,j,k,l=1}^{\nu} b_i a_{i,j} a_{j,k} a_{j,l} = \sum_{i,j=1}^{\nu} b_i a_{i,j} c_j^2$$

$$\Phi(T_7) = \sum_{i,j,k,l=1}^{\nu} b_i a_{i,j} a_{j,k} a_{k,l} = \sum_{i,j,k=1}^{\nu} b_i a_{i,j} a_{j,k} c_k$$

$\Phi(T)$ together with the usual Taylor series coefficients give the coefficients of the elementary differentials when expanding the Runge–Kutta method. In order for the Taylor expansion of the true solution and the Taylor expansion of the Runge–Kutta method to match up to the h^p terms, we need the coefficients of the elementary differentials to be the same. This implies for all trees T with p vertices or less,

$$\Phi(T) = \frac{1}{\gamma(T)},$$

where $\gamma(T)$ is the density of the tree which is defined to be the product of the number of vertices of T with the number of vertices of all possible trees after successively removing roots. This gives the *order conditions*.

$$\sum_{i=1}^{\nu} b_i = 1$$

$$\sum_{i=1}^{\nu} b_i c_i = \frac{1}{2}$$

$$\sum_{i=1}^{\nu} b_i c_i^2 = \frac{1}{3}$$

$$\sum_{i,j=1}^{\nu} b_i a_{i,j} c_j = \frac{1}{6}$$

$$\sum_{i=1}^{\nu} b_i c_i^3 = \frac{1}{4}$$

$$\sum_{i,j=1}^{\nu} b_i c_i a_{i,j} c_j = \frac{1}{8}$$

$$\sum_{i,j=1}^{\nu} b_i a_{i,j} c_j^2 = \frac{1}{12}$$

$$\sum_{i,j,k=1}^{\nu} b_i a_{i,j} a_{j,k} c_k = \frac{1}{24}$$

The number of order conditions explodes with the number of stages.

order p	1	2	3	4	5	6	7	8	...	9
number of conditions	1	2	4	8	17	37	85	200	...	7813

6.10   Revision Exercises

Exercise 6.15. *Consider the scalar ordinary differential $y' = f(y)|$, that is, f is independent of t. It is solved by the following Runge–Kutta method,*

$$\begin{aligned} k_1 &= f(y_n), \\ k_2 &= f(y_n + (1-\alpha)hk_1 + \alpha hk_2), \\ y_{n+1} &= y_n + \frac{h}{2}(k_1 + k_2), \end{aligned}$$

where α is a real parameter.

(a) *Express the first, second, and third derivative of y in terms of f.*

(b) *Perform the Taylor expansion of $y(t_{n+1})$ using the expressions found in the previous part and explain what it means for the method to be of order p.*

(c) *Determine p for the given Runge–Kutta method.*

(d) *Define A-stability, stating explicitly the linear test equation.*

(e) *Suppose the Runge–Kutta method is applied to the linear test equation. Show that then*

$$y_{n+1} = R(h\lambda)y_n$$

and derive $R(h\lambda)$ explicitly.

(f) *Show that the method is A-stable if and only if $\alpha = \frac{1}{2}$.*

Exercise 6.16. *Consider the multistep method for numerical solution of the differential equation $\mathbf{y}' = \mathbf{f}(t, \mathbf{y})$:*

$$\sum_{l=0}^{s} \rho_l \mathbf{y}_{n+l} = h \sum_{l=0}^{s} \sigma_l \mathbf{f}(t_{n+l}, \mathbf{y}_{n+l}), \qquad n = 0, 1, \ldots.$$

(a) *Give the definition of the order of the method.*

(b) *State the test equation, definition of stability region, and A-stability in general.*

(c) *Using Taylor expansions, show that the method is of order p if*

$$\sum_{l=0}^{s} \rho_l = 0, \qquad \sum_{l=0}^{s} l^k \rho_l = k \sum_{l=0}^{s} l^{k-1} \sigma_l, \qquad k = 1, 2, \ldots, p,$$

(d) *State the Dahlqusit equivalence theorem.*

(e) *Determine the order of the two-step method*

$$\mathbf{y}_{n+2} - (1+\alpha)\mathbf{y}_{n+1} + \alpha \mathbf{y}_n = h[\frac{1}{12}(5+\alpha)\mathbf{f}_{n+2} + \frac{2}{3}(1-\alpha)\mathbf{f}_{n+1} - \frac{1}{12}(1+5\alpha)\mathbf{f}_n]$$

for different choices of α.

(f) *For which α is the method convergent?*

Exercise 6.17. *Consider the multistep method for numerical solution of the differential equation $\mathbf{y}' = \mathbf{f}(t, \mathbf{y})$:*

$$\sum_{l=0}^{s} \rho_l \mathbf{y}_{n+l} = h \sum_{l=0}^{s} \sigma_l \mathbf{f}(t_{n+l}, \mathbf{y}_{n+l}), \qquad n = 0, 1, \ldots.$$

(a) *Describe in general what it means that a method is of order p.*

(b) *Define generally the convergence of a method.*

(c) *Define the stability region and A-stability in general.*

(d) *Describe how to determine the stability region of the multistep method.*

(e) Show that the method is of order p if

$$\sum_{l=0}^{s} \rho_l = 0, \qquad \sum_{l=0}^{s} l^k \rho_l = k \sum_{l=0}^{s} l^{k-1} \sigma_l, \qquad k = 1, 2, \ldots, p,$$

(f) Give the conditions on $\rho(w) = \sum_{l=0}^{s} \rho_l w^l$ that ensure convergence.

(g) Hence determine for what values of θ and $\sigma_0, \sigma_1, \sigma_2$ the two-step method

$$\mathbf{y}_{n+2} - (1-\theta)\mathbf{y}_{n+1} - \theta\mathbf{y}_n = h[\sigma_0\mathbf{f}(t_n, \mathbf{y}_n) + \sigma_1\mathbf{f}(t_{n+1}, \mathbf{y}_{n+1}) + \sigma_2\mathbf{f}(t_{n+2}, \mathbf{y}_{n+2})]$$

is convergent and of order 3.

Exercise 6.18. We approximate the solution of the ordinary differential equation

$$\frac{\partial \mathbf{y}}{\partial t} = \mathbf{y}' = \mathbf{f}(t, \mathbf{y}), \qquad t \geq 0,$$

by the s-step method

$$\sum_{l=0}^{s} \rho_l \mathbf{y}_{n+l} = h \sum_{l=0}^{s} \sigma_l \mathbf{f}(t_{n+l}, \mathbf{y}_{n+l}), \qquad n = 0, 1, \ldots, \qquad (6.21)$$

assuming that $\mathbf{y}_n, \mathbf{y}_{n+1}, \ldots, \mathbf{y}_{n+s-1}$ are available. The following complex polynomials are defined:

$$\rho(w) = \sum_{l=0}^{s} \rho_l w^l, \qquad\qquad \sigma(w) = \sum_{l=0}^{s} \sigma_l w^l.$$

(a) When is the method given by (6.21) explicit and when implicit?

(b) Derive a condition involving the polynomials ρ and σ which is equivalent to the s-step method given in (6.21) being of order p.

(c) Define another polynomial and **state** (no proof required) a condition for the method in (6.21) to be A-stable.

(d) Describe the boundary locus method to find the boundary of the stability domain for the method given in (6.21).

(e) What is ρ for the Adams methods and what is the difference between Adams–Bashforth and Adams–Moulton methods?

(f) Let $s = 1$. Derive the Adams–Moulton method of the form

$$\mathbf{y}_{n+1} = \mathbf{y}_n + h\left(Af(t_{n+1}, \mathbf{y}_{n+1}) + Bf(t_n, \mathbf{y}_n)\right).$$

(g) Is the one-step Adams–Moulton method A-stable? Prove your answer.

Exercise 6.19. *We consider the autonomous scalar differential equation*

$$\frac{d}{dt}y(t) = y'(t) = f(y(t)), \qquad y(0) = y_0.$$

Note that f is independent of t.

(a) *Express the second and third derivative of y in terms of f and its derivatives. Write the Taylor expansion of $y(t+h)$ in terms of f and its derivatives up to $O(h^4)$.*

(b) *The differential equation is solved by the Runge–Kutta scheme*

$$
\begin{aligned}
k_1 &= hf(y_n), \\
k_2 &= hf(y_n + k_1), \\
y_{n+1} &= y_n + \tfrac{1}{2}(k_1 + k_2).
\end{aligned}
$$

Show that the scheme is of order 2.

(c) *Define the linear stability domain and A-stability for a general numerical method, stating explicitly the linear test equation on which the definitions are based.*

(d) *Apply the Runge–Kutta scheme given in (b) to the linear test equation from part (c) and find an expression for the linear stability domain of the method. Is the method A-stable?*

(e) *We now modify the Runge–Kutta scheme in the following way*

$$
\begin{aligned}
k_1 &= hf(y_n), \\
k_2 &= hf(y_n + a(k_1 + k_2)), \\
y_{n+1} &= y_n + \tfrac{1}{2}(k_1 + k_2),
\end{aligned}
$$

where $a \in \mathbb{R}$. Apply it to the test equation and find a rational function R such that $y_{n+1} = R(h\lambda)y_n$.

(f) *Explain the maximum modulus principle and use it to find the values of a such that the method given in (e) is A-stable.*

CHAPTER 7

Numerical Differentiation

Numerical differentiation is an ill-conditioned problem in the sense that small perturbations in the input can lead to significant differences in the outcome. It is, however important, since many problems require approximations to derivatives. It introduces the concept of *discretization error*, which is the error occurring when a continuous function is approximated by a set of discrete values. This is also often known as the *local truncation error*. It is different from the *rounding error*, which is the error due to the inherent nature of floating point representation. For a given function $f(x) : \mathbb{R} \to \mathbb{R}$ the *derivative* $f'(x)$ is defined as

$$f'(x) = \lim_{h \to 0} \frac{f(x+h) - f(x)}{h}.$$

Thus $f'(x)$ can be estimated by choosing a small h and letting

$$f'(x) \approx \frac{f(x+h) - f(x)}{h}.$$

This approximation is generally called a *difference quotient*. The important question is how should h be chosen.

Suppose $f(x)$ can be differentiated at least three times, then from Taylor's theorem we can write

$$f(x+h) = f(x) + hf'(x) + \frac{h^2}{2}f''(x) + O(h^3).$$

After rearranging, we have

$$f'(x) = \frac{f(x+h) - f(x)}{h} - \frac{h}{2}f''(x) + O(h^2).$$

The first term is the approximation to the derivative. Thus the absolute value of the *discretization error* or *local truncation error* is approximately $\frac{h}{2}|f''(x)|$.

We now turn to the rounding error. The difference quotient uses the floating point representations

$$f(x+h)^* = f(x+h) + \epsilon_{x+h},$$
$$f(x)^* = f(x) + \epsilon_x.$$

Thus the representation of the approximation is given by

$$\frac{f(x+h) - f(x)}{h} + \frac{\epsilon_{x+h} - \epsilon_x}{h},$$

where we assume for simplicity that the difference and division have been calculated exactly. If $f(x)$ can be evaluated with a relative error of approximately *macheps*, we can assume that

$$|\epsilon_{x+h} - \epsilon_x| \leq macheps \; (|f(x)| + |f(x+h)|).$$

Thus the rounding error is at most *macheps* $(|f(x)| + |f(x+h)|)/h$.

The main point to note is that as h decreases, the discretization error decreases, but the rounding error increases, since we are dividing by h. The ideal choice of h is the one which minimizes the total error. This is the case where the absolute values of the discretization error and the rounding error become the same

$$\frac{h}{2}|f''(x)| = macheps(|f(x)| + |f(x+h)|)/h.$$

However, this involves unknown quantities. Assuming that $\frac{1}{2}|f''(x)|$ and $|f(x)| + |f(x+h)|$ are of order $O(1)$, a good choice for h would satisfy $h^2 \approx macheps$ or in other words $h \approx \sqrt{macheps}$. The total absolute error in the approximation is then of order $O(\sqrt{macheps})$. However, since we assumed that $|f(x)| = O(1)$ in the above analysis, a more realistic choice for h would be $h \approx \sqrt{macheps|f(x)|}$. This, however, does not deal with the assumption on $|f''(x)|$.

Exercise 7.1. *List the assumptions made in the analysis and give an example where at least one of these assumptions does not hold. What does this mean in practice for the approximation of derivatives?*

7.1 Finite Differences

In general, *finite differences* are mathematical expressions of the form $f(x+b) - f(x+a)$. If a finite difference is divided by $b - a$ one gets a *difference quotient*. As we have seen above, these can be used to approximate derivatives. The main basic operators of finite differences are

forward difference operator $\quad \Delta_+ f(x) = f(x+h) - f(x),$

backward difference operator $\quad \Delta_- f(x) = f(x) - f(x-h),$

central difference operator $\quad \delta f(x) = f(x + \frac{1}{2}h) - f(x - \frac{1}{2}h).$

Additionally, we can define

shift operator $\qquad E f(x) = f(x+h),$

averaging operator $\qquad \mu_0 f(x) = \frac{1}{2}(f(x+\frac{1}{2}h) + f(x-\frac{1}{2}h)),$

differential operator $\quad D f(x) = \frac{d}{dx}f(x) = f'(x).$

All operators commute and can be expressed in terms of each other. We can easily see that

$$
\begin{aligned}
\Delta_+ &= E - I, \\
\Delta_- &= I - E^{-1}, \\
\delta &= E^{\frac{1}{2}} - E^{-\frac{1}{2}}, \\
\mu_0 &= \frac{1}{2}(E^{\frac{1}{2}} + E^{-\frac{1}{2}}).
\end{aligned}
$$

Most importantly, it follows from Taylor expansion that

$$E = e^{hD}$$

and thus

$$
\begin{aligned}
\frac{1}{h}\Delta_+ &= \frac{1}{h}(e^{hD} - I) = D + O(h), \\
\frac{1}{h}\Delta_- &= \frac{1}{h}(I - e^{-hD}) = D + O(h).
\end{aligned}
$$

It follows that the forward and backward difference operators approximate the differential operator, or in other words the first derivative with an error of $O(h)$.

Both the averaging and the central difference operator are not well-defined on a grid. However, we have

$$
\begin{aligned}
\delta^2 f(x) &= f(x+h) - 2f(x) + f(x-h), \\
\delta\mu_0 f(x) &= \frac{1}{2}(f(x+h) - f(x-h).
\end{aligned}
$$

Now $\frac{1}{h^2}\delta^2$ approximates the second derivative with an error of $O(h^2)$, because

$$\frac{1}{h^2}\delta^2 = \frac{1}{h^2}(e^{hD} - 2I + e^{-hD}) = D^2 + O(h^2).$$

On the other hand, we have

$$\frac{1}{h}\delta\mu_0 = \frac{1}{2h}(e^{hD} - e^{-hD}) = D + O(h^2).$$

Hence the combination of central difference and averaging operator gives a better approximation to the first derivative. We can also achieve higher accuracy by using the sum of the forward and backward difference

$$\frac{1}{2h}(\Delta_+ + \Delta_-) = \frac{1}{2h}(e^{hD} - I + I - e^{-hD}) = D + O(h^2).$$

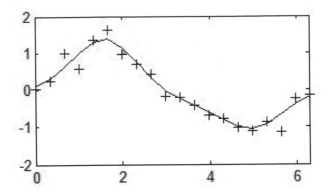

Figure 7.1 Inexact and incomplete data with a fitted curve

Higher-order derivatives can be approximated by applying the difference operators several times. For example the nth order forward, backward, and central difference operators are given by

$$\Delta_+^n f(x) = \sum_{i=0}^n \binom{n}{i} (e^{hD})^{n-i} (-I)^i = \sum_{i=0}^n (-1)^i \binom{n}{i} f(x + (n-i)h),$$

$$\Delta_-^n f(x) = \sum_{i=0}^n \binom{n}{i} (I)^{n-i} (-e^{-hD})^i = \sum_{i=0}^n (-1)^i \binom{n}{i} f(x - ih),$$

$$\delta^n f(x) = \sum_{i=0}^n \binom{n}{i} (e^{h\frac{D}{2}})^{n-i} (-e^{-h\frac{D}{2}})^i = \sum_{i=0}^n (-1)^i \binom{n}{i} f(x + (\frac{n}{2} - i)h).$$

For odd n the nth central difference is again not well-defined on a grid, but this can be alleviated as before by combining one central difference with the averaging operator. After dividing by h^n the nth order forward and backward differences approximate the nth derivative with an error of $O(h)$, while the nth order central difference (if necessary combined with the averaging operator) approximates the nth derivative with an error of $O(h^2)$.

Combination of higher-order differences can also be used to construct better approximations. For example,

$$\frac{1}{h}(\Delta_+ - \frac{1}{2}\Delta_+^2)f(x) = \frac{-1}{2h}(f(x+2h) - 4f(x+h) + 3f(x))$$

can be written with the shift and differential operator as

$$\frac{-1}{2h}(E^2 - 4E + 3I) = \frac{-1}{2h}(e^{2hD} - 4e^{hD} + 3I) = D + O(h^2).$$

The drawback is that more grid points need to be employed. This is called the *bandwidth*. Special schemes at the boundaries are then necessary.

7.2   Differentiation of Incomplete or Inexact Data

If the function f is only known at a finite number of points or if the function values are known to be inexact, it does not make sense to calculate the derivatives from the data as it stands. A much better approach is to fit a curve through the data taking into account the inexactness of the data which can be done, for example, by least squares fitting. Figure 7.1 illustrates this. The derivatives are then approximated by the derivatives of the fitted curve.

No precise statement of the accuracy of the resulting derivatives can be made, since the accuracy of the fitted curve would need to be assessed first.

CHAPTER 8

PDEs

8.1 Classification of PDEs

A partial differential equation (PDE) is an equation of a function of several variables and its partial derivatives. The partial derivatives of a scalar valued function $u : \mathbb{R}^n \to \mathbb{R}$ are often abbreviated as

$$u_{x_i} = \frac{\partial u}{\partial x_i}, \qquad u_{x_i x_j} = \frac{\partial^2 u}{\partial x_i \partial x_j}, \qquad \cdots$$

The PDE has the structure

$$F(x_1, \ldots, x_n, u, u_{x_1}, \ldots, u_{x_n}, u_{x_1 x_1}, u_{x_1 x_2}, \ldots, u_{x_n x_n}, \ldots) = 0. \tag{8.1}$$

Notation can be compacted by using *multi-indices* $\alpha = (\alpha_1, \ldots, \alpha_n) \in \mathbb{N}_0^n$, where $\mathbb{N}_0$ denotes the natural numbers including 0. The notation $|\alpha|$ is used to denote the length $\alpha_1 + \cdots + \alpha_n$. Then

$$D^\alpha u = \frac{\partial^{|\alpha|} u}{\partial x^\alpha} = \frac{\partial^{|\alpha|} u}{\partial x_1^{\alpha_1} \cdots \partial x_n^{\alpha_n}}.$$

Further, let $D^k u = \{D^\alpha u : |\alpha| = k\}$, the construct of all derivatives of degree k. For $k = 1$, $Du = (\frac{\partial u}{\partial x_1}, \ldots, \frac{\partial u}{\partial x_n})$ is the *gradient* of u. For $k = 2$, $D^2 u$ is the *Hessian matrix* of second derivatives given by

$$\begin{pmatrix} \frac{\partial^2 u}{\partial x_1^2} & \frac{\partial^2 u}{\partial x_1 \partial x_2} & \cdots & \frac{\partial^2 u}{\partial x_1 \partial x_n} \\ \frac{\partial^2 u}{\partial x_2 \partial x_1} & \frac{\partial^2 u}{\partial x_2^2} & \cdots & \frac{\partial^2 u}{\partial x_2 \partial x_n} \\ \vdots & \vdots & \ddots & \vdots \\ \frac{\partial^2 u}{\partial x_n \partial x_1} & \frac{\partial^2 u}{\partial x_n \partial x_2} & \cdots & \frac{\partial^2 u}{\partial x_n^2} \end{pmatrix}.$$

Note that the Hessian matrix is symmetric, since it does not matter in which order partial derivatives are taken. (The Hessian matrix of a scalar valued function should not be confused with the Jacobian matrix of a vector valued function which we encountered in the chapter on non-linear systems.) Then

$$F(\mathbf{x}, u(\mathbf{x}), Du(\mathbf{x}), \dots, D^k u(\mathbf{x})) = 0,$$

describes a PDE of order k. We will only consider PDE of order 2 in this chapter, but the principles extend to higher orders.

Definition 8.1 (Linear PDE). *The PDE is called* linear *if it has the form*

$$\sum_{|\alpha| \le k} c_\alpha(\mathbf{x}) D^\alpha u(\mathbf{x}) - f(\mathbf{x}) = 0. \tag{8.2}$$

If $f \equiv 0$, the PDE is homogeneous, *otherwise* inhomogeneous.

Definition 8.2 (Semi-linear PDE). *The PDE is called* semi-linear *if it has the form*

$$\sum_{|\alpha| = k} c_\alpha(\mathbf{x}) D^\alpha u(\mathbf{x}) + G(\mathbf{x}, u(\mathbf{x}), Du(\mathbf{x}), \dots, D^{k-1} u(\mathbf{x})) = 0.$$

In other words, the PDE is linear with regards to the derivatives of highest degree, but nonlinear for lower derivatives.

Definition 8.3 (Quasilinear PDE). *The PDE is called* quasilinear *if it has the form*

$$\sum_{|\alpha| = k} c_\alpha(\mathbf{x}, u(\mathbf{x}), Du(\mathbf{x}), \dots, D^{k-1} u(\mathbf{x})) D^\alpha u(\mathbf{x})$$
$$+ G(\mathbf{x}, u(\mathbf{x}), Du(\mathbf{x}), \dots, D^{k-1} u(\mathbf{x})) = 0.$$

For further classification, we restrict ourselves to quasilinear PDEs of order 2, of the form

$$\sum_{i,j=1}^{n} a_{ij} \frac{\partial^2 u}{\partial x_i \partial x_j} - f = 0, \tag{8.3}$$

where the coefficients a_{ij} and f are allowed to depend on $\mathbf{x}, u$, and the gradient of u. Without loss of generality we can assume that the matrix $A = (a_{ij})$ is symmetric, since otherwise we could rewrite the PDE according to

$$a_{ij} \frac{\partial^2 u}{\partial x_i \partial x_j} + a_{ji} \frac{\partial^2 u}{\partial x_j \partial x_i} = (a_{ij} + a_{ji}) \frac{\partial^2 u}{\partial x_i \partial x_j},$$
$$= \frac{1}{2}(a_{ij} + a_{ji}) \frac{\partial^2 u}{\partial x_i \partial x_j} + \frac{1}{2}(a_{ij} + a_{ji}) \frac{\partial^2 u}{\partial x_j \partial x_i},$$

and the matrix $B = (b_{ij})$ with coefficients $b_{ij} = \frac{1}{2}(a_{ij} + a_{ji})$ is symmetric.

Definition 8.4. *Let $\lambda_1(\mathbf{x}), \dots, \lambda_n(\mathbf{x}) \in \mathbb{R}$ be the eigenvalues of the symmetric coefficient matrix $A = (a_{ij})$ of the PDE given by (8.3) at a point $\mathbf{x} \in \mathbb{R}^n$. The PDE is*

parabolic
> *at* $\mathbf{x}$*, if there exists* $j \in \{1, \dots, n\}$ *for which* $\lambda_j(\mathbf{x}) = 0$,

elliptic
> *at* $\mathbf{x}$*, if* $\lambda_i(\mathbf{x}) > 0$ *for all* $i = 1, \dots, n$,

hyperbolic
> *at* $\mathbf{x}$*, if* $\lambda_j(\mathbf{x}) > 0$ *for one* $j \in \{1, \dots, n\}$ *and* $\lambda_i(\mathbf{x}) < 0$ *for all* $i \neq j$*, or*
> *if* $\lambda_j(\mathbf{x}) < 0$ *for one* $j \in \{1, \dots, n\}$ *and* $\lambda_i(\mathbf{x}) > 0$ *for all* $i \neq j$.

Exercise 8.1. *Consider the PDE*

$$au_{xx} + 2bu_{xy} + cu_{yy} = f,$$

where $a > 0, b, c > 0$, *and* f *are functions of* x, y, u, u_x, *and* u_y. *At* (x, y) *the PDE is*

parabolic, if $b^2 - ac = 0$,

elliptic, if $b^2 - ac < 0$ *and*

hyperbolic, if $b^2 - ac > 0$.

Show that this definition is equivalent to the above definition.

An example of a PDE which changes classification is the *Euler–Tricomi equation*

$$u_{xx} = xu_{yy},$$

which is the simplest model of a transonic flow. It is hyperbolic in the half plane $x > 0$, parabolic at $x = 0$, and elliptic in the half plane $x < 0$.

8.2   Parabolic PDEs

Partial differential equations of evolution are second-order equations with a time component t where only t and the first derivative with regards to t occur in the equation. Second derivatives with regards to t do not feature in the equation and therefore have coefficient zero. Hence the eigenvalue with regards to this component is zero and these are *parabolic* partial differential equations. They describe a wide family of problems in science including heat diffusion, ocean acoustic propagation, and stock option pricing.

In the following sections we focus on the *heat equation* to illustrate the principles. For a function $u(x, y, z, t)$ of three spatial variables (x, y, z) and the time variable t, the heat equation is

$$\frac{\partial u}{\partial t} = \alpha \left(\frac{\partial^2 u}{\partial x^2} + \frac{\partial^2 u}{\partial y^2} + \frac{\partial^2 u}{\partial z^2} \right) = \alpha \nabla^2 u,$$

where α is a positive constant known as the *thermal diffusivity* and where

∇^2 denotes the *Laplace operator* (in three dimensions). For the mathematical treatment it is sufficient to consider the case $\alpha = 1$. We restrict ourselves further and consider the one-dimensional case $u(x,t)$ specified by

$$\frac{\partial u}{\partial t} = \frac{\partial^2 u}{\partial x^2}, \tag{8.4}$$

for $0 \le x \le 1$ and $t \ge 0$. *Initial conditions* $u(x,0)$ describe the state of the system at the beginning and *Dirichlet boundary conditions* $u(0,t)$ and $u(1,t)$ show how the system changes at the boundary over time. The most common example is that of a metal rod heated at a point in the middle. After a long enough time the rod will have constant temperature everywhere.

8.2.1 Finite Differences

We already encountered finite differences in the chapter on numerical differentiation. They are an essential tool in the solution of PDEs. In the following we denote the time step by Δt and the discretization in space by $\Delta x = 1/(M + 1)$; that is, we divide the interval $[0,1]$ into $M + 1$ subintervals. Denote $u(m\Delta x, n\Delta t)$ by u_m^n. Then u_m^0 is given by the initial conditions $u(m\Delta x, 0)$ and we have the known boundary conditions $u_0^n = u(0, n\Delta t)$ and $u_{M+1}^n = u(1, n\Delta t)$. Consider the Taylor expansions

$$
\begin{aligned}
u(x - \Delta x, t) &= u(x,t) - \Delta x \frac{\partial u(x,t)}{\partial x} + \frac{1}{2}(\Delta x)^2 \frac{\partial^2 u(x,t)}{\partial x^2} \\
&\quad - \frac{1}{6}(\Delta x)^3 \frac{\partial^3 u(x,t)}{\partial x^3} + O((\Delta x)^4), \\
u(x + \Delta x, t) &= u(x,t) + \Delta x \frac{\partial u(x,t)}{\partial x} + \frac{1}{2}(\Delta x)^2 \frac{\partial^2 u(x,t)}{\partial x^2} \\
&\quad + \frac{1}{6}(\Delta x)^3 \frac{\partial^3 u(x,t)}{\partial x^3} + O((\Delta x)^4).
\end{aligned}
$$

Adding both together, we deduce

$$\frac{\partial^2 u(x,t)}{\partial x^2} = \frac{1}{(\Delta x)^2}\left[u(x - \Delta x, t) - 2u(x,t) + u(x + \Delta x, t)\right] + O((\Delta x)^2). \tag{8.5}$$

Similarly, in the time direction we have

$$\frac{\partial u(x,t)}{\partial t} = \frac{1}{\Delta t}\left[u(x, t + \Delta t) - u(x,t)\right] + O(\Delta t). \tag{8.6}$$

Inserting the approximations (8.5) and (8.6) into the heat equation (8.4) motivates the numerical scheme

$$u_m^{n+1} = u_m^n + \mu(u_{m-1}^n - 2u_m^n + u_{m+1}^n), \qquad m = 1, \ldots, M, \tag{8.7}$$

where $\mu = \Delta t/(\Delta x)^2$ is the *Courant number* which is kept constant. The method of constructing numerical schemes by approximations to the derivatives is known as *finite differences*. Let $\mathbf{u}^n$ denote the vector $(u_1^n, \ldots, u_M^n)^T$.

The initial condition gives the vector $\mathbf{u}^0$. Using the boundary conditions u_0^n and u_{m+1}^n when necessary, (8.7) can be advanced from $\mathbf{u}^n$ to $\mathbf{u}^{n+1}$ for $n = 0, 1, 2, \ldots$, since from one time step to the next the right-hand side of Equation (8.7) is known. This is an example of a *time marching scheme*.

Keeping μ fixed and letting $\Delta x \to 0$ (which also implies that $\Delta t \to 0$, since $\Delta t = \mu(\Delta x)^2$), the question is: Does for every $T > 0$ the point approximation u_m^n converge uniformly to $u(x, t)$ for $m\Delta x \to x \in [0, 1]$ and $n\Delta t \to t \in [0, T]$? The method here has an extra parameter, μ. It is entirely possible for a method to converge for some choice of μ and diverge otherwise.

Theorem 8.1. $\mu \le \frac{1}{2} \Rightarrow$ *convergence.*

Proof. Define the error as $e_m^n := u_m^n - u(m\Delta x, n\Delta t)$, $m = 0, \ldots, M+1$, $n \ge 0$. Convergence is equivalent to

$$\lim_{\Delta x \to 0} \max_{\substack{m=1,\ldots,M \\ 0 < n \le \lfloor T/\Delta t \rfloor}} |e_m^n| - 0$$

for every constant $T > 0$. Since u satisfies the heat equation, we can equate (8.5) and (8.6), which gives

$$\frac{1}{\Delta t} [u(x, t + \Delta t) - u(x, t)] + O(\Delta t) =$$
$$\frac{1}{(\Delta x)^2} [u(x - \Delta x, t) - 2u(x, t) + u(x + \Delta x, t)] + O((\Delta x)^2).$$

Rearranging yields

$$u(x, t + \Delta t) = u(x, t) + \mu [u(x - \Delta x, t) - 2u(x, t) + u(x + \Delta x, t)] + O((\Delta t)^2) + O(\Delta t(\Delta x)^2).$$

Subtracting this from (8.7) and using $\Delta t = \mu(\Delta x)^2$, it follows that there exists $C > 0$ such that

$$|e_m^{n+1}| \le |e_m^n + \mu(e_{m-1}^n - 2e_m^n + e_{m+1}^n)| + C(\Delta x)^4,$$

where we used the triangle inequality and properties of the O-notation. Let $e_{\max}^n := \max_{m=1,\ldots,M} |e_m^n|$. Then

$$|e_m^{n+1}| \le |\mu e_{m-1}^n + (1 - 2\mu)e_m^n + \mu e_{m+1}^n| + C(\Delta x)^4$$
$$\le (2\mu + |1 - 2\mu|)e_{\max}^n + C(\Delta x)^4,$$

where we used the triangle inequality again. When $\mu \le \frac{1}{2}$, we have $1 - 2\mu \ge 0$ and the modulus sign is not necessary. Hence

$$\|e_m^{n+1}| \le (2\mu + 1 - 2\mu)e_{\max}^n + C(\Delta x)^4 = e_{\max}^n + C(\Delta x)^4.$$

Continuing by induction and using the fact that $e_{\max}^0 = 0$, we deduce

$$e_{\max}^n \leq Cn(\Delta x)^4 \leq C\frac{T}{\Delta t}(\Delta x)^4 = C\frac{T}{\mu}(\Delta x)^2 \to 0$$

as $\Delta x \to 0$. $\qquad\qquad\qquad\qquad\qquad\qquad\qquad\qquad\qquad\qquad\qquad\square$

The restriction on μ has practical consequences, since it follows that $\Delta t \leq \frac{1}{2}(\Delta x)^2$ must hold. Thus the time step Δt has to be tiny compared to Δx in practice. Like the forward Euler method for ODEs, the method given by (8.7) is easily derived, explicit, easy to execute, and simple – but of little use in practice.

In the following we will use finite differences extensively to approximate derivatives occurring in PDEs. To this end we define the *order of accuracy* and *consistency* of a finite difference scheme.

Definition 8.5 (Order of accuracy). *Let $u(\mathbf{x})$ be any sufficiently smooth solution to a linear PDE as given in (8.2). Let F be an operator contructed from finite difference operators using stepsizes Δx and Δt, where Δx is a fixed function of Δt. F has order of accuracy p, if*

$$Fu(\mathbf{x}) = (\Delta t)^p \left[\sum_{|\alpha| \leq k} c_\alpha(\mathbf{x}) D^\alpha u(\mathbf{x}) - f(\mathbf{x}) \right] + O((\Delta t)^{p+1})$$

for all $\mathbf{x} \in \mathbb{R}^n$.

Returning to the previous example, the PDE is

$$(D_t - D_x^2)u(x,t) = 0,$$

while F is given by

$$Fu(x,t) = \left[E_{\Delta t} - I - \mu \left(E_{\Delta x}^{-1} - 2I + E_{\Delta x}, \right) \right] u(x,t),$$

where $E_{\Delta t}$ and $E_{\Delta x}$ denote the shift operators in the x and y direction respectively. Since $\mu = \Delta t/(\Delta x)^2$, $\Delta x = \sqrt{\Delta t/\mu}$ is a fixed funtion of Δt. Denoting the differentiation operator in the t direction by D_t and the differentiation operator in the x direction by D_x, we have

$$E_{\Delta t} = e^{\Delta t D_t} \text{ and } E_{\Delta x} = e^{\Delta x D_x}.$$

Then F becomes

$$\begin{aligned}
Fu(x,t) &= \left[e^{\Delta t D_t} - I - \frac{\Delta t}{(\Delta x)^2} \left(e^{-\Delta x D_x} - 2I + e^{\Delta x D_x} \right) \right] u(x,t) \\
&= \left[\Delta t D_t + \frac{1}{2}(\Delta t D_t)^2 + \ldots - \Delta t \left(D_x^2 + \frac{1}{6}(\Delta x)^2 D_x^4 + \ldots \right) \right] u(x,t) \\
&= \Delta t (D_t - D_x^2)u(x,t) + O((\Delta t)^2),
\end{aligned}$$

since $(\Delta x)^2 = O(\Delta t)$.

Note, since Δx and Δt are linked by a fixed function, the local error can be expressed both in terms of Δx or Δt.

Definition 8.6 (Consistency). *If the order of accuracy p is greater or equal to one, then the finite difference scheme is* consistent.

8.2.2 Stability and Its Eigenvalue Analysis

In our numerical treatment of PDEs we need to ensure that we do not try to achieve the impossible. For example, we describe the solution by point values. However, if the underlying function oscillates, point values can only adequately describe it, if the resolution is small enough to capture the oscillations. This becomes a greater problem if oscillations increase with time. The following defines PDEs where such problems do not arise.

Definition 8.7 (Well-posedness). *A PDE problem is said to be* well-posed *if*

1. *a solution to the problem exists,*

2. *the solution is unique, and*

3. *the solution (in a compact time interval) depends in a uniformly bounded and continuous manner on the initial conditions.*

An example of an ill-posed PDE is the *backward heat equation*. It has the same form as the heat equation, but the thermal diffusivity is negative. Essentially it asks the question: If we know the temperature distribution now, can we work out what it has been? Returning to our example of the metal rod heated at one point: If there is another, but smaller, heat source, this will soon disappear because of diffusion and the solution of the heat equation will be indistinguishable from the solution without the additional heat source. Therefore the solution to the backwards heat equation cannot be unique, since we cannot tell whether we started with one or two heat sources. It can also be shown that the solutions of the backward heat equation blow up at a rate which is unbounded. For more information on the backward heat equation, see Lloyd Nick Trefethen, *The (Unfinished) PDE Coffee Table Book*.

Definition 8.8 (Stability in the context of time marching algorithms for PDEs of evolution). *A numerical method for a well-posed PDE of evolution is stable if for zero boundary conditions it produces a uniformly bounded approximation of the solution in any bounded interval of the form $0 \leq t \leq T$, when $\Delta x \to 0$ and the Courant number (or a generalization thereof) is constant.*

Theorem 8.2 (The Lax equivalence theorem). *For a well-posed underlying PDE which is solved by a* consistent *numerical method, we have stability $\Leftrightarrow$ convergence.*

The above theorem is also known as the *Lax-Richtmyer equivalence theorem*. A proof can be found in J. C. Strikwerda, *Finite Difference Schemes and Partial Differential Equations*, [19].

Stability of (8.7) for $\mu \leq \frac{1}{2}$ is implied by the Lax equivalence theorem, since we have proven convergence. However, it is often easier to prove stability and then deduce convergence. Stability can be proven directly by the *eigenvalue analysis of stability*, as we will show, continuing with our example.

Recall that $\mathbf{u}^n = (u_1^n, \ldots, u_M^n)^T$ is the vector of approximation at time $t = n\Delta t$. The recurrence in (8.7) can be expressed as $\mathbf{u}^{n+1} = A\mathbf{u}^n$, where A is a $M \times M$ *Toeplitz symmetric tri-diagonal (TST) matrix* with entries $1 - 2\mu$ on the main diagonal and μ on the two adjacent diagonals. In general *Toeplitz matrices* are matrices with constant entries along the diagonals. All $M \times M$ TST matrices T have the same set of eigenvectors $\mathbf{q}_1, \ldots, \mathbf{q}_M$. The ith component of $\mathbf{q}_k$ is given by

$$(\mathbf{q}_k)_i = a \sin \frac{\pi i k}{M + 1},$$

where a is a normalization constant which is $\sqrt{2/M}$ for even M and $\sqrt{2/(M + 1)}$ for odd M. The corresponding eigenvalue is

$$\lambda_k = \alpha + 2\beta \cos \frac{\pi k}{M + 1},$$

where α are the diagonal entries and β are the subdiagonal entries. Moreover, the eigenvectors form an orthogonal set. That these are the eigenvectors and eigenvalues can be verified by checking whether the eigenvalue equations

$$T\mathbf{q}_k = \lambda_k \mathbf{q}_k$$

hold for $k = 1, \ldots, M$.

Thus the eigenvalues of A are given by

$$(1 - 2\mu) + 2\mu \cos \frac{\pi k}{M + 1} = 1 - 4\mu(\sin \frac{\pi k}{2M + 2})^2, \qquad k = 1, \ldots, M.$$

Note that

$$0 < (\sin \frac{\pi k}{2M + 2})^2 < 1.$$

For $\mu \leq \frac{1}{2}$, the maximum modulus of eigenvalue is given by

$$|1 - 4\mu(\sin \frac{\pi M}{2M + 2})^2| \leq 1.$$

This is the spectral radius of A and thus the matrix norm $\|A\|$. Hence

$$\|\mathbf{u}^{n+1}\| \leq \|A\|\|\mathbf{u}^n\| \leq \ldots \leq \|A\|^{n+1}\|\mathbf{u}^0\| \leq \|\mathbf{u}^0\|$$

as $n \to \infty$ for every initial condition $\mathbf{u}^0$ and the approximation is uniformly bounded.

For $\mu > \frac{1}{2}$ the maximum modulus of eigenvalue is given by

$$4\mu(\sin \frac{\pi M}{2M + 2})^2 - 1 > 1$$

for M large enough. If the initial condition $\mathbf{u}^0$ happens to be the eigenvector corresponding to the largest (in modulus) eigenvalue λ, then by induction $\mathbf{u}^n = \lambda^n \mathbf{u}^0$, which becomes unbounded as $n \to \infty$.

The *method of lines (MOL)* refers to the construction of schemes for partial differential equations that proceeds by first discretizing the spatial derivatives only and leaving the time variable continuous. This leads to a system of ordinary differential equations to which one of the numerical methods for ODEs we have seen already can be applied. In particular, using the approximation to the second derivative given by (8.5), the *semidiscretization*

$$\frac{du_m(t)}{dt} = \frac{1}{(\Delta x)^2}(u_{m-1}(t) - 2u_m(t) + u_{m+1}(t)), \qquad m = 1, \ldots, M$$

carries a local truncation error of $O((\Delta x)^2)$. This is an ODE system which can be written as

$$\mathbf{u}'(t) = \frac{1}{(\Delta x)^2}\begin{pmatrix} -2 & 1 & 0 & \cdots & 0 \\ 1 & -2 & \ddots & \ddots & \vdots \\ 0 & \ddots & \ddots & \ddots & 0 \\ \vdots & \ddots & \ddots & \ddots & 1 \\ 0 & \cdots & 0 & 1 & -2 \end{pmatrix}\mathbf{u}(t) + \frac{1}{(\Delta x)^2}\begin{pmatrix} u_0(t) \\ 0 \\ \vdots \\ 0 \\ u_{M+1}(t) \end{pmatrix},$$

$$(8.8)$$

where $u_0(t) = u(0,t)$ and $u_{M+1}(t) = u(1,t)$ are the known boundary conditions. The initial condition is given by

$$\mathbf{u}(0) = \begin{pmatrix} u(\Delta x, 0) \\ u(2\Delta x, 0) \\ \vdots \\ u(M\Delta x, 0) \end{pmatrix}.$$

If this system is solved by the forward Euler method the resulting scheme is (8.7), while backward Euler yields

$$u_m^{n+1} - \mu(u_{m-1}^{n+1} - 2u_m^{n+1} + u_{m+1}^{n+1}) = u_m^n.$$

Taking this two-stage approach, first semi-discretization and then applying an ODE solver is conceptually easier than discretizing in both time and space in one step (so-called *full discretization*).

Now, solving the ODE given by (8.8) with the trapezoidal rule gives

$$u_m^{n+1} - \frac{1}{2}\mu(u_{m-1}^{n+1} - 2u_m^{n+1} + u_{m+1}^{n+1}) = u_m^n + \frac{1}{2}\mu(u_{m-1}^n - 2u_m^n + u_{m+1}^n),$$

which is known as the *Crank–Nicolson method*. The error is $O((\Delta t)^3 + (\Delta t)(\Delta x)^2)$ where the $(\Delta t)^3$ term comes from the trapezoidal rule and the

$(\Delta t)(\Delta x)^2$) is inherited from the discretization in space. The Crank–Nicolson scheme has superior qualities, as we will see in the analysis.

The Crank–Nicolson scheme can be written as $Bu^{n+1} = Cu^n$ (assuming zero boundary conditions), where

$$
B = \begin{pmatrix} 1+\mu & -\frac{1}{2}\mu & & \\ -\frac{1}{2}\mu & 1+\mu & & \\ & \ddots & \ddots & -\frac{1}{2}\mu \\ & & -\frac{1}{2}\mu & 1+\mu \end{pmatrix}, C = \begin{pmatrix} 1-\mu & \frac{1}{2}\mu & & \\ \frac{1}{2}\mu & 1-\mu & & \\ & \ddots & \ddots & \frac{1}{2}\mu \\ & & \frac{1}{2}\mu & 1-\mu \end{pmatrix}.
$$

Both B and C are TST matrices. The eigenvalues of B are $\lambda_k^B = 1 + \mu - \mu \cos\frac{\pi k}{M+1} = 1 + 2\mu \sin^2 \frac{\pi k}{2(M+1)}$, while the eigenvalues of C are $\lambda_k^C = 1 - \mu + \mu \cos\frac{\pi k}{M+1} = 1 - 2\mu \sin^2 \frac{\pi k}{2(M+1)}$. Let $Q = (\mathbf{q}_1 \ldots \mathbf{q}_M)$ be the matrix whose columns are the eigenvectors. Thus we can write

$$
Q D_B Q^T \mathbf{u}^{n+1} = Q D_C Q^T \mathbf{u}^n,
$$

where D_B and D_c are diagonal matrices with the eigenvalues of B and C as entries. Multiplying by $Q D_B^{-1} Q^T$ using the fact that $Q^T Q = I$ gives

$$
\mathbf{u}^{n+1} = Q D_B^{-1} D_C Q^T \mathbf{u}^n = Q(D_B^{-1} D_C)^{n+1} Q^T \mathbf{u}^0.
$$

The moduli of the eigenvalues of $D_B^{-1} D_C$ are

$$
\left| \frac{1 - 2\mu \sin^2 \frac{\pi k}{2(M+1)}}{1 + 2\mu \sin^2 \frac{\pi k}{2(M+1)}} \right| \leq 1, \qquad k = 1, \ldots, M.
$$

Thus we can deduce that the Crank–Nicolson method is stable for all $\mu > 0$ and we only need to consider accuracy in our choice of Δt versus Δx.

This technique is the *eigenvalue analysis of stability*. More generally, suppose that a numerical method (for a PDE with zero boundary conditions) can be written in the form

$$
\mathbf{u}_{\Delta x}^{n+1} = A_{\Delta x} \mathbf{u}_{\Delta x}^n,
$$

where $\mathbf{u}_{\Delta x}^n \in \mathbb{R}^M$ and $A_{\Delta x}$ is an $M \times M$ matrix. We use the subscript Δx here to emphasize that the dimensions of $\mathbf{u}^n$ and A depend on Δx, since as Δx decreases M increases. By induction we have $\mathbf{u}_{\Delta x}^n = (A_{\Delta x})^n \mathbf{u}_{\Delta x}^0$. For any vector norm $\| \cdot \|$ and the induced matrix norm $\|A\| = \sup \frac{\|A\mathbf{x}\|}{\|\mathbf{x}\|}$, we have

$$
\|\mathbf{u}_{\Delta x}^n\| = \|(A_{\Delta x})^n \mathbf{u}_{\Delta x}^0\| \leq \|(A_{\Delta x})^n\| \|\mathbf{u}_{\Delta x}^0\| \leq \|(A_{\Delta x})\|^n \|\mathbf{u}_{\Delta x}^0\|.
$$

Stability can now be defined as preserving the boundedness of $\mathbf{u}_{\Delta x}^n$ with respect to the chosen norm $\| \cdot \|$, and it follows from the inequality above that the method is stable if

$$
\|A_{\Delta x}\| \leq 1 \text{ as } \Delta x \to 0.
$$

Usually the norm $\| \cdot \|$ is chosen to be the *averaged Euclidean length*

$$\|\mathbf{u}_{\Delta x}\|_{\Delta x} = \left(\Delta x \sum_{i=1}^{M} |u_i|^2 \right)^{\frac{1}{2}}.$$

Note that the dimensions depend on Δx. The reason for the factor $\Delta x^{1/2}$ is because of

$$\|\mathbf{u}_{\Delta x}\|_{\Delta x} = \left(\Delta x \sum_{i=1}^{M} |u_i|^2 \right)^{\frac{1}{2}} \overset{\Delta x \to 0}{\longrightarrow} \left(\int_0^1 |u(x)|^2 dx \right)^{\frac{1}{2}},$$

where u is a square-integrable function such that $(\mathbf{u}_{\Delta x})_m = u(m\Delta x)$. The sum $\Delta x \sum_{i=1}^{M} |u_i|^2$ is the Riemann sum approximating the integral. The averaging does not affect the induced Euclidean matrix norm (since the averaging factor cancels). For *normal matrices* (i.e., matrices which have a complete set of orthonormal eigenvectors), the Euclidean norm of the matrix equals the spectral radius, i.e., $\|A\| = \rho(A)$, which is the maximum modulus of the eigenvalues, and we are back at the eigenvalue analysis.

Since the Crank–Nicolson method is implicit, we need to solve a system of equations. However, the matrix of the system is TST and its solution by *sparse Cholesky factorization* can be done in $O(M)$ operations. Recall that generally the Cholesky decomposition or Cholesky triangle is a decomposition of a real symmetric, positive-definite matrix A into the unique product of a lower triangular matrix L and its transpose, where L has strictly positive diagonal entries. Thus we want to find $A = LL^T$. The algorithm starts with $i = 1$ and $A^{(1)} := A$. At step i, the matrix $A^{(i)}$ has the form.

$$A^{(i)} = \begin{pmatrix} I_{i-1} & 0 & 0 \\ 0 & b_{ii} & \mathbf{b}_i^T \\ 0 & \mathbf{b}_i & B^{(i)} \end{pmatrix},$$

where I_{i-1} denotes the $(i-1)\times(i-1)$ identity matrix, $B^{(i)}$ is a $(M-i)\times(M-i)$ submatrix, b_{ii} is a scalar, and $\mathbf{b}_i$ is a $(M-i) \times 1$ vector. For $i = 1$, I_0 is a matrix of size 0, i.e., it is not there. We let

$$L_i := \begin{pmatrix} I_{i-1} & 0 & 0 \\ 0 & \sqrt{b_{ii}} & 0 \\ 0 & \frac{1}{\sqrt{b_{ii}}}\mathbf{b}_i & I_{n-i} \end{pmatrix},$$

then $A^{(i)} = L_i A^{(i+1)} L_i^T$, where

$$A^{(i+1)} = \begin{pmatrix} I_{i-1} & 0 & 0 \\ 0 & 1 & 0 \\ 0 & 0 & B^{(i)} - \frac{1}{\sqrt{b_{ii}}}\mathbf{b}_i\mathbf{b}_i^T \end{pmatrix}.$$

Since $\mathbf{b}_i \mathbf{b}_i^T$ is an outer product, this algorithm is also called the *outer product version* (other names are the *Cholesky–Banachiewicz algorithm* and the *Cholesky–Crout algorithm*). After M steps, we get $A^{(M+1)} = I$, since $B^{(M)}$ and $\mathbf{b}_M$ are of size 0. Hence, the lower triangular matrix L of the factorization is $L := L_1 L_2 \ldots L_M$.

For TST matrices we have

$$A^{(1)} = \begin{pmatrix} a_{11} & \mathbf{b}_1^T \\ \mathbf{b}_1 & B^{(1)} \end{pmatrix},$$

where $\mathbf{b}_1^T = (a_{12}, 0, \ldots, 0)$. Note that $B^{(1)}$ is a tridiagonal symmetric matrix. So

$$L_1 = \begin{pmatrix} \sqrt{a_{11}} & 0 & 0 \\ \frac{a_{12}}{\sqrt{a_{11}}} & 1 & 0 \\ 0 & 0 & I_{n-2} \end{pmatrix} \text{ and } A^{(2)} = \begin{pmatrix} 1 & 0 \\ 0 & B^{(1)} - \frac{1}{\sqrt{a_{11}}} \mathbf{b}_1 \mathbf{b}_1^T \end{pmatrix}.$$

Now the matrix formed by $\mathbf{b}_1 \mathbf{b}_1^T$ has only one non-zero entry in the top left corner, which is a_{12}^2. It follows that $B^{(1)} - \frac{1}{\sqrt{a_{11}}} \mathbf{b}_1 \mathbf{b}_1^T$ is again a tridiagonal symmetric positive-definite matrix. Thus the algorithm calculates successively smaller tridiagonal symmetric positive-definite matrices where only the top left element has to be updated. Additionally the matrices L_i differ from the identity matrix in only two entries. Thus the factorization can be calculated in $O(M)$ operations. We will see later when discussing the *Hockney algorithm* how the structure of the eigenvectors can be used to obtain solutions to this system of equations using the *fast Fourier transform*.

8.2.3 Cauchy Problems and the Fourier Analysis of Stability

So far we have looked at PDEs where explicit boundary conditions are given. We now consider PDEs where the spatial variable extends over the whole range of $\mathbb{R}$. Again, we have to ensure that a solution exists, and therefore restrict ourselves to so-called Cauchy problems.

Definition 8.9 (Cauchy problem). *A PDE is known as a Cauchy problem, if there are no explicit boundary conditions but the initial condition $u(x, 0)$ must be square-integrable in $(-\infty, \infty)$; that is, the integral of the square of the absolute value is finite.*

To solve a Cauchy problem, we assume a recurrence of the form

$$\sum_{k=-r}^{s} \alpha_k u_{m+k}^{n+1} = \sum_{k=-r}^{s} \beta_k u_{m+k}^n, \qquad m \in \mathbb{Z}, n \in \mathbb{Z}^+, \tag{8.9}$$

where the coefficients α_k and β_k are independent of m and n, but typically depend on μ. For example, for the Crank–Nicolson method we have

$$r = 1, \quad s = 1 \quad \alpha_0 = 1 + \mu, \quad \alpha_{\pm 1} = -\mu/2, \quad \beta_0 = 1 - \mu, \quad \beta_{\pm 1} = \mu/2.$$

The stability of (8.9) is investigated by Fourier analysis. This is independent of the underlying PDE. The numerical stability is a feature of the algebraic recurrences.

Definition 8.10 (Fourier transform). *Let* $\mathbf{v} = (v_m)_{m \in \mathbb{Z}}$ *be a sequence of numbers in* $\mathbb{C}$. *The Fourier transform of this sequence is the function*

$$\hat{v}(\theta) = \sum_{m \in \mathbb{Z}} v_m e^{-im\theta}, \qquad -\pi \le \theta \le \pi. \tag{8.10}$$

The elements of the sequence can be retrieved from $\hat{v}(\theta)$ *by calculating*

$$v_m = \frac{1}{2\pi} \int_{\pi}^{\pi} \hat{v}(\theta) e^{im\theta} d\theta$$

for $m \in \mathbb{Z}$, *which is the inverse transform.*

The sequences and functions are equipped with the following norms

$$\|\mathbf{v}\| = \left(\sum_{m \in \mathbb{Z}} |v_m|^2 \right)^{\frac{1}{2}} \qquad \text{and} \qquad \|\hat{v}\| = \left(\frac{1}{2\pi} \int_{-\pi}^{\pi} |\hat{v}(\theta)|^2 d\theta \right)^{\frac{1}{2}}.$$

Lemma 8.1 (Parseval's identity). *For any sequence* $\mathbf{v}$, *we have the identity* $\|\mathbf{v}\| = \|\hat{v}\|$.

Proof. We have

$$\int_{-\pi}^{\pi} e^{-il\theta} d\theta = \left\{ \begin{array}{ll} 2\pi, & l = 0 \\ 0, & l \ne 0. \end{array} \right\}.$$

So by definition,

$$
\begin{aligned}
\|\hat{v}\|^2 &= \frac{1}{2\pi} \int_{-\pi}^{\pi} |\sum_{m \in \mathbb{Z}} v_m e^{-im\theta}|^2 d\theta = \frac{1}{2\pi} \int_{-\pi}^{\pi} \sum_{m \in \mathbb{Z}} \sum_{k \in \mathbb{Z}} v_m \bar{v}_k e^{-i(m-k)\theta} d\theta \\
&= \frac{1}{2\pi} \sum_{m \in \mathbb{Z}} \sum_{k \in \mathbb{Z}} v_m \bar{v}_k \int_{-\pi}^{\pi} e^{-i(m-k)\theta} d\theta = \sum_{m \in \mathbb{Z}} \sum_{k \in \mathbb{Z}} v_m \bar{v}_k \delta_{mk} = \|\mathbf{v}\|^2.
\end{aligned}
$$

$\square$

The implication of Parseval's identity is that the Fourier transform is an *isometry* of the Euclidean norm, which is an important reason for its many applications.

For the *Fourier analysis of stability* we have for every time step n a sequence $\mathbf{u}^n = (u_m^n)_{m \in \mathbb{Z}}$. For $\theta \in [-\pi, \pi]$, let $\hat{u}^n(\theta) = \sum_{m \in \mathbb{Z}} u_m^n e^{-im\theta}$ be the

Fourier transform of the sequence $\mathbf{u}^n$. Multiplying (8.9) by $e^{-im\theta}$, and summing over $m \in \mathbb{Z}$, gives for the left-hand side

$$\sum_{m \in \mathbb{Z}} e^{-im\theta} \sum_{k=-r}^{s} \alpha_k u_{m+k}^{n+1} = \sum_{k=-r}^{s} \alpha_k \sum_{m \in \mathbb{Z}} e^{-im\theta} u_{m+k}^{n+1}$$

$$= \sum_{k=-r}^{s} \alpha_k \sum_{m \in \mathbb{Z}} e^{-i(m-k)\theta} u_m^{n+1}$$

$$= \left(\sum_{k=-r}^{s} \alpha_k e^{ik\theta} \right) \sum_{m \in \mathbb{Z}} e^{-im\theta} u_m^{n+1}$$

$$= \left(\sum_{k=-r}^{s} \alpha_k e^{ik\theta} \right) \hat{u}^{n+1}(\theta).$$

Similarly, the right-hand side is $\left(\sum_{k=-r}^{s} \beta_k e^{ik\theta} \right) \hat{u}^n(\theta)$ and we deduce

$$\hat{u}^{n+1}(\theta) = H(\theta)\hat{u}^n(\theta) \qquad \text{where} \qquad H(\theta) = \frac{\sum_{k=-r}^{s} \beta_k e^{ik\theta}}{\sum_{k=-r}^{s} \alpha_k e^{ik\theta}}. \qquad (8.11)$$

The function H is called the *amplification factor* of the recurrence given by (8.9).

Theorem 8.3. *The method given by (8.9) is stable if and only if $|H(\theta)| \le 1$ for all $\theta \in [-\pi, \pi]$.*

Proof. Since we are solving a Chauchy problem and m ranges over the whole of $\mathbb{Z}$, the equations are identical for all Δx and there is no need to insist explicitly that $\|\mathbf{u}^n\|$ remains uniformly bounded for $\Delta x \to 0$. The definition of stability says that there exists $C > 0$ such that $\|\mathbf{u}^n\| \le C$ for all $n \in \mathbb{Z}$. Since the Fourier transform is an isometry, this is equivalent to $\|\hat{u}^n\| \le C$ for all $n \in \mathbb{Z}$. Iterating (8.11), we deduce

$$\hat{u}^{n+1}(\theta) = [H(\theta)]^{n+1} \hat{u}^0(\theta), \qquad \theta \in [-\pi, \pi], n \in \mathbb{Z}^+.$$

To prove the first direction, let's assume $|H(\theta)| \le 1$ for all $\theta \in [-\pi, \pi]$. Then by the above equation $|\hat{u}^n(\theta)| \le |\hat{u}^0(\theta)|$ it follows that

$$\|\hat{u}^n\|^2 = \frac{1}{2\pi} \int_{-\pi}^{\pi} |\hat{u}^n(\theta)|^2 d\theta \le \frac{1}{2\pi} \int_{-\pi}^{\pi} |H(\theta)|^{2n} |\hat{u}^0(\theta)|^2 d\theta$$

$$\le \frac{1}{2\pi} \int_{-\pi}^{\pi} |\hat{u}^0(\theta)|^2 d\theta = \|\hat{u}^0\|^2,$$

and hence we have stability.

The other direction is more technical, since we have to construct an example of an initial condition where the solution will become unbounded. Suppose that there exists $\theta^* \in (-\pi, \pi)$ such that $|H(\theta^*)| = 1 + \epsilon > 1$. Since H is rational, it is continuous everywhere apart from where the denominator vanishes, which cannot be at θ^*, since it takes a finite value there. Hence there exist $\theta^- < \theta^* < \theta^+ \in [-\pi, \pi]$ such that $|H(\theta)| \geq 1 + \frac{1}{2}\epsilon$ for all $\theta \in [\theta^-, \theta^+]$. Let

$$\hat{u}^0(\theta) = \left\{ \begin{array}{cc} \sqrt{\frac{2\pi}{\theta^+ - \theta^-}}, & \theta^- \leq \theta \leq \theta^+, \\ 0, & \text{otherwise.} \end{array} \right\}$$

This is a step function over the interval $[\theta^-, \theta^+]$. We can calculate the sequence which gives rise to this step function by

$$u_m^0 = \frac{1}{2\pi} \int_{-\pi}^{\pi} \hat{u}^0(\theta) e^{im\theta} d\theta = \frac{1}{2\pi} \int_{\theta^-}^{\theta^+} \sqrt{\frac{2\pi}{\theta^+ - \theta^-}} e^{im\theta} d\theta.$$

For $m = 0$ we have $e^{-im\theta} = 1$ and therefore

$$u_0^0 = \sqrt{\frac{\theta^+ - \theta^-}{2\pi}}.$$

For $m \neq 0$ we get

$$u_m^0 = \frac{1}{\sqrt{2\pi(\theta^+ - \theta^-)}} \left[\frac{1}{im} e^{im\theta} \right]_{\theta^-}^{\theta^+} = \frac{1}{im\sqrt{2\pi(\theta^+ - \theta^-)}} (e^{im\theta^+} - e^{im\theta^-}).$$

The complex-valued function

$$u(x) = \left\{ \begin{array}{cc} \sqrt{\frac{\theta^+ - \theta^-}{2\pi}}, & x = 0 \\ \frac{1}{ix\sqrt{2\pi(\theta^+ - \theta^-)}} (e^{ix\theta^+} - e^{ix\theta^-}), & x \neq 0 \end{array} \right\}$$

is well defined and continuous, since for $x \to 0$ it tends to the value for $x = 0$ (which can be verified by using the expansions of the exponentials). It is also square integrable, since it tends to zero for $x \to \infty$ due to x being in the denominator. Therefore it is a suitable choice for an initial condition.

On the other hand,

$$\begin{aligned} \|\hat{u}^n\| &= \frac{1}{\sqrt{2\pi}} \left(\int_{-\pi}^{\pi} |H(\theta)|^{2n} |\hat{u}^0(\theta)|^2 d\theta \right)^{\frac{1}{2}} \\ &= \frac{1}{\sqrt{2\pi}} \left(\int_{\theta^-}^{\theta^+} |H(\theta)|^{2n} |\hat{u}^0(\theta)|^2 d\theta \right)^{\frac{1}{2}} \\ &\geq \frac{1}{\sqrt{2\pi}} (1 + \frac{1}{2}\epsilon)^n \left(\int_{\theta^-}^{\theta^+} 2\pi(\theta^+ - \theta^-)^{-1} d\theta \right)^{\frac{1}{2}} \\ &= (1 + \frac{1}{2}\epsilon)^n \stackrel{n \to \infty}{\longrightarrow} \infty, \end{aligned}$$

since the last integral equates to 1. Thus the method is then unstable. $\qquad \square$

Consider the Cauchy problem for the heat equation and recall the first method based on solving the semi-discretized problem with forward Euler.

$$u_m^{n+1} = u_m^n + \mu(u_{m-1}^n - 2u_m^n + u_{m+1}^n).$$

Putting this into the new context, we have

$$r = 1, \quad s = 1 \quad \alpha_0 = 1, \quad \alpha_{\pm 1} = 0, \quad \beta_0 = 1 - 2\mu, \quad \beta_{\pm 1} = \mu.$$

Therefore

$$H(\theta) = 1 + \mu(e^{-i\theta} - 2 + e^{i\theta}) = 1 - 4\mu \sin^2 \frac{\theta}{2}, \qquad \theta \in [-\pi, \pi]$$

and thus $1 \geq H(\theta) \geq H(\pi) = 1 - 4\mu$. Hence the method is stable if and only if $\mu \leq \frac{1}{2}$.

On the other hand, for the backward Euler method we have

$$u_m^{n+1} - \mu(u_{m-1}^{n+1} - 2u_m^{n+1} + u_{m+1}^{n+1}) = u_m^n$$

and therefore

$$H(\theta) = [1 - \mu(e^{-i\theta} - 2 + e^{i\theta})]^{-1} = [1 + 4\mu \sin^2 \frac{\theta}{2}]^{-1} \in (0, 1],$$

which implies stability for all $\mu > 0$.

The Crank–Nicolson scheme given by

$$u_m^{n+1} - \frac{1}{2}\mu(u_{m-1}^{n+1} - 2u_m^{n+1} + u_{m+1}^{n+1}) = u_m^n + \frac{1}{2}\mu(u_{m-1}^n - 2u_m^n + u_{m+1}^n)$$

results in

$$H(\theta) = \frac{1 + \frac{1}{2}\mu(e^{-i\theta} - 2 + e^{i\theta})}{1 - \frac{1}{2}\mu(e^{-i\theta} - 2 + e^{i\theta})} = \frac{1 - 2\mu \sin^2 \frac{\theta}{2}}{1 + 2\mu \sin^2 \frac{\theta}{2}},$$

which lies in $(-1, 1]$ for all $\theta \in [-\pi, \pi]$ and all $\mu > 0$.

Exercise 8.2. *Apply the Fourier stability test to the difference equation*

$$u_m^{n+1} = \frac{1}{2}(2 - 5\mu + 6\mu^2)u_m^n + \frac{2}{3}\mu(2 - 3\mu)(u_{m-1}^n + u_{m+1}^n)$$
$$- \frac{1}{12}\mu(1 - 6\mu)(u_{m-2}^n + u_{m+2}^n).$$

Deduce that the test is satisfied if and only if $0 \leq \mu \leq \frac{2}{3}$.

The eigenvalue stability analysis and the Fourier stability analysis are tackling two fundamentally different problems. In the eigenvalue framework, boundaries are incorporated, while in the Fourier analysis we have $m \in \mathbb{Z}$, which corresponds to $x \in \mathbb{R}$ in the underlying PDE. It is no trivial task to translate Fourier analysis to problems with boundaries. When either $r \geq 2$ or

$s \geq 2$ there are not enough boundary values to satisfy the recurrence equations near the boundary. This means the discretized equations need to be amended near the boundary and the identity (8.11) is no longer valid. It is not enough to extend the values u_m^n with zeros for $m \notin \{1, 2, \ldots, M\}$. In general a great deal of care needs to be taken to combine Fourier analysis with boundary conditions. How to treat the problem at the boundaries has to be carefully considered to avoid the instability which then propagates from the boundary inwards.

With many parabolic PDEs, e.g., the heat equation, the Euclidean norm of the exact solution decays (for zero boundary conditions) and good methods share this behaviour. Hence they are robust enough to cope with inwards error propagation from the boundary into the solution domain, which might occur when discretized equations are amended there. The situation is more difficult for many hyperbolic equations, e.g., the wave equation, since the exact solution keeps the energy (a.k.a. the Euclidean norm) constant and so do many good methods. In that case any error propagation from the boundary delivers a false result. Additional mathematical techniques are necessary in this situation.

Exercise 8.3. *The Crank–Nicolson formula is applied to the heat equation* $u_t = u_{xx}$ *on a rectangular mesh* $(m\Delta x, n\Delta t)$, $m = 0, 1, \ldots, M + 1$, $n = 0, 1, 2, \ldots$, *where* $\Delta x = 1/(M + 1)$. *We assume zero boundary conditions* $u(0, t) = u(1, t) = 0$ *for all* $t \geq 0$. *Prove that the estimates* $u_n^m \approx u(m\Delta x, n\Delta t)$ *satisfy the equation*

$$\sum_{m=1}^{M} [(u_m^{n+1})^2 - (u_m^n)^2] = -\frac{1}{2}\mu \sum_{m=1}^{M+1} (u_m^{n+1} + u_m^n - u_{m-1}^{n+1} - u_{m-1}^n)^2.$$

This shows that $\sum_{m=1}^{M}(u_m^n)^2$ *is monotonically decreasing with increasing* n *and the numerical solution mimics the decaying behaviour of the exact solution. (Hint: Substitute the value of* $u_m^{n+1} - u_m^n$ *that is given by the Crank–Nicolson formula into the elementary equation* $\sum_{m=1}^{M}[(u_m^{n+1})2 - (u_m^n)2] = \sum_{m=1}^{M}(u_m^{n+1} - u_m^n)(u_m^{n+1} + u_m^n)$. *It is also helpful occasionally to change the index* m *of the summation by one.)*

8.3 Elliptic PDEs

To move from the one-dimensional case to higher dimensional cases, let's first look at *elliptic partial differential equations*. Since the heat equation in two dimensions is given by $u_t = \nabla^2 u$, let's look at the *Poisson equation*

$$\nabla^2 u(x, y) = u_{xx}(x, y) + u_{yy}(x, y) = f(x, y) \text{ for all } (x, y) \in \Omega,$$

where Ω is an open connected domain in $\mathbb{R}^2$ with boundary $\partial\Omega$. For all $(x, y) \in \partial\Omega$ we have the Dirichlet boundary condition $u(x, y) = \phi(x, y)$. We assume that f is continuous in Ω and that ϕ is twice differentiable. We lay a square grid over Ω with uniform spacing of Δx in both the x and y direction. Further, we assume that $\partial\Omega$ is part of the grid. For our purposes Ω is a rectangle, but the results hold for other domains.

8.3.1 Computational Stencils

The finite difference given by Equation (8.5) gives an $O(\Delta x^2)$ approximation to the second derivative. We employ this in both the x and y direction to obtain the approximation

$$\nabla^2 u(x,y) \approx \frac{1}{(\Delta x)^2}[u(x-\Delta x,y)+u(x+\Delta x,y)+u(x,y-\Delta x)$$
$$+u(x,y+\Delta x)-4u(x,y)],$$

which produces a local error of $O((\Delta x)^2)$. This gives rise to the *five-point method*

$$u_{l-1,m}+u_{l+1,m}+u_{l,m-1}+u_{l,m+1}-4u_{l,m}=(\Delta x)^2 f_{l,m}, \qquad (l\Delta x, m\Delta x)\in\Omega,$$

where $f_{l,m}=f(l\Delta x, m\Delta x)$ and $u_{l,m}$ approximates $u(l\Delta x, m\Delta x)$. A compact notation is the *computational stencil* (also known as *computational molecule*)

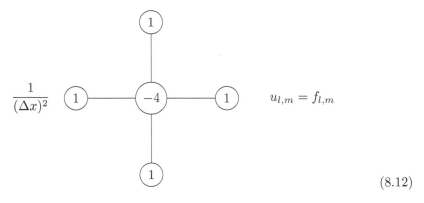

$$(8.12)$$

Whenever $(l\Delta x, m\Delta x) \in \partial\Omega$, we substitute the appropriate value $\phi(l\Delta x, m\Delta x)$. This procedure leads to a (sparse) set of linear algebraic equations, whose solution approximates the true solution at the grid points.

By using the approximation

$$\frac{\partial^2 u(x,y)}{\partial x^2} = \frac{1}{(\Delta x)^2}[-\frac{1}{12}u(x-2\Delta x,y)+\frac{4}{3}u(x-\Delta x,y)-\frac{5}{2}u(x,y)$$
$$+\frac{4}{3}u(x+\Delta x,y)-\frac{1}{12}u(x+2\Delta x,y)]+O((\Delta x)^4)$$

we obtain the computational stencil

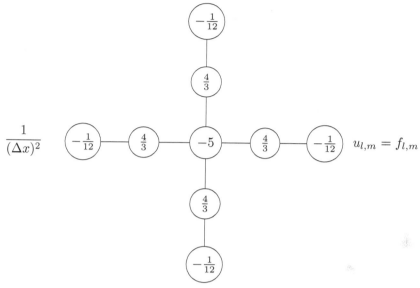

$$(8.13)$$

which produces a local error of $O((\Delta x)^4)$. However, the implementation of this method is more complicated, since at the boundary, values of points outside the boundary are needed. These values can be approximated by nearby values. For example, if we require an approximation to $u(l\Delta x, m\Delta x)$, where $m\Delta x$ lies outside the boundary, we can set

$$u(l\Delta x, m\Delta x) \approx \frac{1}{4}u((l+1)\Delta x, (m-1)\Delta x) + \frac{1}{2}u(l\Delta x, (m-1)\Delta x)$$
$$+ \frac{1}{4}u((l-1)\Delta x, (m-1)\Delta x).$$

$$(8.14)$$

The set of linear algebraic equations has to be modified accordingly to take these adjustments into account.

Now consider the approximation

$$u(x+\Delta x, y) + u(x-\Delta x, y) = \left(e^{\Delta x D_x} + e^{-\Delta x D_x}\right) u(x,y)$$
$$= 2u(x,y) + (\Delta x)^2 \frac{\partial^2 u}{\partial x^2}(x,y) + O((\Delta x)^4),$$

where D_x denotes the differential operator in the x-direction. Applying the same principle in the y direction, we have

$$u(x + \Delta x, y + \Delta x) + u(x - \Delta x, y + \Delta x)$$
$$+u(x + \Delta x, y - \Delta x) + u(x - \Delta x, y - \Delta x) =$$
$$\left(e^{\Delta x D_x} + e^{-\Delta x D_x}\right)\left(e^{\Delta x D_y} + e^{-\Delta x D_y}\right)u(x, y) =$$
$$[2 + (\Delta x)^2 \frac{\partial^2}{\partial x^2} + O((\Delta x)^4)] \times [2 + (\Delta x)^2 \frac{\partial^2}{\partial y^2} + O((\Delta x)^4)]u(x, y) =$$
$$4u(x, y) + 2(\Delta x)^2 \frac{\partial^2 u}{\partial x^2}(x, y) + 2(\Delta x)^2 \frac{\partial^2 u}{\partial y^2}(x, y) + O((\Delta x)^4).$$

This motivates the computational stencil

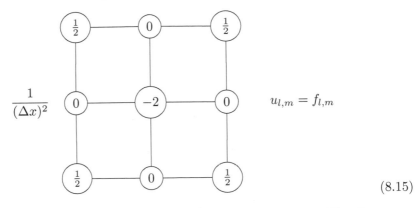

(8.15)

Both (8.12) and (8.15) give $O((\Delta x)^2)$ approximations to $\nabla^2 u$. Computational stencils can be added: $2\times$ (8.12) + (8.15) and dividing by 3 results in the *nine-point method*:

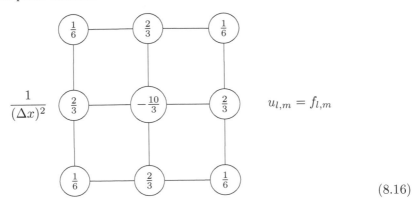

(8.16)

This method has advantages, as we will see in the following.

Exercise 8.4. *We have seen that the nine-point method provides a numerical solution to the Poisson equation with local error $O((\Delta x)^2)$. Show that for the Laplace equation $\nabla^2 u = 0$ the local error is $O((\Delta x)^4)$.*

Generally the nine-point formula is an $O((\Delta x)^2)$ approximation. However, this can be increased to $O((\Delta x)^4)$. In the above exercise the $(\Delta x)^2$ error term is

$$\frac{1}{12}(\Delta x)^2 \left(\frac{\partial^4 u}{\partial x^4} + 2\frac{\partial^4 u}{\partial x^2 \partial y^2} + \frac{\partial^4 u}{\partial y^4} \right),$$

which vanishes for $\nabla^2 u = 0$. This can be rewritten as

$$\frac{1}{12}(\Delta x)^2 \left[\frac{\partial^2}{\partial x^2}(\frac{\partial^2 u}{\partial x^2} + \frac{\partial^2 u}{\partial y^2}) + \frac{\partial^2}{\partial y^2}(\frac{\partial^2 u}{\partial x^2} + \frac{\partial^2 u}{\partial y^2}) \right] =$$
$$\frac{1}{12}(\Delta x)^2 \left[\frac{\partial^2 f}{\partial x^2} + \frac{\partial^2 f}{\partial y^2} \right] = \frac{1}{12}(\Delta x)^2 \nabla^2 f.$$

The Laplacian of f can be approximated by the five-point method (8.12) and adding this to the right-hand side of (8.16) gives the scheme

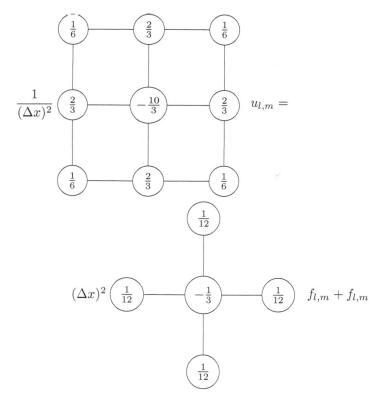

which has a local error of $O((\Delta x)^4)$, since the $(\Delta x)^2$ error term is canceled.

Exercise 8.5. *Determine the order of the local error of the finite difference* *approximation to $\partial^2 u / \partial x \partial y$, which is given by the computational stencil*

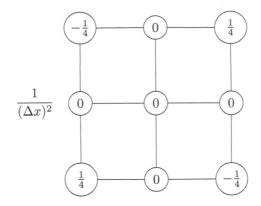

We have seen that the first error term in the nine-point method is

$$\frac{1}{12}(\Delta x)^2 \left(\frac{\partial^4 u}{\partial x^4}(x,y) + 2\frac{\partial^4 u}{\partial x^2 \partial y^2}(x,y) + \frac{\partial^4 u}{\partial y^4}(x,y) \right) = \frac{1}{12}(\Delta x)^2 \nabla^4 u(x,y),$$

while the method given by (8.13) has no $(\Delta x)^2$ error term. Now we can combine these two methods to generate an approximation for ∇^4. In particular, let the new method be given by $12 \times (8.16) - 12 \times (8.13)$, and dividing by $(\Delta x)^2$ gives

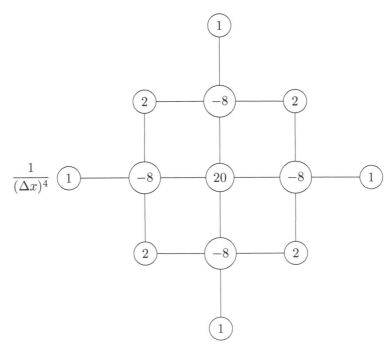

Exercise 8.6. *Verify that the above approximation of ∇^4 has a local error of $O((\Delta x)^2)$ and identify the first error term.*

Thus, by knowing the first error term, we can combine different finite difference schemes to obtain new approximations to different partial differential equations.

So far, we have only looked at equispaced square grids. The boundary, however, often fails to fit exactly into a square grid. Thus we sometimes need to approximate derivatives using non-equispaced points at the boundary. In the interior the grid remains equispaced. For example, suppose we try to approximate the second directional derivative and that the grid points have the spacing Δx in the interior and $\alpha \Delta x$ at the boundary, where $0 < \alpha \le 1$. Using the Taylor expansion, one can see that

$$\frac{1}{(\Delta x)^2}\left[\frac{2}{\alpha+1}g(x-\Delta x) - \frac{2}{\alpha}g(x) + \frac{2}{\alpha(\alpha+1)}g(x+\alpha\Delta x)\right]$$
$$= g''(x) + \frac{1}{2}(\alpha-1)g'''(x)\Delta x + O((\Delta x)^2),$$

with error of $O(\Delta x)$. Note that $\alpha = 1$ recovers the finite difference with error $O((\Delta x)^2)$ that we have already used. Better approximation can be obtained by taking two equispaced points on the interior side.

$$\frac{1}{(\Delta x)^2}\left[\frac{\alpha-1}{\alpha+2}g(x-2\Delta x) - \frac{2(\alpha-2)}{\alpha+1}g(x-\Delta x)\right.$$
$$\left. +\frac{\alpha-3}{\alpha}g(x) + \frac{6}{\alpha(\alpha+1)(\alpha+2)}g(x+\alpha\Delta x)\right]$$
$$= g''(x) + O((\Delta x)^2).$$

8.3.2 Sparse Algebraic Systems Arising from Computational Stencils

For simplicity we restrict our attention to the case of Ω being a *square* where the sides are divided into pieces of length Δx by $M+1$. Thus we need to estimate M^2 unknown function values $u(l\Delta x, m\Delta x)$, $l,m = 1,\dots,M$. Let $N = M^2$. The computational stencils then yield an $N \times N$ system of linear equations. How this is represented in matrix form depends of course on the way the grid points are arranged into a one-dimensional array. In the *natural ordering* the grid points are arranged by columns. Using this ordering, the five-point method gives the linear system $(\Delta x)^{-2}A\mathbf{u} = \mathbf{b}$ where A is the block tridiagonal matrix

$$A = \begin{pmatrix} B & I & & & \\ I & B & I & & \\ & \ddots & \ddots & \ddots & \\ & & I & B & I \\ & & & I & B \end{pmatrix} \text{ and } B = \begin{pmatrix} -4 & 1 & & & \\ 1 & -4 & 1 & & \\ & \ddots & \ddots & \ddots & \\ & & 1 & -4 & 1 \\ & & & 1 & -4 \end{pmatrix}.$$

Note that B is a TST matrix. The vector $\mathbf{b}$ is given by the right-hand sides $f_{l,m}$ (following the same ordering) and the boundary conditions. More specifically, if $u_{l,m}$ is such that for example $u_{l,m+1}$ lies on the boundary, we let $u_{l,m+1} = \phi(l\Delta x, (m+1)\Delta x)$ and the right-hand side $b_{l,m} = f_{l,m} - \phi(l\Delta x, (m+1)\Delta x)$.

The method specified by (8.13) has the associated matrix

$$A = \begin{pmatrix} B & \frac{4}{3}I & -\frac{1}{12}I & & & & \\ \frac{4}{3}I & B & \frac{4}{3}I & -\frac{1}{12}I & & & \\ -\frac{1}{12}I & \frac{4}{3}I & B & \frac{4}{3}I & -\frac{1}{12}I & & \\ & \ddots & \ddots & \ddots & \ddots & \ddots & \\ & & -\frac{1}{12}I & \frac{4}{3}I & B & \frac{4}{3}I & -\frac{1}{12}I \\ & & & -\frac{1}{12}I & \frac{4}{3}I & B & \frac{4}{3}I \\ & & & & -\frac{1}{12}I & \frac{4}{3}I & B \end{pmatrix}$$

where

$$B = \begin{pmatrix} -5 & \frac{4}{3} & -\frac{1}{12} & & & \\ \frac{4}{3} & -5 & \frac{4}{3} & -\frac{1}{12} & & \\ -\frac{1}{12} & \frac{4}{3} & -5 & \frac{4}{3} & -\frac{1}{12} & \\ & \ddots & \ddots & \ddots & \ddots & \ddots \\ & & -\frac{1}{12} & \frac{4}{3} & -5 & \frac{4}{3} & -\frac{1}{12} \\ & & & -\frac{1}{12} & \frac{4}{3} & -5 & \frac{4}{3} \\ & & & & -\frac{1}{12} & \frac{4}{3} & -5 \end{pmatrix}.$$

Again the boundary conditions need to be incorporated into the right-hand side, using the approximation given by (8.14) for points lying outside Ω.

Now the nine-point method has the associated block tridiagonal matrix

$$A = \begin{pmatrix} B & C & & & \\ C & B & C & & \\ & \ddots & \ddots & \ddots & \\ & & C & B & C \\ & & & C & B \end{pmatrix} \tag{8.17}$$

where the blocks are given by

$$B = \begin{pmatrix} -\frac{10}{3} & \frac{2}{3} & & & \\ \frac{2}{3} & -\frac{10}{3} & \frac{2}{3} & & \\ & \ddots & \ddots & \ddots & \\ & & \frac{2}{3} & -\frac{10}{3} & \frac{2}{3} \\ & & & \frac{2}{3} & -\frac{10}{3} \end{pmatrix} \quad \text{and} \quad C = \begin{pmatrix} \frac{2}{3} & \frac{1}{6} & & & \\ \frac{1}{6} & \frac{2}{3} & \frac{1}{6} & & \\ & \ddots & \ddots & \ddots & \\ & & \frac{1}{6} & \frac{2}{3} & \frac{1}{6} \\ & & & \frac{1}{6} & \frac{2}{3} \end{pmatrix}.$$

With these examples, we see that the symmetry of computational stencils yields symmetric block matrices with constant blocks only on the main diagonal and one or two subdiagonals where the blocks themselves are symmetric Toeplitz matrices with only the main diagonal, and one or two subdiagonals nonzero. This special structure lends itself to efficient algorithms to solve the large system of equations.

8.3.3 Hockney Algorithm

For simplicity we restrict ourselves to tridiagonal systems, which results in the *Hockney algorithm*. Recall that the normalized eigenvectors and the eigenvalues for a general $M \times M$ tridiagonal symmetric Toeplitz (TST) matrix given by

$$B = \begin{pmatrix} \alpha & \beta & & & \\ \beta & \alpha & \beta & & \\ & \ddots & \ddots & \ddots & \\ & & \beta & \alpha & \beta \\ & & & \beta & \alpha \end{pmatrix}$$

are

$$\lambda_k^B = \alpha + 2\beta \cos \frac{k\pi}{M+1}, \qquad (\mathbf{q}_k)_j = a \sin \frac{jk\pi}{M+1}, \qquad k, j = 1, \ldots, M,$$

where a is a normalization constant. Let $Q = (\mathbf{q}_1 \ldots \mathbf{q}_M)$ be the matrix whose columns are the eigenvectors. We consider the matrices B and C occurring in the nine-point method. The matrix A can then be written as

$$A = \begin{pmatrix} QD_BQ^T & QD_CQ^T & & & \\ QD_CQ^T & QD_BQ^T & QD_CQ^T & & \\ & \ddots & \ddots & \ddots & \\ & & QD_CQ^T & QD_BQ^T & QD_CQ^T \\ & & & QD_CQ^T & QD_BQ^T \end{pmatrix}.$$

Setting $\mathbf{v}_k = (\Delta x)^{-2} Q^T \mathbf{u}_k$ and multiplying by a diagonal block matrix where each block equals Q^T, our system becomes

$$\begin{pmatrix} D_B & D_C & & & \\ D_C & D_B & D_C & & \\ & \ddots & \ddots & \ddots & \\ & & D_C & D_B & D_C \\ & & & D_C & D_B \end{pmatrix} \begin{pmatrix} \mathbf{v}_1 \\ \mathbf{v}_2 \\ \vdots \\ \mathbf{v}_{M-1} \\ \mathbf{v}_M \end{pmatrix} = \begin{pmatrix} \mathbf{c}_1 \\ \mathbf{c}_2 \\ \vdots \\ \mathbf{c}_{M-1} \\ \mathbf{c}_M \end{pmatrix}$$

where $\mathbf{c}_k = Q^T \mathbf{b}_k$. At this stage we reorder the gird by rows instead of columns. In other words, we permute $\mathbf{v} \mapsto \hat{\mathbf{v}} = P\mathbf{v}, \mathbf{c} \mapsto \hat{\mathbf{c}} = P\mathbf{c}$, such that the portion $\hat{\mathbf{v}}_1$ is made out of the first components of each of the portions $\mathbf{v}_1, \ldots, \mathbf{v}_M$, the portion $\hat{\mathbf{v}}_2$ is made out of the second components of each of the portions $\mathbf{v}_1, \ldots, \mathbf{v}_M$, and so on. Permutations come essentially at no computational cost since in practice we store $\mathbf{c}, \mathbf{v}$ as 2D arrays (which are addressed accordingly) and not in one long vector. This yields a new system

$$
\begin{pmatrix}
\Lambda_1 & & & \\
& \Lambda_2 & & \\
& & \ddots & \\
& & & \Lambda_M
\end{pmatrix}
\begin{pmatrix}
\hat{\mathbf{v}}_1 \\
\hat{\mathbf{v}}_2 \\
\vdots \\
\hat{\mathbf{v}}_M
\end{pmatrix}
=
\begin{pmatrix}
\hat{\mathbf{c}}_1 \\
\hat{\mathbf{c}}_2 \\
\vdots \\
\hat{\mathbf{c}}_M
\end{pmatrix},
$$

where

$$
\Lambda_k =
\begin{pmatrix}
\lambda_k^B & \lambda_k^C & & & \\
\lambda_k^C & \lambda_k^B & \lambda_k^C & & \\
& \ddots & \ddots & \ddots & \\
& & \lambda_k^C & \lambda_k^B & \lambda_k^C \\
& & & \lambda_k^C & \lambda_k^B
\end{pmatrix}.
$$

These are M uncoupled systems, $\Lambda_k \hat{\mathbf{v}}_k = \hat{\mathbf{c}}_k$ for $k = 1, \ldots, M$. Since these systems are tridiagonal, each can be solved fast in $O(M)$ operations. Hence the steps of the algorithm and their computational cost are as follows:

- Form the products $\mathbf{c}_k = Q^T \mathbf{b}_k$ for $k = 1, \ldots, M$ taking $O(M^3)$ operations.

- Solve the $M \times M$ tridiagonal system $\Lambda_k \hat{\mathbf{v}}_k = \hat{\mathbf{c}}_k$ for $k = 1, \ldots, M$ taking $O(M^2)$ operations.

- Form the products $\mathbf{u}_k = Q \mathbf{v}_k$ for $k = 1, \ldots, M$ taking $O(M^3)$ operations.

The computational bottleneck consists of the $2M$ matrix-vector products by the matrices Q and Q^T. The elements of Q are $(\mathbf{q}_k)_j = a \sin \frac{jk\pi}{M+1}$. Now $\sin \frac{jk\pi}{M+1}$ is the imaginary part of $\exp \frac{ijk\pi}{M+1}$. This observation lends itself to a considerable speedup.

Definition 8.11 (Discrete Fourier Transform). *Let* $\mathbf{x} = (x_l)_{l \in \mathbb{Z}}$ *be a sequence of complex numbers with period* n. *i.e.,* $x_{l+n} = x_l$. *Set* $\omega_n = \exp \frac{2\pi i}{n}$, *the primitive root of unity of degree* n. *The* discrete Fourier transform (DFT) $\hat{\mathbf{x}}$ *of* $\mathbf{x}$ *is given by*

$$
\hat{x}_j = \frac{1}{n} \sum_{l=0}^{n-1} \omega_n^{-jl} x_l, \qquad j = 0, \ldots, n - 1.
$$

It can be easily proven that the discrete Fourier transform is an isomorphism and that the inverse is given by

$$x_l = \sum_{j=0}^{n-1} \omega_n^{jl} \hat{x}_j, \qquad l = 0, \dots, n-1.$$

For $l = 1, \dots, M$ the lth component of the product $Q\mathbf{y}$ is

$$(Q\mathbf{y})_l = a \sum_{j=1}^{M} y_j \sin \frac{lj\pi}{M+1} = a \sum_{j=0}^{M} y_j \sin \frac{lj\pi}{M+1} = a\text{Im} \sum_{j=0}^{M} y_j \exp \frac{ilj\pi}{M+1},$$

since $\sin \frac{l0\pi}{M+1} = 0$. Here Im denotes the imaginary part of the expression. To obtain a formula similar to the discrete Fourier transform we need a factor of 2 multiplying π. Subsequently, we need to extend the sum, which we can do by setting $y_{M+1} = \dots = y_{2M+1} = 0$.

$$\begin{aligned}
(Q\mathbf{y})_l &= a\text{Im} \sum_{j=0}^{M} y_j \exp \frac{ilj2\pi}{2M+2} = a\text{Im} \sum_{j=0}^{2M+1} y_j \exp \frac{ilj2\pi}{2M+2} \\
&= a\text{Im} \sum_{j=0}^{2M+1} y_j \omega_{2M+2}^{lj}.
\end{aligned}$$

Thus, multiplication by Q can be reduced to calculating an inverse DFT.

The calculation can be sped up even more by using the *fast Fourier transform (FFT)*. Suppose that n is a power of 2, i.e., $n = 2^L$, and denote by

$$\hat{\mathbf{x}}^E = (\hat{x}_{2l})_{l \in \mathbb{Z}} \text{ and } \hat{\mathbf{x}}^O = (\hat{x}_{2l+1})_{l \in \mathbb{Z}}$$

the even and odd portions of $\hat{\mathbf{x}}$. Note that both $\hat{\mathbf{x}}^E$ and $\hat{\mathbf{x}}^O$ have period $n/2 = 2^{L-1}$. Suppose we already know the inverse DFT of both the short sequences $\mathbf{x}^E$ and $\mathbf{x}^O$. Then it is possible to assemble $\mathbf{x}$ in a small number of operations. Remembering $w_n^n = 1$,

$$\begin{aligned}
x_l &= \sum_{j=0}^{2^L-1} w_{2^L}^{jl} \hat{x}_j = \sum_{j=0}^{2^{L-1}-1} w_{2^L}^{2jl} \hat{x}_{2j} + \sum_{j=0}^{2^{L-1}-1} w_{2^L}^{(2j+1)l} \hat{x}_{2j+1} \\
&= \sum_{j=0}^{2^{L-1}-1} w_{2^{L-1}}^{jl} \hat{x}_j^E + w_{2^L}^l \sum_{j=0}^{2^{L-1}-1} w_{2^{L-1}}^{jl} \hat{x}_j^O = x_l^E + w_{2^L}^l x_l^O.
\end{aligned}$$

Thus it only costs n products to calculate $\mathbf{x}$, provided that $\mathbf{x}^E$ and $\mathbf{x}^O$ are known. This can be reduced even further to $n/2$, since the second half of the sequence can be calculated as

$$x_{l+2^{L-1}} = x_{l+2^{L-1}}^E + w_{2^L}^{l+2^{L-1}} x_{l+2^{L-1}}^O = x_l^E + w_{2^L}^{2^{L-1}} w_{2^L}^l x_l^O = x_l^E - w_{2^L}^l x_l^O.$$

Thus the products $\omega_{2L}^l x_l^O$ only need to be evaluated for $l = 0, \ldots, n/2 - 1$.

To execute the FFT we start with vectors with only one element and in the ith stage, $i = 1, \ldots, L$, assemble 2^{L-i} vectors of length 2^i from vectors of length 2^{i-1}. Altogether the cost of the FFT is $\frac{1}{2}n \log_2 n$ products. For example, if $n = 1024 = 2^{10}$, the cost is $\approx 5 \times 10^3$ compared to $\approx 10^6$ for the matrix multiplication. For $n = 2^{20}$ the numbers become $\approx 1.05 \times 10^7$ for the FFT compared to $\approx 1.1 \times 10^{12}$, which represents a factor of more than 10^5.

The following schematic shows the structure of the FFT algorithm for $n = 8$. The bottom row shows the indices forming the vectors consisting of only one element. The branches show how the vectors are combined with which factors $\pm \omega_{2L}^l$.

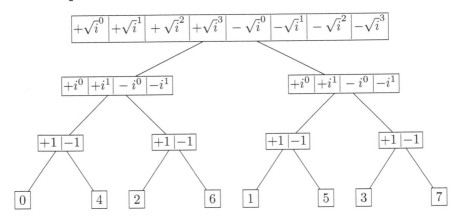

The FFT was discovered by Gauss (and forgotten), rediscovered by Lanczos (and forgotten), and, finally, rediscovered by Cooley and Tuckey under which name it is implemented in many packages. It is a classic divide-and-conquer algorithm.

Exercise 8.7. *Let* $(\hat{x}_0, \hat{x}_1, \hat{x}_2, \hat{x}_3, \hat{x}_4, \hat{x}_5, \hat{x}_6, \hat{x}_7) = (2, 0, 6, -2, 6, 0, 6, 2)$. *By applying the inverse of the FFT algorithm, calculate* $x_l = \sum_{j=0}^{7} \omega_8^{jl} \hat{x}_j$ *for* $l = 0, 2, 4, 6$, *where* $\omega_8 = \exp \frac{2i\pi}{8}$.

8.3.4 Multigrid Methods

In the previous section, we have exploited the special structure in the matrices of computational stencils we encountered so far. Another approach are multigrid methods which were developed from the computational observation that when the Gauss–Seidel method is applied to solve the five-point formula on a square $M \times M$ grid, the error decays substantially in each iteration for the first few iterations. Subsequently, the rate slows down and the method settles to its slow asymptotic rate of convergence. Note that this is the error associated with the Gauss–Seidel solution to the linear equations, not the error in the solution of the original Laplacian equation.

Recall that for a system of linear equations given by $A\mathbf{x} = (D+L+U)\mathbf{x} = \mathbf{b}$, where D, L, U are the diagonal, strictly lower triangular, and strictly upper triangular parts of A, respectively, the Gauss–Seidel and Jacobi methods are:

Gauss–Seidel
$$(D + L)\mathbf{x}^{(k+1)} = -U\mathbf{x}^{(k)} + \mathbf{b},$$

Jacobi
$$D\mathbf{x}^{(k+1)} = -(U + L)\mathbf{x}^{(k)} + \mathbf{b}.$$

Specifically for the five-point method, this results in (using the natural ordering)

Gauss–Seidel
$$u_{l-1,m}^{(k+1)} + u_{l,m-1}^{(k+1)} - 4u_{l,m}^{(k+1)} = -u_{l+1,m}^{(k)} - u_{l,m+1}^{(k)} + (\Delta x)^2 f_{l,m},$$

Jacobi
$$-4u_{l,m}^{(k+1)} = -u_{l-1,m}^{(k)} - u_{l,m-1}^{(k)} - u_{l+1,m}^{(k)} - u_{l,m+1}^{(k)} + (\Delta x)^2 f_{l,m}.$$

So why does the rate of convergence change so dramatically? In our analysis we consider the Jacobi method. The iteration matrix is given by

$$H = \frac{1}{4} \begin{pmatrix} B & I & & & \\ I & B & I & & \\ & \ddots & \ddots & \ddots & \\ & & I & B & I \\ & & & I & B \end{pmatrix} \quad \text{where } B = \begin{pmatrix} 0 & 1 & & & \\ 1 & 0 & 1 & & \\ & \ddots & \ddots & \ddots & \\ & & 1 & 0 & 1 \\ & & & 1 & 0 \end{pmatrix}.$$

We know from our previous results that the eigenvalues of B are $\lambda_i^B = 2\cos\frac{i\pi}{M+1}$, $i = 1, \ldots, M$. Changing the ordering from columns to rows, we obtain the matrix

$$\frac{1}{4} \begin{pmatrix} \Lambda_1 & & & \\ & \Lambda_2 & & \\ & & \ddots & \\ & & & \Lambda_M \end{pmatrix} \quad \text{where } \Lambda_i = \begin{pmatrix} \lambda_i^B & 1 & & & \\ 1 & \lambda_i^B & 1 & & \\ & \ddots & \ddots & \ddots & \\ & & 1 & \lambda_i^B & 1 \\ & & & 1 & \lambda_i^B \end{pmatrix}.$$

The eigenvalues of Λ_i are $\lambda_i^B + 2\cos\frac{j\pi}{M+1}$, $j = 1, \ldots, M$. We deduce that the eigenvalues of the system are

$$\lambda_{i,j} = \frac{1}{4}\left(2\cos\frac{i\pi}{M+1} + 2\cos\frac{j\pi}{M+1}\right) = 1 - \left(\sin^2\frac{i\pi}{2(M+1)} + \sin^2\frac{j\pi}{2(M+1)}\right).$$

Hence all the eigenvalues are smaller than 1 in modulus, since i and j range from 1 to M, guaranteeing convergence; however, the spectral radius is close to 1, being $1 - 2\sin^2\frac{\pi}{2(M+1)} \approx 1 - \frac{\pi^2}{2M^2}$. The larger M, the closer the spectral radius is to 1.

Let $\mathbf{e}^{(k)}$ be the error in the kth iteration and let $\mathbf{v}_{i,j}$ be the orthonormal eigenvectors. We can expand the error with respect to this basis.

$$\mathbf{e}^{(k)} = \sum_{i,j=1}^{M} e_{i,j}^{(k)} \mathbf{v}_{i,j}.$$

Iterating, we have

$$\mathbf{e}^{(k)} = H^k \mathbf{e}^{(0)} \Rightarrow |e_{i,j}^{(k)}| = |\lambda_{i,j}|^k |e_{i,j}^{(0)}|.$$

Thus the components of the error (with respect to the basis of eigenvectors) decay at a different rate for different values of i, j, which are the frequencies. We separate those into low frequencies (LF) where both i and j lie in $[1, \frac{M+1}{2})$, which results in the angles lying between zero and $\pi/4$, high frequencies (HF) where both i and j lie in $[\frac{M+1}{2}, M]$, which results in the angles lying between $pi/4$ and $\pi/2$, and mixed frequencies (MF) where one of i and j lies in $[1, \frac{M+1}{2})$ and the other lies in $[\frac{M+1}{2}, M]$. Let us determine the least factor by which the amplitudes of the mixed frequencies are damped. Either i or j is at least $\frac{M+1}{2}$ while the other is at most $\frac{M+1}{2}$ and thus $\sin^2 \frac{i\pi}{2(M+1)} \in [0, \frac{1}{2}]$ while $\sin^2 \frac{j\pi}{2(M+1)} \in [\frac{1}{2}, 1]$. It follows that

$$1 - (\sin^2 \frac{i\pi}{2(M+1)} + \sin^2 \frac{j\pi}{2(M+1)}) \in [-\frac{1}{2}, \frac{1}{2}]$$

for i, j in the mixed frequency range. Hence the amplitudes are damped by at least a factor of $\frac{1}{2}$, which corresponds to the observations. This explains the observations, but how can it be used to speed up convergence?

Firstly, let's look at the high frequencies. Good damping of those is achieved by using the Jacobi method with *relaxation* (also known as *damped Jacobi*) which gives a damping factor of $\frac{3}{5}$. The proof of this is left to the reader in the following exercise.

Exercise 8.8. *Find a formula for the eigenvalues of the iteration matrix of the Jacobi with relaxation and an optimal value for the relaxation parameter for the MF and HF components combined.*

From the analysis and the results from the exercise, we can deduce that the damped Jacobi method converges fast for HF and MF. This is also true for the damped Gauss–Seidel method. For the low frequencies we note that these are the high frequencies if we consider a grid with spacing $2\Delta x$ instead of Δx.

To examine this further we restrict ourselves to one space dimension. The matrix arising from approximating the second derivative is given by

$$\begin{pmatrix} -2 & 1 & & & & \\ 1 & -2 & 1 & & & \\ & & \ddots & \ddots & \ddots & \\ & & & 1 & -2 & 1 \\ & & & & 1 & -2 \end{pmatrix}.$$

For the Jacobi method, the corresponding iteration matrix is

$$\frac{1}{2}\begin{pmatrix} 0 & 1 & & & & \\ 1 & 0 & 1 & & & \\ & \ddots & \ddots & \ddots & & \\ & & 1 & 0 & 1 \\ & & & 1 & 0 \end{pmatrix}. \tag{8.18}$$

Assume that both matrices have dimension $(n-1) \times (n-1)$ and that n is even. For $k = 1, \ldots, n-1$, the eigenvalues of the iteration matrix are

$$\lambda_k = \frac{1}{2}(0 + 2\cos\frac{k\pi}{n}) = \cos\frac{k\pi}{n},$$

and the corresponding eigenvector $\mathbf{v}_k$ has the components

$$(\mathbf{v}_k)_j = \sin\frac{kj\pi}{n}.$$

$|\lambda_k|$ is largest for $k = 1$ and $k = n-1$, while for $k = n/2$ we have $\lambda_k = 0$, since $\cos\pi/2 = 0$. We deduce that we have a fast reduction of error for $k \in (n/4, 3n/4)$, while for $k \in [1, n/4] \cup [3n/4, n-1]$ the error reduces slowly. For $k \in [3n/4, n-1]$ a faster reduction in error is usually obtained via relaxation. To consider $k \in [1, n/4]$, take every second component of $\mathbf{v}_k$ forming a new vector $\hat{\mathbf{v}}_k$ with components

$$(\hat{\mathbf{v}}_k)_j = \sin\frac{k2j\pi}{n} = \sin\frac{kj\pi}{n/2}$$

for $j = 1, \ldots, n/2 - 1$. This is now the eigenvector of a matrix with the same form as in (8.18) but with dimension $(n/2 - 1) \times (n/2 - 1)$. The corresponding eigenvalue is

$$\lambda_k = \cos\frac{k\pi}{n/2}.$$

For $k = n/4$ where we had a slow reduction in error on the fine grid, we now have

$$\lambda_k = \cos\frac{n/4\pi}{n/2} = \cos\frac{\pi}{2} = 0,$$

the fastest reduction possible.

The idea of the multigrid method is to cover the square domain by a range of nested grids, of increasing coarseness, say,

$$\Omega_{\Delta x} \subset \Omega_{2\Delta x} \subset \cdots,$$

so that on each grid we remove the contribution of high frequencies relative to this grid. A typical *multigrid sweep* starts at the finest grid, travels to the coarsest (where the number of variables is small and we can afford to solve the equations directly), and back to the finest. Whenever we coarsen the grid, we compute the residual

$$\mathbf{r}_{\Delta x} = \mathbf{b}_{\Delta x} - A_{\Delta x}\mathbf{x}_{\Delta x},$$

where Δx is the size of the grid we are coarsening and $\mathbf{x}_{\Delta x}$ is the current solution on this grid. The values of the residual are restricted to the coarser grid by combining nine fine values by the restriction operator R

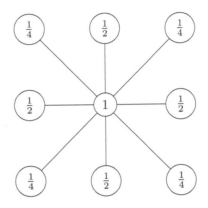

Thus the new value on the coarse grid is an average of the fine value at this point and its eight neighbouring fine values. Then we solve for the residual, i.e., we iterate to solve the equations

$$A_{2\Delta x}\mathbf{x}_{2\Delta x} = \mathbf{r}_{2\Delta x} = R\mathbf{r}_{\Delta x}. \tag{8.19}$$

(Note that the residual becomes the new $\mathbf{b}$.)

On the way back, refinement entails a prolongation P by linear interpolation, which is the exact opposite of restriction. The values of the coarse grid are distributed to the fine grid such that

- if the points coincide, the value is the same,

- if the point has two coarse neighbours at each side, the value is the average of those two,

- if the point has four coarse neighbours (top left, top right, bottom left, bottom right), the value is the average of those four.

We then correct $\mathbf{x}_{\Delta x}$ by $P\mathbf{x}_{2\Delta x}$.

Usually only a moderate number of iterations (3 to 5) is employed in each restriction to solve (8.19). At each prolongation one or two iterations are necessary to remove high frequencies which have been reintroduced by the prolongation. We check for convergence at the end of each sweep. We repeat the sweeps until convergence has occurred.

Before the first multigrid sweep, however, we need to obtain good starting values for the finest grid. This is done by starting from the coarsest grid solving the system of equations there and prolonging the solution to the finest grid in a zig-zag fashion. That means we do not go directly to the finest grid, but return after each finer grid to the coarsest grid to obtain better initial values for the solution.

Exercise 8.9. *The function $u(x,y) = 18x(1-x)y(1-y), 0 \leq x, y \leq 1$, is the solution of the Poisson equation $u_{xx} + u_{yy} = 36(x^2 + y^2 - x - y) = f(x,y)$, subject to zero boundary conditions. Let $\Delta x = 1/6$ and seek the solution of the five-point method*

$$u_{m-1,n} + u_{m+1,n} + u_{m,n-1} + u_{m,n+1} - 4u_{m,n} = (\Delta x)^2 f(mh, nh), \quad 1 \leq m, n \leq 5,$$

where $u_{m,n}$ is zero if one of m, n is 0 or 6. Let the multigrid method be applied, using only this fine grid and a coarse grid of mesh size $1/3$, and let every $u_{m,n}$ be zero initially. Calculate the 25 residuals of the starting vector on the fine grid. Then, following the restriction procedure, find the residuals for the initial calculation on the coarse grid. Solve the equations on the coarse grid exactly. The resultant estimates of u at the four interior points of the coarse grid all have the value $5/6$. By applying the prolongation operator to these estimates, find the 25 starting values of $u_{m,n}$ for the subsequent iterations of Jacobi on the fine grid. Further, show that if one Jacobi iteration is performed, then $u_{3,3} = 23/24$ occurs, which is the estimate of $u(1/2, 1/2) = 9/8$.

8.4   Parabolic PDEs in Two Dimensions

We are now combining the analysis of elliptic partial differential equations from the previous chapter with the analysis of parabolic partial differential equations. Let's look at the heat equation on the unit square

$$\frac{\partial u}{\partial t} = \nabla^2 u, \qquad 0 \leq x, y \leq 1, t \geq 0,$$

where $u = u(x, y, t)$. We are given initial condition at $t = 0$ and zero boundary conditions on $\partial \Omega$, where $\Omega = [0, 1]^2 \times [0, \infty)$. We generalize the method of lines which was introduced earlier to derive an algorithm. To this purpose we lay a square grid over the unit square with mesh size Δx. Let $u_{l,m}(t) \approx u(l\Delta x, m\Delta x, t)$ and let $u_{l,m}^n \approx u_{l,m}(n\Delta t)$. The five-point method approximating the right-hand side of the PDE results in

$$u_{l,m}' = \frac{1}{(\Delta x)^2}(u_{l-1,m} + u_{l+1,m} + u_{l,m-1} + u_{l,m+1} - 4u_{l,m}).$$

Using the previous analysis, in matrix form this is

$$\mathbf{u}' = \frac{1}{(\Delta x)^2} A_* \mathbf{u}, \tag{8.20}$$

where A_* is the block TST matrix

$$A_* = \begin{pmatrix} B & I & & & \\ I & B & I & & \\ & \ddots & \ddots & \ddots & \\ & & I & B & I \\ & & & I & B \end{pmatrix} \quad \text{and } B = \begin{pmatrix} -4 & 1 & & & \\ 1 & -4 & 1 & & \\ & \ddots & \ddots & \ddots & \\ & & 1 & -4 & 1 \\ & & & 1 & -4 \end{pmatrix}.$$

Using the forward Euler method to discretize in time yields

$$u_{l,m}^{n+1} = u_{l,m}^n + \mu(u_{l-1,m}^n + u_{l+1,m}^n + u_{l,m-1}^n + u_{l,m+1}^n - 4u_{l,m}^n), \qquad (8.21)$$

where $\mu = \frac{\Delta t}{(\Delta x)^2}$. Again, in matrix form this is

$$\mathbf{u}^{n+1} = A\mathbf{u}^n, \qquad A = I + \mu A_*.$$

The local error is $O((\Delta t)^2 + \Delta t(\Delta x)^2) = O((\Delta x)^4)$, when μ is held constant, since the forward Euler method itself carries an error of $O((\Delta t)^2)$ and the discretization in space carries an error of $O((\Delta x)^2)$, which gets multiplied by Δt when the forward Euler method is applied.

As we have seen, the eigenvalues of A_* are given by

$$\begin{aligned} \lambda_{i,j}(A_*) &= -4 + 2\cos\frac{i\pi}{M+1} + 2\cos\frac{j\pi}{M+1} \\ &= -4\left(\sin^2\frac{i\pi}{2(M+1)} + \sin^2\frac{j\pi}{2(M+1)}\right) \end{aligned}$$

and thus the eigenvalues of A are

$$\lambda_{i,j}(A) = 1 - 4\mu\left(\sin^2\frac{i\pi}{2(M+1)} + \sin^2\frac{j\pi}{2(M+1)}\right).$$

To achieve stability, the spectral radius of A has to be smaller or equal to one. This is satisfied if and only if $\mu \leq \frac{1}{4}$. Thus, compared to the one-dimensional case where $\mu \leq \frac{1}{2}$, we are further restricted in our choice of $\Delta t \leq \frac{1}{4}(\Delta x)^2$. Generally, using the same discretization in space for all space dimensions and employing the forward Euler method, we have $\mu \leq \frac{1}{2d}$ where d is the number of space dimensions.

The Fourier analysis of stability generalizes equally to two dimensions when the space dimensions are an infinite plane. Of course, the range and indices have to be extended to two dimensions. For a given sequence of numbers $\mathbf{u} = (u_{l,m})_{l,m \in \mathbb{Z}}$, the 2-D Fourier transform of this sequence is

$$\hat{u}(\theta, \psi) = \sum_{l,m \in \mathbb{Z}} u_{l,m} e^{-i(l\theta + m\psi)}, \qquad -\pi \leq \theta, \psi \leq \pi.$$

All the previous results generalize. In particular, the Fourier transform is an isometry.

$$\left(\sum_{l,m\in\mathbb{Z}}|u_{l,m}|^2\right)^{1/2} =: \|\mathbf{u}\| = \|\hat{u}\| := \left(\frac{1}{4\pi^2}\int_{-\pi}^{\pi}\int_{-\pi}^{\pi}|\hat{u}(\theta,\psi)|^2 d\theta d\phi\right)^{1/2}.$$

Assume our numerical method takes the form

$$\sum_{k,j=-r}^{s}\alpha_{k,j}u_{l+k,m+j}^{n+1} = \sum_{k,j=-r}^{s}\beta_{k,j}u_{l+k,m+j}^{n}.$$

We can, analogous to (8.11), define the *amplification factor*

$$H(\theta,\psi) = \frac{\sum_{k,j=-r}^{s}\beta_{k,j}e^{i(k\theta+j\psi)}}{\sum_{k,j=-r}^{s}\alpha_{k,j}e^{i(k\theta+j\psi)}},$$

and the method is stable if and only if $|H(\theta,\psi)| \le 1$ for all $\theta,\psi\in[-\pi,\pi]$.

For the method given in (8.21) we have

$$H(\theta,\psi) = 1 + \mu(e^{-i\theta} + e^{i\theta} + e^{-i\psi} + e^{i\psi} - 4) = 1 - 4\mu\left(\sin^2\frac{\theta}{2} + \sin^2\frac{\psi}{2}\right),$$

and we again deduce stability if and only if $\mu \le \frac{1}{4}$.

If we apply the trapezoidal rule instead of the forward Euler method to our semi-discretization (8.20), we obtain the *two-dimensional Crank–Nicolson method*

$$u_{l,m}^{n+1} - \tfrac{1}{2}\mu(u_{l-1,m}^{n+1} + u_{l+1,m}^{n+1} + u_{l,m-1}^{n+1} + u_{l,m+1}^{n+1} - 4u_{l,m}^{n+1}) =$$
$$u_{l,m}^{n} + \tfrac{1}{2}\mu(u_{l-1,m}^{n} + u_{l+1,m}^{n} + u_{l,m-1}^{n} + u_{l,m+1}^{n} - 4u_{l,m}^{n}),$$

or in matrix form

$$(I - \frac{1}{2}\mu A_*)\mathbf{u}^{n+1} = (I + \frac{1}{2}\mu A_*)\mathbf{u}^n. \tag{8.22}$$

The local error is $O((\Delta t)^3 + \Delta t(\Delta x)^2)$, since the trapezoidal rule carries an error of $O((\Delta t)^3)$. Similarly to the one-dimensional case, the method is stable if and only if the moduli of the eigenvalues of $A = (I - \frac{1}{2}\mu A_*)^{-1}(I + \frac{1}{2}\mu A_*)$ are less than or equal to 1. Both matrices $I - \frac{1}{2}\mu A_*$ and $I + \frac{1}{2}\mu A_*$ are block TST matrices and share the same set of eigenvectors with A_*. Thus the eigenvalue of A corresponding to a particular eigenvector is easily calculated as the inverse of the eigenvalue of $I - \frac{1}{2}\mu A_*$ times the eigenvalue of $I + \frac{1}{2}\mu A_*$,

$$|\lambda_{i,j}(A)| = \left|\frac{1 + \frac{1}{2}\mu\lambda_{i,j}(A_*)}{1 - \frac{1}{2}\mu\lambda_{i,j}(A_*)}\right|.$$

This is always less than one, because the numerator is always less than the denominator, since all the eigenvalues of A_* lie in $(-8,0)$.

Exercise 8.10. *Deduce the amplification factor for the two-dimensional Crank–Nicolson method.*

8.4.1 Splitting

Solving parabolic equations with explicit methods typically leads to restrictions of the form $\Delta t \sim \Delta x^2$, and this is generally not acceptable. Instead, implicit methods are used, for example, Crank–Nicolson. However, this means that in each time step a system of linear equations needs to be solved. This can become very costly for several space dimensions. The matrix $I - \frac{1}{2}\mu A_*$ is in structure similar to A_*, so we can apply the Hockney algorithm.

However, since the two-dimensional Crank–Nicolson method already carries a local truncation error of $O((\Delta t)^3 + \Delta t(\Delta x)^2) = O((\Delta t)^2)$ (because of $\Delta t = (\mu \Delta x)^2$), the system does not need to be solved exactly. It is enough to solve it within this error. Using the following operator notation (central difference operator applied twice),

$$\delta_x^2 u_{l,m} = u_{l-1,m} - 2u_{l,m} + u_{l+1,m}, \qquad \delta_y^2 u_{l,m} = u_{l,m-1} - 2u_{l,m} + u_{l,m+1},$$

the Crank–Nicolson method becomes

$$[I - \frac{1}{2}\mu(\delta_x^2 + \delta_y^2)]u_{l,m}^{n+1} = [I + \frac{1}{2}\mu(\delta_x^2 + \delta_y^2)]u_{l,m}^n.$$

We have, however, the same magnitude of local error if this formula is replaced by

$$[I - \frac{1}{2}\mu\delta_x^2][I - \frac{1}{2}\mu\delta_y^2]u_{l,m}^{n+1} = [I + \frac{1}{2}\mu\delta_x^2][I + \frac{1}{2}\mu\delta_y^2]u_{l,m}^n,$$

which is called the *split version of Crank–Nicolson*. Note that this modification decouples the x and y direction, since operator multiplication means that they can be applied after each other. Therefore this technique is called dimensional splitting. We will see below what practical impact that has, but we first examine the error introduced by the modification.

Multiplying the split version of Crank–Nicolson out,

$$[I - \frac{1}{2}\mu(\delta_x^2 + \delta_y^2) + \frac{1}{4}\mu^2\delta_x^2\delta_y^2]u_{l,m}^{n+1} = [I + \frac{1}{2}\mu(\delta_x^2 + \delta_y^2) + \frac{1}{4}\mu^2\delta_x^2\delta_y^2]u_{l,m}^n,$$

we see that on each side a term of the form $\frac{1}{4}\mu^2\delta_x^2\delta_y^2$ is introduced. The extra error introduced into the modified scheme is therefore

$$e = \frac{1}{4}\mu^2\delta_x^2\delta_y^2(u_{l,m}^{n+1} - u_{l,m}^n).$$

Now the term in the bracket is an approximation to the first derivative in the time direction times Δt. In fact, the difference between the two schemes is

$$e = \frac{1}{4}\mu^2\delta_x^2\delta_y^2\left(\Delta t\frac{\partial}{\partial t}u_{l,m}(t) + O((\Delta t)^2)\right) = \frac{1}{4}\mu^2\Delta t\delta_x^2\delta_y^2\frac{\partial}{\partial t}u_{l,m}(t) + O((\Delta t)^2).$$

We know that $\delta_x^2/(\Delta x)^2$ and $\delta_y^2/(\Delta x)^2$ are approximations to the second partial derivatives in the space directions carrying an error of $O(\Delta x)^2$. Thus we can write, using $\mu = \Delta t/(\Delta x)^2$,

$$
\begin{aligned}
e &= \frac{(\Delta t)^2}{4} \frac{1}{(\Delta x)^2} \delta_x^2 \frac{1}{(\Delta x)^2} \delta_y^2 \Delta t \frac{\partial}{\partial t} u_{l,m}(t) + O((\Delta t)^2) \\
&= \frac{(\Delta t)^3}{4} \frac{\partial^2}{\partial x^2} \frac{\partial^2}{\partial y^2} \frac{\partial}{\partial t} u(x,y,t) + O((\Delta t)^3 (\Delta x)^2) + O((\Delta t)^2) = O((\Delta t)^2).
\end{aligned}
$$

In matrix form, the new method is equivalent to splitting the matrix A_* into the sum of two matrices, A_x and A_y, where

$$
A_x = \begin{pmatrix} -2I & I & & & \\ I & \ddots & \ddots & & \\ & \ddots & \ddots & I & \\ & & I & -2I \end{pmatrix}, \qquad A_y = \begin{pmatrix} B & & & \\ & B & & \\ & & \ddots & \\ & & & B \end{pmatrix},
$$

where

$$
B = \begin{pmatrix} -2 & 1 & & \\ 1 & \ddots & \ddots & \\ & \ddots & \ddots & 1 \\ & & 1 & -2 \end{pmatrix}.
$$

We then solve the *uncoupled system*

$$
(I - \frac{1}{2}\mu A_x)(I - \frac{1}{2}\mu A_y)\mathbf{u}^{n+1} = (I + \frac{1}{2}\mu A_x)(I + \frac{1}{2}\mu A_y)\mathbf{u}^n
$$

in two steps, first solving

$$
(I - \frac{1}{2}\mu A_x)\mathbf{u}^{n+1/2} = (I + \frac{1}{2}\mu A_x)(I + \frac{1}{2}\mu A_y)\mathbf{u}^n
$$

then solving

$$
(I - \frac{1}{2}\mu A_y)\mathbf{u}^{n+1} = \mathbf{u}^{n+1/2}.
$$

The matrix $I - \frac{1}{2}\mu A_y$ is block diagonal, where each block is $I - \frac{1}{2}\mu B$. Thus solving the above system is equivalent to solving the same tridiagonal system $(I - \frac{1}{2}\mu B)\mathbf{u}_i^{n+1} = \mathbf{u}_i^{n+1/2}$ for different right-hand sides M times, for which the same method can be reused. Here the vector $\mathbf{u}$ has been divided into vectors $\mathbf{u}_i$ of size M for $i = 1,\ldots,M$. The matrix $I - \frac{1}{2}\mu A_x$ is of the same form after a reordering of the grid which changes the right hand sides. Thus we first have to calculate $(I + \frac{1}{2}\mu A_x)(I + \frac{1}{2}\mu A_y)\mathbf{u}^n$, then reorder, solve the first system, then reorder and solve the second system.

Speaking more generally, suppose the method of lines results after discretization in space in the linear system of ODEs given by

$$
\mathbf{u}' = A\mathbf{u}, \qquad \mathbf{u}(0) = \mathbf{u}_0,
$$

where $\mathbf{u}_0$ is derived from the initial condition.

We formally define the *matrix exponential* by Taylor series

$$e^B = \sum_{k=0}^{\infty} \frac{1}{k!} B^k.$$

Using this definition to differentiate e^{tA}, we have

$$\frac{de^{tA}}{dt} = \sum_{k=0}^{\infty} \frac{1}{k!} \frac{d((tA)^k)}{dt} = A \sum_{k=0}^{\infty} \frac{1}{k!} (tA)^k = Ae^{tA},$$

and the solution to the system of ODEs is $\mathbf{u}(t) = e^{(tA)}\mathbf{u}_0$, or at time t_{n+1},

$$\mathbf{u}(t_{n+1}) = e^{\Delta t A}\mathbf{u}(t_n).$$

Many methods for ODEs are actually approximations to the matrix exponential. For example, applying the forward Euler method to the ODE results in

$$\mathbf{u}^{n+1} = (I + \Delta t A)\mathbf{u}^n.$$

The corresponding approximation to the exponential is $1 + z = e^z + O(z^2)$. On the other hand, if the trapezoidal rule is used instead, we have

$$\mathbf{u}^{n+1} = \left[(I - \frac{1}{2}\Delta t A)^{-1}(I + \frac{1}{2}\Delta t A) \right]$$

with the corresponding approximation to the exponential

$$\frac{1 + \frac{1}{2}z}{1 - \frac{1}{2}z} = e^z + O(z^3).$$

Now we apply dimensional splitting by $A = A_x + A_y$, where A_x and A_y contain the contributions in the x and y direction. If A_x and A_y commute, then $e^{(tA)} = e^{t(A_x+A_y)} = e^{tA_x}e^{tA_y}$ and we could solve the system of ODEs by approximating each exponential independently. Using the approximation to the exponential given by the forward Euler method gives

$$\mathbf{u}^{n+1} = (I + \Delta t A_x)(I + \Delta t A_y)\mathbf{u}^n,$$

while the trapezoidal approximation yields

$$\mathbf{u}^{n+1} = \left[(I - \frac{1}{2}\Delta t A_x)^{-1}(I + \frac{1}{2}\Delta t A_x)(I - \frac{1}{2}\Delta t A_y)^{-1}(I + \frac{1}{2}\Delta t A_y) \right] \mathbf{u}^n.$$

The advantage is that up to reordering all matrices involved are tridiagonal (if only neighbouring points are used) and the system of equations can be solved cheaply.

However, the assumption $e^{t(A_x+A_y)} = e^{tA_x}e^{tA_y}$ is generally false. Taking the first few terms of the definition of the matrix exponential, we have

$$e^{t(A_x+A_y)} = I + t(A_x + A_y) + \frac{1}{2}t^2(A_x^2 + A_xA_y + A_yA_x + A_y^2) + O(t^3)$$

while

$$\begin{aligned} e^{tA_x}e^{tA_y} &= [I + tA_x + \tfrac{1}{2}t^2A_x^2 + O(t^3)] \times [I + tA_y + \tfrac{1}{2}t^2A_y^2 + O(t^3)] \\ &= I + t(A_x + A_y) + \tfrac{1}{2}t^2(A_x^2 + 2A_xA_y + A_y^2) + O(t^3). \end{aligned}$$

Hence the difference is $\frac{1}{2}t^2(A_xA_y - A_yA_x) + O(t^3)$, which does not vanish if the matrices A_x and A_y do not commute. Still, splitting is, when suitably implemented, a powerful technique to drastically reduce computational expense. Common splitting techniques are

Beam and Warming's splitting
$$e^{t(A_x+A_y)} = e^{tA_x}e^{tA_y} + O(t^2),$$

Strang's splitting
$$e^{t(A_x+A_y)} = e^{\frac{1}{2}tA_x}e^{tA_y}e^{\frac{1}{2}tA_x} + O(t^3),$$

Parallel splitting
$$e^{t(A_x+A_y)} = \frac{1}{2}e^{tA_x}e^{tA_y} + \frac{1}{2}e^{tA_y}e^{tA_x} + O(t^3).$$

Let $r = p/q$ be a rational function where p and q are polynomials, which approximates the exponential function. As long as the order of the error in this approximation is the same as the order of the error in the splitting, approximating $e^{\Delta tA_x}$ and $e^{\Delta tA_y}$ by $r(\Delta tA_x)$ and $r(\Delta tA_y)$ results in the same local error. Note that if $r(B)$ for a matrix B is evaluated, we calculate $q(B)^{-1}p(B)$ where $q(B)^{-1}$ is the inverse of the matrix formed by computing $q(B)$. The ordering of the multiplication is important, since matrices generally do not commute.

For example, if $r(z) = e^z + O(z^2)$, and employing Beam and Warming's splitting, we can construct the time stepping method

$$\mathbf{u}^{n+1} = r(\Delta tA_x)r(\Delta tA_y)\mathbf{u}^n$$

which produces an error of $O((\Delta t)^2)$. The choice $r(z) = (1 + \frac{1}{2}z)/(1 - \frac{1}{2}z)$ produces the split Crank–Nicolson scheme. On the other hand, as long as $r(z) = e^z + O(z^3)$, using Strang's splitting to obtain

$$\mathbf{u}^{n+1} = r(\frac{1}{2}\Delta tA_x)r(\Delta tA_y)r(\frac{1}{2}\Delta tA_x)\mathbf{u}^n$$

carries an error of $O((\Delta t)^3)$.

Stability depends on the eigenvalues of A_x and A_y as well as on the the properties of r. In the case of the split Crank–Nicolson we examined above, A_x and A_y are (up to a re-ordering) TST matrices with eigenvalues $(\Delta x)^{-2}(-2 +$

$\cos \frac{k\pi}{M+1}$) for $k = 1, \ldots, M$, where each eigenvalue is M-fold. It is easy to see that these eigenvalues are nonpositive. So as long as r specifies an A-stable method, that is, $|r(z)| < 1$ for all $z \in \mathbb{C}^-$, we have stability.

Exercise 8.11. *Let $F(t) = e^{tA}e^{tB}$ be the first order Beam–Warming splitting of $e^{t(A+B)}$. Generally the splitting error is of the form $t^2 C$ for some matrix C. If C has large eigenvalues the splitting error can be large even for small t. Show that*

$$F(t) = e^{t(A+B)} + \int_0^t e^{(t-\tau)(A+B)} \left(e^{\tau A} B - B e^{\tau A} \right) e^{\tau B} d\tau.$$

(Hint: Find explicitly $G(t) = F'(t) - (A+B)F(t)$ and use variation of constants to find the solution of the linear matrix ODE $F' = (A+B)F + G$, $F(0) = I$.)

Suppose that a matrix norm $\| \cdot \|$ is given and that there exist real constants c_A, c_B and c_{A+B} such that

$$\|e^{tA}\| \le e^{c_A t}, \qquad \|e^{tB}\| \le e^{c_B t}, \qquad \|e^{t(A+B)}\| \le e^{c_{A+B} t}.$$

Prove that

$$\|F(t) - e^{t(A+B)}\| \le 2\|B\| \frac{e^{(c_A + c_B)t} - e^{c_{A+B} t}}{c_A + c_B - c_{A+B}}.$$

Hence, for $c_A, c_B \le 0$, the splitting error remains relatively small even for large t. ($e^{c_{A+B} t}$ is an intrinsic error.)

So far we have made it easy for ourselves by assuming zero boundary conditions. We now consider the *splitting of inhomogeneous systems* where the boundary conditions are also allowed to vary over time. In general, the linear ODE system is of the form

$$\mathbf{u}'(t) = (A_x + A_y)\mathbf{u}(t) + \mathbf{b}(t), \qquad \mathbf{u}(0) = \mathbf{u}_0, \qquad (8.23)$$

where $\mathbf{b}$ originates in the boundary conditions and also possibly a forcing term in the original PDE. We have seen that the solution to the homogeneous system (where $\mathbf{b}(t) = 0$) is

$$\mathbf{u}(t) = e^{t(A_x + A_y)}\mathbf{u}_0.$$

We assume that the solution of the inhomogeneous system is of the form

$$\mathbf{u}(t) = e^{t(A_x + A_y)}\mathbf{c}(t).$$

Inserting this into the inhomogeneous differential equation, we obtain

$$(A_x + A_y)e^{t(A_x + A_y)}\mathbf{c}(t) + e^{t(A_x + A_y)}\mathbf{c}'(t) = (A_x + A_y)e^{t(A_x + A_y)}\mathbf{c}(t) + \mathbf{b}(t)$$

and thus

$$\mathbf{c}'(t) = e^{-t(A_x + A_y)}\mathbf{b}(t) \quad \Rightarrow \quad \mathbf{c}(t) = \int_0^t e^{-\tau(A_x + A_y)}\mathbf{b}(\tau)d\tau + \mathbf{c}_0.$$

Using the initial condition $\mathbf{u}(0) = \mathbf{u}_0$, the exact solution of (8.23) is provided by

$$
\begin{aligned}
\mathbf{u}(t) &= e^{t(A_x + A_y)} \left[\mathbf{u}_0 + \int_0^t e^{-\tau(A_x + A_y)} \mathbf{b}(\tau) d\tau \right] \\
&= e^{t(A_x + A_y)} \mathbf{u}_0 + \int_0^t e^{(t-\tau)(A_x + A_y)} \mathbf{b}(\tau) d\tau, \qquad t \geq 0.
\end{aligned}
$$

This technique of deriving the solution to an inhomogeneous differential equation is called *variation of constants*.

In particular, for each time step $n = 0, 1, \ldots$

$$
\mathbf{u}((n+1)\Delta t) = e^{\Delta t(A_x + A_y)} \mathbf{u}(n\Delta t) + \int_{n\Delta t}^{(n+1)\Delta t} e^{((n+1)\Delta t - \tau)(A_x + A_y)} \mathbf{b}(\tau) d\tau.
$$

Often, we can evaluate the integral explicitly; for example, when $\mathbf{b}(t)$ is a linear combination of exponential and polynomial terms. If $\mathbf{b}$ is constant, then

$$
\mathbf{u}((n+1)\Delta t) = e^{\Delta t(A_x + A_y)} \mathbf{u}(n\Delta t) + (A_x + A_y)^{-1} \left(e^{\Delta t(A_x + A_y)} - I \right) \mathbf{b}.
$$

However, this observation does not get us any further, since, even if we split the exponential, an equivalent technique to split $(A_x + A_y)^{-1}$ does not exist. The solution is not to compute the integral explicitly but to use *quadrature rules* instead. One of those is the trapezium rule given by

$$
\int_0^h g(\tau) d\tau = \frac{1}{2} h[g(0) + g(h)] + O(h^3).
$$

This gives

$$
\begin{aligned}
\mathbf{u}((n+1)\Delta t) = \; & e^{\Delta t(A_x + A_y)} \mathbf{u}(n\Delta t) + \\
& \frac{1}{2} \Delta t \left[e^{\Delta t(A_x + A_y)} \mathbf{b}(n\Delta t) + \mathbf{b}((n+1)\Delta t) \right] + O((\Delta t)^3).
\end{aligned}
$$

We then gather the exponentials together and replace them by their splittings and use an approximation r to the exponential. For example, using Strang's splitting results in

$$
\mathbf{u}^{n+1} = r(\frac{1}{2} \Delta t A_x) r(\Delta t A_y) r(\frac{1}{2} \Delta t A_x) \left[\mathbf{u}^n + \frac{1}{2} \Delta t \mathbf{b}^n \right] + \frac{1}{2} \Delta t \mathbf{b}^{n+1}.
$$

Again, we have tridiagonal systems which are inexpensive to solve.

As an example, lets look at the general diffusion equation in two space dimensions

$$
\frac{\partial u}{\partial t} = \nabla^T (a \nabla u) + f = \frac{\partial}{\partial x}(a u_x) + \frac{\partial}{\partial y}(a u_y) + f, \qquad -1 \leq x, y \leq 1,
$$

where $a(x, y) > 0$ and $f(x, y)$ are given, as are initial conditions on the unit square and Dirichlet boundary conditions on $\partial[0, 1]^2 \times [0, \infty)$. Every space derivative is replaced by a central difference at the midpoint, for example

$$
\begin{aligned}
\frac{\partial}{\partial x} u(x, y) &= \frac{1}{\Delta x} \delta_x u(x, y) + O((\Delta x)^2) \\
&= \frac{1}{\Delta x} \left(u(x + \frac{1}{2}\Delta x, y) - u(x - \frac{1}{2}\Delta x, y) \right) + O((\Delta x)^2).
\end{aligned}
$$

This yields the ODE system

$$
\begin{aligned}
u'_{l,m} = \frac{1}{(\Delta x)^2} \Big[& a_{l-\frac{1}{2},m} u_{l-1,m} + a_{l+\frac{1}{2}} u_{l+1,m} + a_{l,m-\frac{1}{2}} u_{l,m-1} \\
& + a_{l,m+\frac{1}{2}} u_{l,m+1} + (a_{l-\frac{1}{2},m} + a_{l+\frac{1}{2},m} + a_{l,m-\frac{1}{2}} + a_{l,m+\frac{1}{2}}) u_{l,m} \Big] + f_{l,m}.
\end{aligned}
$$

The resulting matrix A is split in such a way that A_x consists of all the $a_{l\pm\frac{1}{2},m}$ terms, while A_y includes the remaining $a_{l,m\pm\frac{1}{2}}$ terms. Again, if the grid is ordered by columns, A_y is tridiagonal; if it is ordered by rows, A_x is tridiagonal. The vector $\mathbf{b}$ consists of $f_{l,m}$ and incorporates the boundary conditions.

What we have looked at so far is known as *dimensional splitting*. In addition there also exists *operational splitting*, to resolve non-linearities. As an example, consider the *reaction-diffusion equation* in one space dimension

$$
\frac{\partial u}{\partial t} = \frac{\partial^2 u}{\partial x^2} + \alpha u (1 - u).
$$

For simplicity we assume zero boundary conditions at $x = 0$ and $x = 1$. Discretizing in space, we arrive at

$$
u'_m = \frac{1}{(\Delta x)^2} (u_{m-1} - 2u_m + u_{m+1}) + \alpha u_m (1 - u_m).
$$

We separate the diffusion from the reaction part by keeping one part constant and advancing the other by half a time step. We add the superscript n to the part which is kept constant. In particular we advance by $\frac{1}{2}\Delta t$ solving

$$
u'_m = \frac{1}{(\Delta x)^2} (u_{m-1} - 2u_m + u_{m+1}) + \alpha u_m^n (1 - u_m^n),
$$

i.e., keeping the reaction part constant. This can be done, for example, by Crank–Nicolson. Then we advance another half time step solving

$$
u'_m = \frac{1}{(\Delta x)^2} (u_{m-1}^{n+\frac{1}{2}} - 2u_m^{n+\frac{1}{2}} + u_{m+1}^{n+\frac{1}{2}}) + \alpha u_m (1 - u_m),
$$

this time keeping the diffusion part constant. The second ODE is a linear Riccati equation, i.e., the right-hand side is a quadratic in u_m which can be solved explicitly (see for example [21] D. Zwillinger, *Handbook of Differential Equations*).

8.5 Hyperbolic PDEs

Hyperbolic PDEs are qualitatively different from elliptic and parabolic PDEs. A perturbation of the initial or boundary data of an elliptic or parabolic PDE changes the solution at all points in the domain instantly. Solutions of hyperbolic PDEs are, however, wave-like. If the initial data of a hyperbolic PDE is disturbed, the effect is not felt at once at every point in the domain. Disturbances travel along the characteristics of the equation with finite propagation speed. Good numerical methods mimic this behaviour. Hyperbolic PDEs have been studied extensively, since they are part of many engineering and scientific areas. As an example see [14] R. J. LeVeque, *Finite Volume Methods for Hyperbolic Problems*. We have already studied how to construct algorithms based on finite differences and the order of error these carry, as well as stability. We consider these principles examining the advection and wave equation as examples.

8.5.1 Advection Equation

A useful example of hyperbolic PDEs is the *advection equation*

$$\frac{\partial}{\partial t}u(x,t) = -c\frac{\partial}{\partial x}u(x,t),$$

with initial condition $u(x,0) = \phi(x)$, where $\phi(x)$ has finite support, that is, it is zero outside a finite interval. By the chain rule an exact solution of the advection equation is given by

$$u(x,t) = \phi(x - ct).$$

As time passes, the initial condition retains its shape while shifting with velocity c to the right or left depending on the sign of c (to the right for positive c, to the left for negative c). This has been likened to a wind blowing from left to right or vice versa. For simplicity let $c = -1$, which gives $u_t(x,t) = u_x(x,t)$, and let the support of ϕ lie in $[0,1]$. We restrict ourselves to the interval $[0,1]$ by imposing the boundary condition $u(1,t) = \phi(t+1)$.

Let $\Delta x = \frac{1}{M+1}$. We start by semidiscretizing the right-hand side by the sum of the forward and backward difference

$$\frac{\partial}{\partial t}u_m(t) = \frac{1}{2\Delta x}[u_{m+1}(t) - u_{m-1}(t)] + O((\Delta x)^2). \tag{8.24}$$

Solving the resulting ODE $u'_m(t) = (2\Delta x)^{-1}[u_{m+1}(t) - u_{m-1}(t)]$ by forward Euler results in

$$u_m^{n+1} = u_m^n + \frac{1}{2}\mu(u_{m+1}^n - u_{m-1}^n), \qquad m = 1,\ldots,M, \qquad n \geq 0,$$

where $\mu = \frac{\Delta t}{\Delta x}$ is the Courant number. The overall local error is $O((\Delta t)^2 +$

$\Delta t (\Delta x)^2)$. In matrix form this is $\mathbf{u}^{n+1} = A\mathbf{u}^n$ where

$$A = \begin{pmatrix} 1 & \frac{1}{2}\mu & & \\ -\frac{1}{2}\mu & 1 & \ddots & \\ & \ddots & \ddots & \frac{1}{2}\mu \\ & & -\frac{1}{2}\mu & 1 \end{pmatrix}.$$

Now the matrix A is tridiagonal and Toeplitz. However, it is not symmetric, but skew-symmetric. Similar to TST matrices, the eigenvalues and eigenvectors of

$$\begin{pmatrix} \alpha & \beta & & \\ -\beta & \alpha & \ddots & \\ & \ddots & \ddots & \beta \\ & & -\beta & \alpha \end{pmatrix}$$

are given by $\lambda_k = \alpha + 2i\beta \cos \frac{k\pi}{M+1}$, with corresponding eigenvector $\mathbf{v}_k$, which has as j^{th} component $i^j \sin \frac{jk\pi}{M+1}$, $j = 1, \dots, M$. So for A, $\lambda_k = 1 + i\mu \cos \frac{k\pi}{M+1}$ and $|\lambda_k|^2 = 1 + \mu^2 \cos^2 \frac{k\pi}{M+1} > 1$. Hence we have instability for any μ.

It is, however, sufficient to have a local error of $O(\Delta x)$ when discretizing in space, since it is multiplied by Δt, which is then $O((\Delta x)^2)$ for a fixed μ. Thus if we discretized in space by the forward difference

$$\frac{\partial}{\partial t} u_m(t) = \frac{1}{\Delta x}[u_{m+1}(t) - u_m(t)] + O(\Delta x)$$

and solve the resulting ODE again by the forward Euler method, we arrive at

$$u_m^{n+1} = u_m^n + \mu(u_{m+1}^n - u_m^n), \qquad m = 1, \dots M, \qquad n \geq 0. \qquad (8.25)$$

This method is known as the *upwind method*. It takes its name because we are taking additional information from the point $m + 1$ which is against the wind, which is blowing from right to left since c is negative. It makes logical sense to take information from the direction the wind is blowing from. This implies that the method has to be adjusted for positive c to use u_{m-1}^n instead of u_{m+1}^n. It also explains the instability of the first scheme we constructed, since there information was taken from both sides of u_m in the form of u_{m-1} and u_{m+1}.

In matrix form, the upwind method is $\mathbf{u}^{n+1} = A\mathbf{u}^n$ where

$$A = \begin{pmatrix} 1-\mu & \mu & & \\ & 1-\mu & \ddots & \\ & & \ddots & \mu \\ & & & 1-\mu \end{pmatrix}.$$

Now the matrix A is not normal and thus its 2-norm is not equal to its spectral radius, but equal to the square root of the spectral radius of AA^T. Now

$$AA^T = \begin{pmatrix} (1-\mu)^2 + \mu^2 & \mu(1-\mu) & & & \\ \mu(1-\mu) & (1-\mu)^2 + \mu^2 & \ddots & & \\ & \ddots & \ddots & \ddots & \\ & & \ddots & (1-\mu)^2 + \mu^2 & \mu(1-\mu) \\ & & & \mu(1-\mu) & (1-\mu)^2 \end{pmatrix},$$

which is not TST, since the entry in the bottom right corner differs. The eigenvalues can be calculated solving a three term recurrence relation (see for example [18]). However, defining $\|\mathbf{u}^n\|_\infty = \max_m |u_m^n|$, it follows from (8.25) that

$$\|\mathbf{u}^{n+1}\|_\infty - \max_m |u_m^{n+1}| \leq \max_m \{|1-\mu||u_m^n| + \mu|u_{m+1}^n|\} \leq (|1-\mu| + \mu)\|\mathbf{u}^n\|_\infty.$$

As long as $\mu \in (0,1]$ we have $\|\mathbf{u}^{n+1}\|_\infty \leq \|\mathbf{u}^n\|_\infty \leq \cdots \leq \|\mathbf{u}^0\|_\infty$ and hence stability.

If we keep with our initial space discretization as in (8.24), but now solve the resulting ODE with the second order mid-point rule

$$\mathbf{y}_{n+1} = \mathbf{y}_{n-1} + 2\Delta t\mathbf{f}(t_n, \mathbf{y}_n),$$

the outcome is the *two-step leapfrog method*

$$u_m^{n+1} = \mu(u_{m+1}^n - u_{m-1}^n) + u_m^{n-1}.$$

The local truncation error is $O((\Delta t)^3 + \Delta t(\Delta x)^2) = O((\Delta x)^3)$, because $\Delta t = \mu\Delta x$. Since it is a two-step method another method has to be chosen to compute the first time step.

We analyze the stability by the Fourier technique. Since it is a two-step method, we have

$$\hat{u}^{n+1}(\theta) = \mu(e^{i\theta} - e^{-i\theta})\hat{u}^n(\theta) + \hat{u}^{n-1}(\theta) = 2i\mu\sin\theta\hat{u}^n(\theta) + \hat{u}^{n-1}(\theta).$$

Thus we are looking for solutions to the recurrence relation

$$\hat{u}^{n+1}(\theta) - 2i\mu\sin\theta\hat{u}^n(\theta) - \hat{u}^{n-1}(\theta) = 0.$$

Generally for a recurrence relation given by $ax_{n+1} + bx_n + cx_{n-1} = 0$, $n \geq 1$, we let $x_\pm$ be the roots of the characteristic equation $ax^2 + bx + c = 0$. For $x_- \neq x_+$ the general solution is $x_n = \alpha x_+^n + \beta x_-^n$, where α and β are constants derived from the initial values x_0 and x_1. For $x_- = x_+$, the solution is $x_n = (\alpha + \beta n)x_+^n$, where again α and β are constants derived from the initial values x_0 and x_1. In our case

$$\hat{u}_\pm(\theta) = \frac{1}{2}\left(2i\mu\sin\theta \pm \sqrt{4i^2\mu^2\sin^2\theta - 4*(-1)}\right) = i\mu\sin\theta \pm \sqrt{1 - \mu^2\sin^2\theta}.$$

We have stability if $|\hat{u}_{\pm}(\theta)| \leq 1$ for all $\theta \in [-\pi, \pi]$ and we do not have a double root on the unit circle. For $\mu > 1$ the square root is imaginary at $\theta = \pm\pi/2$ and then

$$|\hat{u}_+(\pi/2)| = \mu + \sqrt{\mu - 1} > 1 \text{ and } |\hat{u}_-(-\pi/2)| = |-\mu - \sqrt{\mu - 1}| > 1.$$

For $\mu = 1$ we have a double root for both $\theta = \pi/2$ and $\theta = -\pi/2$, since the square root vanishes. In this case $\hat{u}_{\pm}(\pm\pi/2) = \pm i$, which lies on the unit circle. Thus we have instability for $\mu \geq 1$. For $|\mu| < 1$ we have stability, because in this case

$$|\hat{u}_{\pm}(\theta)|^2 = \mu^2 \sin^2 \theta + 1 - \mu^2 \sin^2 \theta = 1.$$

The leapfrog method is a good example of how instability can be introduced from the boundary. Calculating u_m^{n+1} for $m = 0$ we are lacking the value u_{-1}^n. Setting $u_{-1}^n = 0$ introduces instability which propagates inwards. However, stability can be recovered by letting $u_0^{n+1} = u_1^n$.

8.5.2 The Wave Equation

Once we have seen solutions for the advection equation, it is easy to derive solutions for the wave equation

$$\frac{\partial^2 u}{\partial t^2} = \frac{\partial^2 u}{\partial x^2}$$

in an appropriate domain of $\mathbb{R} \times \mathbb{R}^+$ with boundary conditions and initial conditions for u and $\frac{\partial u}{\partial t}$, since they are fundamentally linked. Specifically, let (v, w) be solutions to the system of advection equations given by

$$\frac{\partial v}{\partial t} = \frac{\partial w}{\partial x},$$
$$\frac{\partial w}{\partial t} = \frac{\partial v}{\partial x}.$$

Then

$$\frac{\partial^2 v}{\partial t^2} = \frac{\partial^2 w}{\partial t \partial x} = \frac{\partial^2 w}{\partial x \partial t} = \frac{\partial^2 v}{\partial x^2}.$$

Then imposing the correct initial and boundary conditions on v and w, we have $u = v$.

More generally speaking, once we have a method for the advection equation, this can be generalized to the system of advection equations

$$\frac{\partial}{\partial t}\mathbf{u} = A\frac{\partial}{\partial x}\mathbf{u},$$

where all the eigenvalues of A are real and nonzero to ensure the system of equations is hyperbolic. For the wave equation, A is given by

$$A = \begin{pmatrix} 0 & 1 \\ 1 & 0 \end{pmatrix}.$$

If the original method for the advection equation is stable for all $\mu \in [a, b]$ where $a < 0 < b$, then the method for the system of advection equations is stable as long as $a \leq \lambda\mu \leq b$ for all eigenvalues λ of A. Again for the wave equation the eigenvalues are ± 1 with corresponding eigenvectors $(1, 1)^T$ and $(1, -1)^T$. Thus using the upwind method (8.25) (for which we have the condition $|\mu| \leq 1$) we calculate v_m^n and w_m^n according to

$$v_m^{n+1} = v_m^n + \mu(w_{m+1}^n - w_m^n), \qquad w_m^{n+1} = w_m^n + \mu(v_{m+1}^n - v_m^n).$$

Eliminating the w_m^ns and letting $u_m^n = v_m^n$, we obtain

$$u_m^{n+1} - 2u_m^n + u_m^{n-1} = \mu^2(u_{m+1}^n - 2u_m^n + u_{m-1}^n),$$

which is the *leapfrog scheme*. Note that we could also have obtained the method by using the usual finite difference approximating the second derivative.

Since this is intrinsically a two-step method in the time direction, we need to calculate u_m^1. One possibility is to use the forward Euler method and let $u_m^1 = u(m\Delta x, 0) + \Delta t u_t(m\Delta x, 0)$ where both terms on the right-hand side are given by the initial conditions. This carries an error of $O((\Delta t)^2)$. However, considering the Taylor expansion,

$$
\begin{aligned}
u(m\Delta x, \Delta t) \;=\;& u(m\Delta x, 0) + \Delta t u_t(m\Delta x, 0) \\
& + \frac{1}{2}(\Delta t)^2 u_{tt}(m\Delta x, 0) + O((\Delta t)^3) \\
\;=\;& u(m\Delta x, 0) + \Delta t u_t(m\Delta x, 0) \\
& + \frac{1}{2}(\Delta t)^2 u_{xx}(m\Delta x, 0) + O((\Delta t)^3) \\
\;=\;& u(m\Delta x, 0) + \Delta t u_t(m\Delta x, 0) \\
& + \frac{1}{2}\mu(u((m-1)\Delta x, 0) - 2u(m\Delta x, 0) + u((m+1)\Delta x, 0)) \\
& + O((\Delta t)^2(\Delta x)^2 + (\Delta t)^3).
\end{aligned}
$$

We see that approximating according to the last line has better accuracy.

The Fourier stability analysis of the leapfrog method for the Cauchy problem yields

$$\hat{u}^{n+1}(\theta) - 2\hat{u}^n(\theta) + \hat{u}^{n-1}(\theta) = \mu(e^{i\theta} - 2 + e^{-i\theta})\hat{u}^n(\theta) = -4\mu\sin^2\frac{\theta}{2}\hat{u}^n(\theta).$$

This recurrence relation has the characteristic equation

$$x^2 - 2(1 - 2\mu\sin^2\frac{\theta}{2})x + 1 = 0$$

with roots $x_\pm = (1 - 2\mu\sin^2\frac{\theta}{2}) \pm \sqrt{(1 - 2\mu\sin^2\frac{\theta}{2})^2 - 1}$. The product of the roots is 1. For stability we require the moduli of both roots to be less than or equal to 1 and if a root lies at 1 it has to be a single root. Thus the roots must be a complex conjugate pair and this leads to the inequality

$$(1 - 2\mu \sin^2 \frac{\theta}{2})^2 \leq 1.$$

This condition is fulfilled if and only if $\mu = (\Delta t / \Delta x)^2 \leq 1$.

8.6   Spectral Methods

Let f be a function on the interval $[-1, 1]$. Its *Fourier series* is given by

$$f(x) = \sum_{n=-\infty}^{\infty} \hat{f}_n e^{i\pi n x}, \qquad x \in [-1, 1],$$

where the *Fourier coefficients* $\hat{f}_n$ are given by

$$\hat{f}_n = \frac{1}{2} \int_{-1}^{1} f(\tau) e^{-i\pi n \tau} d\tau, \qquad , n \in \mathbb{Z}.$$

Letting $\theta = \pi \tau$, which implies $d\tau = d\theta / \pi$, the Fourier coefficients can also be calculated by the following formula

$$\hat{f}_n = \frac{1}{2\pi} \int_{-\pi}^{\pi} f\left(\frac{\theta}{\pi}\right) e^{-in\theta} d\theta, \qquad , n \in \mathbb{Z}.$$

This formula is also widely used.

As examples for Fourier series take $\cos(\pi x)$ and $\sin(\pi x)$. Both have only two non-zero coefficients in their Fourier expansion:

$$\cos(\pi x) = \frac{1}{2} \left(e^{i\pi x} + e^{-i\pi x}\right),$$
$$\sin(\pi x) = \frac{1}{2i} \left(e^{i\pi x} - e^{-i\pi x}\right).$$

We define the *N-point truncated Fourier approximation* ϕ_N by

$$\phi_N(x) = \sum_{n=-N/2+1}^{N/2} \hat{f}_n e^{i\pi n x}, \qquad x \in [-1, 1],$$

where here and elsewhere in this section $N \geq 2$ is an even integer.

Theorem 8.4 (The de la Valleé Poussin theorem). *If the function f is Riemann integrable and the coefficients $\hat{f}_n$ are of order $O(n^{-1})$ for large n, then $\phi_N(x) = f(x) + O(N^{-1})$ as $N \to \infty$ for every point x in the open interval $(-1, 1)$, where f is Lipschitz.*

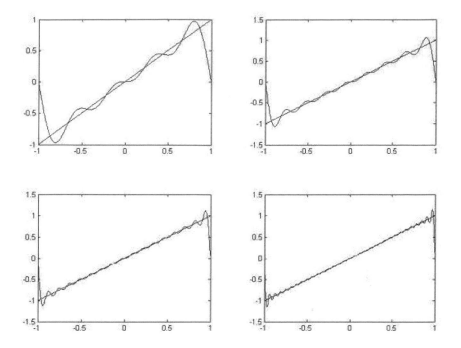

Figure 8.1 Gibbs effect when approximating the line $y = x$ for the choices $N = 4, 8, 16, 32$

Note that the above theorem explicitly excludes the endpoints of the interval. This is due to the *Gibbs phenomenon*. Figure 8.1 illustrates this. The Gibbs effect involves both the fact that Fourier sums overshoot at a discontinuity, and that this overshoot does not die out as the frequency increases. With increasing N the point where that overshoot happens moves closer and closer to the discontinuity. So once the overshoot has passed by a particular x, convergence at the value of x is possible.However, convergence at the endpoints -1 and 1 cannot be guaranteed. It is possible to show (as a consequence of the Dirichlet–Jordan theorem) that

$$\phi_N(\pm 1) \to \frac{1}{2}[f(-1) + f(1)] \qquad \text{as} \qquad N \to \infty$$

and hence there is no convergence unless f is periodic.

For proofs and more in-depth analysis, see [13] T. W. Körner, *Fourier Analysis*.

Theorem 8.5. *Let f be an analytic function in $[-1, 1]$, which can be extended analytically to a complex domain Ω and which is periodic with period 2, i.e., $f^{(m)}(1) = f^{(m)}(-1)$ for all $m = 1, 2, \ldots$. Then the Fourier coefficients $\hat{f}_n$ are*

$O(n^{-m})$ *for any* $m = 1, 2, \ldots$. *Moreover, the Fourier approximation* ϕ_N *is of infinite order, that is,* $|\phi_N - f| = O(N^{-p})$ *for any* $p = 1, 2, \ldots$.

Proof. We only sketch the proof. Using integration by parts we can deduce

$$
\begin{aligned}
\hat{f}_n &= \frac{1}{2} \int_{-1}^{1} f(\tau) e^{-i\pi n \tau} d\tau \\
&= \left[\frac{1}{2} f(\tau) \frac{e^{-i\pi n \tau}}{(-i\pi n)} \right]_{-1}^{1} - \frac{1}{2} \int_{-1}^{1} f'(\tau) \frac{e^{-i\pi n \tau}}{(-i\pi n)} d\tau.
\end{aligned}
$$

The first term vanishes, since $f(-1) = f(1)$ and $e^{\pm i\pi n} = \cos n\pi \pm i \sin n\pi = \cos n\pi$. Thus

$$
\hat{f}_n = \frac{1}{\pi i n} \widehat{f'}_n.
$$

Using $f^{(m)}(1) = f^{(m)}(-1)$ for all $m = 1, 2, \ldots$ and multiple integration by parts gives

$$
\hat{f}_n = \frac{1}{\pi i n} \widehat{f'}_n = \left(\frac{1}{\pi i n} \right)^2 \widehat{f''}_n = \left(\frac{1}{\pi i n} \right)^3 \widehat{f'''}_n = \ldots.
$$

Hence

$$
\hat{f}_n = \left(\frac{1}{\pi i n} \right)^m \widehat{f^{(m)}}_n, \qquad m = 1, 2, \ldots.
$$

Now using Cauchy's theorem of complex analysis $|\widehat{f^{(m)}}_n|$ can be bounded by $c m! \alpha^m$ for some constants $c, \alpha > 0$. Then

$$
\begin{aligned}
|\phi_N(x) - f(x)| &= \left| \sum_{n=-N/2+1}^{N/2} \hat{f}_n e^{i\pi n x} - \sum_{n=-\infty}^{\infty} \hat{f}_n e^{i\pi n x} \right| \\
&\leq \sum_{-\infty}^{-N/2} |\hat{f}_n| + \sum_{N/2+1}^{\infty} |\hat{f}_n| \\
&= \sum_{-\infty}^{-N/2} \frac{|\widehat{f^{(m)}}_n|}{(-\pi n)^m} + \sum_{N/2+1}^{\infty} \frac{|\widehat{f^{(m)}}_n|}{(\pi n)^m} \\
&\leq c m! \left(\frac{\alpha}{\pi} \right)^m \left[\frac{1}{(N/2)^m} + 2 \sum_{n=N/2+1}^{\infty} \frac{1}{n^m} \right] \\
&\leq c m! \left(\frac{\alpha}{\pi} \right)^m \left[\frac{1}{(N/2)^m} + 2 \int_{N/2+1}^{\infty} \frac{d\tau}{\tau^m} \right] \\
&= c m! \left(\frac{\alpha}{\pi} \right)^m \left[\frac{1}{(N/2)^m} + \frac{2}{(m-1)(N/2+1)^{m-1}} \right] \\
&\leq C_m N^{-m+1}.
\end{aligned}
$$

□

Definition 8.12 (Convergence at spectral speed). *An N-point approximation ϕ_N of a function f converges to f at spectral speed if $|\phi_N - f|$ decays pointwise faster than $O(N^{-p})$ for any $p = 1, 2, \ldots$.*

The fast convergence of Fourier approximations rests on two properties of the underlying function: analyticity and periodicity. If one is not satisfied the rate of convergence in general drops to polynomial. In general, the speed of convergence of the truncated Fourier series of a function f depends on the smoothness of the function. In fact, the smoother the function the higher the convergence, i.e., for $f \in C^p(-1; 1)$ (i.e., the derivatives up to p exist and are continuous), we receive an $O(N^{-p})$ order of convergence.

We now consider the heat equation $u_t = u_{xx}$ on the interval $[-1, 1]$ with given initial condition $u(x, 0) = g(x)$, periodic boundary conditions, and a *normalization condition*, i.e.,

$$u(-1, t) = u(1, t), \quad u_x(-1, t) = u_x(1, t), \quad t \geq 0$$

$$\int_{-1}^{1} u(x, t) dx = 0, \qquad\qquad t \geq 0.$$

We approximate u by its Nth order Fourier expansion in x

$$u(x, t) \approx \sum_{n=-N/2+1}^{N/2} \hat{u}_n(t) e^{i\pi n x}.$$

Differentiating once with respect to t and on the other hand differentiating twice with respect to x gives

$$u_t \approx \sum_{n=-N/2+1}^{N/2} \hat{u}_n'(t) e^{i\pi n x},$$

$$u_{xx} \approx \sum_{n=-N/2+1}^{N/2} \hat{u}_n(t)(i\pi n)^2 e^{i\pi n x}.$$

Equating yields for each coefficient $\hat{u}_n$ the ODE

$$\hat{u}_n'(t) = -\pi^2 n^2 \hat{u}_n(t), \qquad n = -N/2 + 1, \ldots, N/2.$$

This has the exact solution $\hat{u}_n(t) = c_n e^{-\pi^2 n^2 t}$ for $n \neq 0$. The constant coefficients c_n are given by the initial condition, since

$$g(x) = u(x, 0) \approx \sum_{n=-N/2+1}^{N/2} \hat{u}_n(0) e^{i\pi n x} = \sum_{n=-N/2+1}^{N/2} c_n e^{i\pi n x}.$$

Approximating g by its Nth order Fourier expansion, we see that $c_n = \hat{g}_n$ is the appropriate choice.

For $n = 0$ the ODE simplifies to $\hat{u}_0'(t) = 0$ and thus $\hat{u}_0(t) = c_0$. The constant c_0 is determined by the normalization condition

$$\int_{-1}^{1} \sum_{n=-N/2+1}^{N/2} c_n e^{-\pi^2 n^2 t} e^{i\pi n x} dx = \sum_{n=-N/2+1}^{N/2} c_n e^{-\pi^2 n^2 t} \int_{-1}^{1} e^{i\pi n x} dx = 2c_0,$$

since the integral is zero for $n \neq 0$. Thus $\hat{u}_0(t) = 0$.

For more general problems we need to examine the *algebra of Fourier expansions*. The set of all analytic functions $f : [-1, 1] \to \mathbb{C}$, which are periodic with period 2 and which can be extended analytically into the complex plane, form a linear space. The Fourier series of sums and scalar products of such functions are well-defined. In particular for

$$f(x) = \sum_{n=-\infty}^{\infty} \hat{f}_n e^{i\pi n x}, \qquad g(x) = \sum_{n=-\infty}^{\infty} \hat{g}_n e^{i\pi n x}$$

we have

$$f(x) + g(x) = \sum_{n=-\infty}^{\infty} (\hat{f}_n + \hat{g}_n) e^{i\pi n x} \qquad af(x) = \sum_{n=-\infty}^{\infty} a\hat{f}_n e^{i\pi n x}.$$

Moreover,

$$f(x)g(x) = \sum_{n=-\infty}^{\infty} \left(\sum_{m=-\infty}^{\infty} \hat{f}_{n-m} \hat{g}_m \right) e^{i\pi n x} = \sum_{n=-\infty}^{\infty} \left(\{\hat{f}_n\} * \{\hat{g}_n\} \right) e^{i\pi n x},$$

where $*$ denotes the convolution operator acting on sequences. In addition, the derivative of f is given by

$$f'(x) = i\pi \sum_{n=-\infty}^{\infty} n\hat{f}_n e^{i\pi n x}.$$

Since $\{\hat{f}_n\}$ decays faster than $O(n^{-m})$ for all $m \in \mathbb{Z}^+$, it follows that all derivatives of f have rapidly convergent Fourier expansions.

We now have the tools at our disposal to solve the heat equation with non-constant coefficient $\alpha(x)$. In particular,

$$\frac{\partial u(x,t)}{\partial t} = \frac{\partial}{\partial x} \left(\alpha(x) \frac{\partial u(x,t)}{\partial x} \right)$$

for $-1 \leq x \leq 1, t \geq 0$ and given initial conditions $u(x,0)$. Letting

$$u(x,t) = \sum_{n=-\infty}^{\infty} \hat{u}_n(t) e^{i\pi n x} \quad \text{and} \quad \alpha(x) = \sum_{n=-\infty}^{\infty} \hat{\alpha}_n e^{i\pi n x}$$

we have

$$\frac{\partial u(x,t)}{\partial x} = \sum_{n=-\infty}^{\infty} \hat{u}_n(t) i\pi n e^{i\pi nx}$$

and by convolution,

$$\alpha(x)\frac{\partial u(x,t)}{\partial x} = \sum_{n=-\infty}^{\infty} \left(\sum_{m=-\infty}^{\infty} \hat{\alpha}_{n-m} i\pi m \hat{u}_m(t) \right) e^{i\pi nx}.$$

Differentiating again and truncating, we deduce the following system ODEs for the coefficients $\hat{u}_n$

$$\hat{u}'_n(t) = -\pi^2 \sum_{m=-N/2+1}^{N/2} nm\hat{\alpha}_{n-m}\hat{u}_m(t), \qquad n = -N/2+1, \ldots, N/2.$$

We can now proceed to discretize in time by applying a suitable ODE solver. For example, the forward Euler method results in

$$\hat{u}_n^{k+1} = \hat{u}_n^k - \Delta t\pi^2 \sum_{m=-N/2+1}^{N/2} nm\hat{\alpha}_{n-m}\hat{u}_m^k,$$

or in vector form,

$$\hat{\mathbf{u}}^{k+1} = (I + \Delta t\hat{A})\hat{\mathbf{u}}^k,$$

where $\hat{A}$ has elements $\hat{A}_{nm} = -\pi^2 nm\hat{\alpha}_{n-m}$, $m,n = -N/2+1, \ldots, N/2$. Note that every row and every column in $\hat{A}$ has a common factor. If $\alpha(x)$ is constant, i.e., $\alpha(x) \equiv \alpha_0$, then $\hat{\alpha}_n = 0$ for all $n \neq 0$ and $\hat{A}$ is a diagonal matrix

$$-\pi^2 \begin{pmatrix} (-N/2+1)^2\alpha_0 & 0 & \cdots & & & & 0 \\ 0 & \ddots & & & & & \\ & & \alpha_0 & & & & \\ & & & 0 & & & \\ & & & & \alpha_0 & & \\ & & & & & \ddots & 0 \\ 0 & & \cdots & & & 0 & (N/2)^2\alpha_0 \end{pmatrix}.$$

It is easy to see that the eigenvalues of $(I + \Delta t\hat{A})$ are given by $1, 1 - \Delta t\pi^2\alpha_0, \ldots, 1 - \Delta t\pi^2(N/2-1)^2\alpha_0, 1 - \Delta t\pi^2(N/2)^2\alpha_0$. Thus for the method to be numerically stable, Δt must scale like N^{-2}.

Speaking more generally, the maximum eigenvalue of the matrix approximating the kth derivative is $(N/2)^k$. This means that Δt in spectral approximations for linear PDEs with constant coefficients must scale like N^{-k}, where k is the maximal order of differentiation.

In the analysis of the spectral speed of convergence we have seen that analyticity and $f^{(m)}(1) = f^{(m)}(-1)$ for all $m = 1, 2, \ldots$ is crucial. What to do, however, if the latter does not hold? We can force values at the endpoints to be equal. Consider the function

$$g(x) = f(x) - \frac{1}{2}(1 - x)f(-1) - \frac{1}{2}(1 + x)f(1),$$

which satisfies $g(\pm 1) = 0$. Now the Fourier coefficients $\hat{g}_n$ are $O(n^{-1})$. According to the de la Vallée Poussin theorem, the rate of convergence of the N-terms truncated Fourier expansion of g is hence $O(N^{-1})$. This idea can be iterated. By letting

$$h(x) = g(x) + a(1 - x)(1 + x) + b(1 - x)(1 + x)^2,$$

which already satisfies $h(\pm 1) = g(\pm 1) = 0$, and choosing a and b appropriately, we achieve $h'(\pm 1) = 0$. Here the Fourier coefficients $\hat{h}_n$ are $O(n^{-2})$. However, the values of the derivatives at the boundaries need to be known and with every step the degree of the polynomial which needs to be added to achieve zero boundary conditions increases by 2.

Another possibility to deal with the lack of periodicity is the use of *Chebyshev polynomials (of the first kind)*, which are defined by $T_n(x) = \cos(n \arccos x)$, $n \geq 0$. Each T_n is a polynomial of degree n, i.e.,

$$T_0(x) = 1, \quad T_1(x) = x, \quad T_2(x) = 2x^2 - 1, \quad T_3(x) = 4x^3 - 3x, \quad \ldots$$

The sequence T_n obeys the three-term recurrence relation

$$T_{n+1}(x) = 2x T_n(x) - T_{n-1}(x), \qquad n = 1, 2, \ldots.$$

Moreover, they form a sequence of orthogonal polynomials, which are orthogonal with respect to the weight function $(1 - x^2)^{-\frac{1}{2}}$ in $(-1, 1)$. More specifically,

$$\int_{-1}^{1} T_m(x) T_n(x) \frac{dx}{\sqrt{1 - x^2}} = \left\{ \begin{array}{ll} \pi & m = n = 0 \\ \frac{\pi}{2} & m = n \geq 1 \\ 0 & m \neq n. \end{array} \right\}. \tag{8.26}$$

This can be proven by letting $x = \cos\theta$ and using the identity $T_n(\cos\theta) = \cos n\theta$.

Now since the Chebyshev polynomials are mutually orthogonal, a general integrable function f can be expanded in

$$f(x) = \sum_{n=0}^{\infty} \check{f}_n T_n(x). \tag{8.27}$$

The formulae for the coefficients can be derived by multiplying (8.27) by $T_m(x)(1 - x^2)^{-\frac{1}{2}}$ and integrating over $(-1, 1)$. Further using the orthogonality property (8.26) results in the explicit expression for the coefficients

$$\check{f}_0 = \frac{1}{\pi} \int_{-1}^{1} f(x) \frac{dx}{\sqrt{1 - x^2}}, \quad \check{f}_n = \frac{2}{\pi} \int_{-1}^{1} f(x) T_n(x) \frac{dx}{\sqrt{1 - x^2}}, \quad n = 1, 2, \ldots.$$

The connection to Fourier expansions can be easily seen by letting $x = \cos\theta$,

$$
\begin{aligned}
\int_{-1}^{1} f(x) T_n(x) \frac{dx}{\sqrt{1-x^2}} &= \int_{0}^{\pi} f(\cos\theta) T_n(\cos\theta) d\theta \\
&= \frac{1}{2} \int_{-\pi}^{\pi} f(\cos\theta) \cos n\theta \, d\theta \\
&= \frac{1}{2} \int_{-\pi}^{\pi} f(\cos\theta) \frac{1}{2}(e^{in\theta} + e^{-in\theta}) d\theta \\
&= \frac{\pi}{2} (\widehat{f(\cos\theta)}_{-n} + \widehat{f(\cos\theta)}_{n}),
\end{aligned}
$$

since the general Fourier transform of a function g defined in the interval $[a, b]$, $a < b$, and which is periodic with period $b - a$, is given by the sequence

$$
\hat{g} = \frac{1}{b-a} \int_{a}^{b} g(\tau) e^{-\frac{2\pi i n \tau}{b-a}} d\tau.
$$

In particular, letting $g(x) = f(\cos x)$ and $[a, b] = [-\pi, \pi]$, we have

$$
\hat{g} = \frac{1}{2\pi} \int_{-\pi}^{\pi} g(\tau) e^{-in\tau} d\tau
$$

and the result follows. Thus we can deduce

$$
\check{f}_n = \left\{ \begin{array}{ll} \widehat{f(\cos\theta)}_0, & n = 0, \\ \widehat{f(\cos\theta)}_{-n} + \widehat{f(\cos\theta)}_{n}, & n = 1, 2, \ldots \end{array} \right\}.
$$

Thus for a general integrable function f the computation of its Chebyshev expansion is equivalent to the Fourier expansion of the function $f(\cos\theta)$. The latter is periodic with period 2π. In particular, if f can be extended analytically, then $\check{f}_n$ decays spectrally fast for large n. So the Chebyshev expansion inherits the rapid convergence of spectral methods without ever assuming that f is periodic.

Similar to Fourier expansions, we have an *algebra of Chebyshev expansions*. The Chebyshev expansions of sums and scalar products are well-defined. In particular for

$$
f(x) = \sum_{n=0}^{\infty} \check{f}_n T_n(x), \qquad g(x) = \sum_{n=0}^{\infty} \check{g}_n T_n(x)
$$

we have

$$
f(x) + g(x) = \sum_{n=0}^{\infty} (\check{f}_n + \check{g}_n) T_n(x) \qquad af(x) = \sum_{n=0}^{\infty} a\check{f}_n T_n(x).
$$

Moreover, since

$$
\begin{aligned}
T_m(x) T_n(x) &= \cos(m \arccos x) \cos(n \arccos x) \\
&= \frac{1}{2} \left[\cos((m-n) \arccos x) + \cos((m+n) \arccos x) \right] \\
&= \frac{1}{2} \left[T_{|m-n|}(x) + T_{m+n}(x) \right],
\end{aligned}
$$

we have

$$
\begin{aligned}
f(x)g(x) &= \sum_{m=0}^{\infty} \check{f}_m T_m(x) \sum_{n=0}^{\infty} \check{g}_n T_n(x) \\
&= \frac{1}{2} \sum_{m,n=0}^{\infty} \check{f}_m \check{g}_n [T_{|m-n|}(x) + T_{m+n}(x)] \\
&= \frac{1}{2} \sum_{m,n=0}^{\infty} \check{f}_m (\check{g}_{m+n} + \check{g}_{|m-n|}) T_n(x),
\end{aligned}
$$

where the last equation is due to a change in summation.

Lemma 8.2 (Derivatives of Chebyshev polynomials). *The derivatives of Chebyshev polynomials can be explicitly expressed as the linear combinations of Chebyshev polynomials*

$$
T'_{2n}(x) = 4n \sum_{l=0}^{n-1} T_{2l+1}(x), \tag{8.28}
$$

$$
T'_{2n+1}(x) = (2n+1)T_0(x) + 2(2n+1) \sum_{l=1}^{n} T_{2l}(x). \tag{8.29}
$$

Proof. We only proof (8.28), since the proof of (8.29) follows a similar argument. Since $T_{2n}(x) = \cos(2n \arccos x)$, we have

$$
T'_{2n}(x) = 2n \sin(2n \arccos x) \frac{1}{\sqrt{1-x^2}}.
$$

Letting $x = \cos\theta$ and rearranging, it follows that

$$
\sin\theta \, T'_{2n}(\cos\theta) = 2n \sin(2n\theta).
$$

Multiplying the right-hand side of (8.28) by $\sin\theta$, we have

$$
\begin{aligned}
4n \sin\theta \sum_{l=0}^{n-1} T_{2l+1}(\cos\theta) &= 4n \sum_{l=0}^{n-1} \cos((2l+1)\theta) \sin\theta \\
&= 2n \sum_{l=0}^{n-1} (\sin((2l+2)\theta) - \sin 2l\theta) = 2n \sin 2n\theta.
\end{aligned}
$$

This concludes the proof. $\qquad\square$

We can use this Lemma to express the coefficients of the Chebyshev expansion of the derivative in terms of the original Chebyshev coefficients $\check{f}_n$

$$
\check{f}'_n = \frac{2}{c_n} \sum_{\substack{p=n+1 \\ n+p \text{ odd}}}^{\infty} p \check{f}_p, \tag{8.30}
$$

where c_n is 2 for $n = 0$ and 1 for $n \geq 1$. We now have the tools at our hand to use Chebyshev expansions in a similar way as Fourier expansions. However, this results in much more complicated relations, since (8.30) shows that a Chebyshev coefficient of the derivative is linked to infinitely many original Chebyshev coefficients, while the equivalent relation between Fourier coefficients is one-to-one.

8.6.1 Spectral Solution to the Poisson Equation

We consider the Poisson equation

$$\nabla^2 u = f, \qquad -1 \leq x, y \leq 1,$$

where f is analytic and obeys periodic boundary conditions

$$f(-1, y) = f(1, y), \quad -1 \leq y \leq 1, \quad f(x, -1) = f(x, 1), \quad -1 \leq x \leq 1.$$

Additionally we impose the following periodic boundary conditions on u

$$\begin{aligned} u(-1, y) = u(1, y), \quad u_x(-1, y) = u_x(1, y), \quad -1 \leq y \leq 1, \\ u(x, -1) = u(x, 1), \quad u_y(x, -1) = u_y(x, 1), \quad -1 \leq y \leq 1. \end{aligned} \qquad (8.31)$$

These conditions alone only define the solution up to an additative constant. As we have already seen in the spectral solution to the heat equation, we add a normalization condition:

$$\int_{-1}^{1} \int_{-1}^{1} u(x, y) dx dy = 0. \qquad (8.32)$$

Since f is analytic, its Fourier expansion

$$f(x, y) = \sum_{k,l=-\infty}^{\infty} \hat{f}_{k,l} e^{i\pi(kx+ly)}$$

is spectrally convergent. From the Fourier expansion of f we can calculate the coefficients of the Fourier expansion of u

$$u(x, y) = \sum_{k,l=-\infty}^{\infty} \hat{u}_{k,l} e^{i\pi(kx+ly)}.$$

From the normalization condition we obtain

$$0 = \int_{-1}^{1} \int_{-1}^{1} u(x, y) dx dy = \sum_{k,l=-\infty}^{\infty} \hat{u}_{k,l} \int_{-1}^{1} \int_{-1}^{1} e^{i\pi(kx+ly)} dx dy = 4\hat{u}_{0,0}.$$

For the other coefficients,

$$\nabla^2 u(x, y) = -\pi^2 \sum_{k,l=-\infty}^{\infty} (k^2 + l^2) \hat{u}_{k,l} e^{i\pi(kx+ly)} = \sum_{k,l=-\infty}^{\infty} \hat{f}_{k,l} e^{i\pi(kx+ly)},$$

thus summarizing

$$
\hat{u}_{k,l} = \begin{cases} -\dfrac{1}{\pi^2(k^2 + l^2)}\hat{f}_{k,l}, & k, l \in \mathbb{Z}\backslash\{0,0\} \\ 0 & (k,l) = (0,0). \end{cases}
$$

This solution is not representative for its application to general PDEs. The reason is the special structure of the Poisson equation, because $\phi_{k,l} = e^{i\pi(kx+ly)}$ are the eigenfunctions of the Laplace operator with eigenvalue $-\pi^2(k^2 + l^2)$, since

$$
\nabla^2 \phi_{k,l} = -\pi^2(k^2 + l^2)\phi_{k,l},
$$

and they obey periodic boundary conditions.

The concept can be extended to general second-order elliptic PDEs specified by the equation

$$
\nabla^T(a\nabla u) = f, \qquad -1 \le x, y \le 1,
$$

where a is a positive analytic function and f is an analytic function, and both a and f are periodic. We again impose the boundary conditions (8.31) and the normalization condition (8.32). Writing

$$
\begin{aligned}
a(x,y) &= \sum_{k,l=-\infty}^{\infty} \hat{a}_{k,l} e^{i\pi(kx+ly)}, \\
f(x,y) &= \sum_{k,l=-\infty}^{\infty} \hat{f}_{k,l} e^{i\pi(kx+ly)}, \\
u(x,y) &= \sum_{k,l=-\infty}^{\infty} \hat{u}_{k,l} e^{i\pi(kx+ly)},
\end{aligned}
$$

and rewriting the PDE using the fact that the Laplacian ∇^2 is the divergence ∇^T of the gradient ∇u as

$$
\nabla^T(a\nabla u) = \frac{\partial}{\partial x}(au_x) + \frac{\partial}{\partial y}(au_y) = a\nabla^2 u + a_x u_x + a_y u_y,
$$

we get

$$
\begin{aligned}
-\pi^2 &\left(\sum_{k,l=-\infty}^{\infty} \hat{a}_{k,l} e^{i\pi(kx+ly)} \right) \left(\sum_{m,n=-\infty}^{\infty} (m^2 + n^2)\hat{u}_{m,n} e^{i\pi(mx+ny)} \right) \\
-\pi^2 &\left(\sum_{k,l=-\infty}^{\infty} k\hat{a}_{k,l} e^{i\pi(kx+ly)} \right) \left(\sum_{m,n=-\infty}^{\infty} m\hat{u}_{m,n} e^{i\pi(mx+ny)} \right) \\
-\pi^2 &\left(\sum_{k,l=-\infty}^{\infty} l\hat{a}_{k,l} e^{i\pi(kx+ly)} \right) \left(\sum_{m,n=-\infty}^{\infty} n\hat{u}_{m,n} e^{i\pi(mx+ny)} \right) \\
= &\sum_{k,l=-\infty}^{\infty} \hat{f}_{k,l} e^{i\pi(kx+ly)}.
\end{aligned}
$$

Again the normalization condition yields $\hat{u}_{0,0} = 0$. Replacing the products by convolutions using the bivariate version

$$\left(\sum_{k,l=-\infty}^{\infty} \hat{f}_{k,l} e^{i\pi(kx+ly)} \right) \left(\sum_{m,n=-\infty}^{\infty} \hat{g}_{m,n} e^{i\pi(mx+ny)} \right) =$$
$$\sum_{k,l=-\infty}^{\infty} \left(\sum_{m,n=-\infty}^{\infty} \hat{f}_{k-m,l-n} \hat{g}_{m,n} \right) e^{i\pi(kx+ly)}$$

and truncating the sums to $-N/2 + 1 \le k, l \le N/2$, we get a system of $N^2 - 1$ linear algebraic equations in the unknowns $\hat{u}_{k,l}, -N/2 + 1 \le k, l \le N/2, (k, l) \ne (0, 0)$

$$-\pi^2 \sum_{m,n=-N/2+1}^{N/2} \left[\hat{a}_{k-m,l-n}(m^2 + n^2) + (k - m)a_{k-m,l-n}m + \right.$$
$$\left. (l - n)\hat{a}_{k-m,l-n}n \right] \hat{u}_{m,n} = \hat{f}_{k,l}.$$

The main difference between methods arising from computational stencils and spectral methods is that the former leads to large sparse matrices, while the latter leads to small but dense matrices.

8.7 Finite Element Method

So far we have calculated estimates of the values of the solution to a differential equation at discrete grid points. We now take a different approach by choosing an element from a finite dimensional space of functions as an approximation to the true solution. For example, if the true solution is a function of one variable, this space may be made up of piecewise polynomials such as linear or cubic splines. There are analogous piecewise polynomials of several variables.

This technique was developed because many differential equations arise from variational calculations. For example, let u describe a physical system and let $I(u)$ be an integral that gives the total energy in the system. The system is in a steady state if u is the function that minimizes $I(u)$. Often it can then be deduced that u is the solution to a differential equation. Solving this differential equation might be the best way of calculating u.

On the other hand, it is sometimes possible to show that the solution u of a differential equation is the function in an infinite dimensional space which minimizes a certain integral $I(u)$. The task is to choose a finite dimensional subspace $\mathbf{V}$ and then to find the element $\tilde{u}$ in this subspace that minimizes $I(\tilde{u})$.

We will not go into detail, but only give an introduction to *finite elements*. Ern and Guermond cover the topic extensively in [8].

We will consider a one-dimensional example. Let u be the solution to the differential equation

$$u''(x) = f(x), \qquad 0 \le x \le 1,$$

subject to the boundary conditions $u(0) = u_0$ and $u(1) = u_1$, where f is a smooth function from $[0, 1]$ to $\mathbb{R}$. Moreover, let w be the solution of the minimization of the integral

$$I(w) = \int_0^1 [w'(x)]^2 + 2w(x)f(x)dx$$

subject to the same boundary conditions $w(0) = u_0$ and $w(1) = u_1$. We will show that $u = w$. That is, the solution to the differential equation is also the minimizer of $I(w)$. In other words, we prove $v = w - u = 0$.

We have

$$
\begin{aligned}
I(w) = I(u + v) &= \int_0^1 [u'(x) + v'(x)]^2 + 2[u(x) + v(x)]f(x)dx \\
&= \int_0^1 [u'(x)]^2 + 2u(x)f(x)dx \\
&\quad + 2\int_0^1 u'(x)v'(x) + v(x)f(x)dx + \int_0^1 [v'(x)]^2 dx.
\end{aligned}
$$

The first integral is $I(u)$ and the last integral is always non-negative. For the second integral integrating by parts, we obtain

$$
\begin{aligned}
\int_0^1 & u'(x)v'(x) + v(x)f(x)dx = \\
&= [u'(x)v(x)]_0^1 - \int_0^1 u''(x)v(x)dx + \int_0^1 v(x)f(x)dx \\
&= u'(1)v(1) - u'(0)v'(0) + \int_0^1 v(x)(f(x) - u''(x))dx = 0,
\end{aligned}
$$

since $v(0) = v(1) = 0$, since u and w have the same boundary conditions, and since $u''(x) = f(x)$. Hence

$$I(w) = I(u) + \int_0^1 [v'(x)]^2 dx.$$

Since w minimizes $I(w)$ we have $I(w) \le I(u)$. It follows that

$$\int_0^1 [v'(x)]^2 dx = 0$$

and thus $v(x) \equiv 0$, since $v(0) = 0$.

The above solution can be adjusted for any boundary condition $u(0) = a$ and $u(1) = b$, because the linear polynomial $a - u_0 + (b - u_1 - a + u_0)x$ can be added to the solution. This does not change the second derivative, but achieves the desired boundary conditions. Thus we can assume zero boundary conditions $w(0) = 0$ and $w(1) = 0$ when minimizing $I(w)$.

This example can be extended to two dimensions. We consider the *Poisson equation*

$$\nabla^2 u(x, y) = u_{xx}(x, y) + u_{yy}(x, y) = f(x, y)$$

on the square $[0, 1] \times [0, 1]$ with zero boundary conditions. We assume that f is continuous on $[0, 1] \times [0, 1]$. The solution to the Poisson equation is equivalent to minimizing

$$I(w) = \int_0^1 \int_0^1 (w_x(x, y))^2 + (w_y(x, y))^2 + 2w(x, y)f(x, y)dxdy.$$

As before, we let $v = w - u$ and hence

$$
\begin{aligned}
I(w) \;=\; & I(u + v) \\
=\; & \int_0^1 \int_0^1 (u_x(x, y) + v_x(x, y))^2 + (u_y(x, y) + v_y(x, y))^2 \\
& + \int_0^1 \int_0^1 2(u(x, y) + v(x, y))f(x, y)dxdy \\
=\; & \int_0^1 \int_0^1 [u_x(x, y)]^2 + [u_y(x, y)]^2 + 2u(x, y)f(x, y)dxdy \\
& + 2\int_0^1 \int_0^1 u_x(x, y)v_x(x, y) + u_y(x, y)v_y(x, y) + v(x, y)f(x, y)dxdy \\
& + \int_0^1 \int_0^1 [v_x(x, y)]^2 + [v_y(x, y)]^2 dxdy.
\end{aligned}
$$

Simplifying the second integral

$$
\int_0^1 \int_0^1 u_x(x, y)v_x(x, y) + u_y(x, y)v_y(x, y) + 2v(x, y)f(x, y)dxdy =
$$

$$
= \int_0^1 [u_x(x, y)v(x, y)]_0^1 \, dy - \int_0^1 \int_0^1 u_{xx}(x, y)v(x, y)dxdy
$$

$$
+ \int_0^1 [u_y(x, y)v(x, y)]_0^1 \, dx - \int_0^1 \int_0^1 u_{yy}(x, y)v(x, y)dydx
$$

$$
+ \int_0^1 \int_0^1 v(x, y)f(x, y)dxdy
$$

$$
= \int_0^1 u_x(1, y)v(1, y) - u_x(0, y)v(0, y)dy
$$

$$
+ \int_0^1 u_y(x, 1)v(x, 1) - u_y(x, 0)v(x, 0)dx
$$

$$
+ \int_0^1 \int_0^1 v(x, y)(f(x, y) - u_{xx}(x, y) - u_{yy}(x, y))dxdy = 0,
$$

since $v(x, 0) = v(x, 1) = v(0, y) = v(1, y) = 0$ for all $x, y \in [0, 1]$ and $f(x, y) =$

$u_{xx}(x, y) + u_{yy}(x, y)$ for all $(x, y) \in [0, 1] \times [0, 1]$. Hence

$$I(w) = I(u) + \int_0^1 \int_0^1 [v_x(x, y)]^2 + [v_y(x, y)]^2 dx dy.$$

On the other hand, $I(w)$ is minimal and thus

$$\int_0^1 \int_0^1 [v_x(x, y)]^2 + [v_y(x, y)]^2 dx dy = 0.$$

This implies $v(x, y) \equiv 0$ and thus the solution to the Poisson equation also minimizes the integral.

We have seen two examples of the first step of the finite element method. The first step is to rephrase the problem as a variational problem. The true solution lies in an infinite dimensional space of functions and is minimal with respect to a certain functional. The next step is to choose a finite dimensional subspace S and find the element of that subspace which minimizes the functional.

To be more formal, let the functional be of the form

$$I(u) = \mathcal{A}(u, u) + 2\langle u, f \rangle, \tag{8.33}$$

where $\langle u, f \rangle$ is a scalar product. In the first example

$$\mathcal{A}(u, v) = \int_0^1 u'(x) v'(x) dx$$

and the scalar product is

$$\langle u, f \rangle = \int_0^1 u(x) f(x) dx.$$

In the second example, on the other hand,

$$\mathcal{A}(u, v) = \int_0^1 \int_0^1 u_x(x, y) v_x(x, y) + u_y(x, y) v_y(x, y) dx dy$$

and the scalar product is

$$\langle u, f \rangle = \int_0^1 \int_0^1 u(x, y) f(x, y) dx dy.$$

In the following we assume that the functional $\mathcal{A}(u, v)$ satisfies the following properties:

Symmetry
$$\mathcal{A}(u, v) = \mathcal{A}(v, u).$$

Bi-linearity

$\mathcal{A}(u, v)$ is a linear function of u for fixed v:

$$\mathcal{A}(\lambda u + \mu w, v) = \lambda \mathcal{A}(u, v) + \mu \mathcal{A}(w, v).$$

Due to symmetry, $\mathcal{A}(u, v)$ is also a linear functional of v for fixed u.

Non-negativity

$\mathcal{A}(u, u) \geq 0$ for all u and $\mathcal{A}(u, u) = 0$ if and only if $u \equiv 0$.

Theorem 8.6. *If the above assumptions hold, then $u \in S$ minimizes $I(u)$ as in (8.33) if and only if for all non-zero functions $v \in S$,*

$$\mathcal{A}(u, v) + \langle v, f \rangle = 0.$$

Proof. Let $u \in S$ minimize $I(u)$, then for any non-zero $v \in S$ and scalars λ,

$$
\begin{aligned}
I(u) &\leq I(u + \lambda v) = \mathcal{A}(u + \lambda v, u + \lambda v) + 2\langle u + \lambda v, f \rangle \\
&= I(u) + 2\lambda \left[\mathcal{A}(u, v) + \langle v, f \rangle \right] + \lambda^2 \mathcal{A}(v, v).
\end{aligned}
$$

It follows that $2\lambda \left[\mathcal{A}(u, v) + \langle v, f \rangle \right] + \lambda^2 \mathcal{A}(v, v)$ has to be non-negative for all λ. This is only possible if the linear term in λ vanishes, that is,

$$\mathcal{A}(u, v) + \langle u, f \rangle = 0.$$

To proof the other direction let $u \in S$ such that $\mathcal{A}(u, v) + \langle v, f \rangle = 0$ for all non-zero functions $v \in S$. Then for any $v \in S$,

$$
\begin{aligned}
I(u + v) &= \mathcal{A}(u + v, u + v) + 2\langle u + v, f \rangle \\
&= I(u) + 2 \left[\mathcal{A}(u, v) + \langle v, f \rangle \right] + \mathcal{A}(v, v) \\
&= I(u) + \mathcal{A}(v, v) \geq I(u).
\end{aligned}
$$

Thus u minimizes $I(u)$ in S. $\qquad\square$

We are looking for the element u in the finite dimensional space S that minimizes $I(u)$ as in (8.33). Let the dimension of S be n and let v_i, $i = 1, \ldots, n$ be a basis of S. Remember that S is a space of functions; for example, the elements of S could be piecewise polynomials and the basis could be B-splines. We can express u as a linear combination of the basis functions

$$u = \sum_{i=1}^{n} a_i v_i.$$

We know that $\mathcal{A}(u, v) + \langle v, f \rangle = 0$ for all non-zero functions in S. In particular, the equation holds for all basis functions,

$$\mathcal{A}(u, v_j) + \langle v_j, f \rangle = \mathcal{A}\left(\sum_{i=1}^{n} a_i v_i, v_j \right) + \langle v_j, f \rangle = \sum_{i=1}^{n} \mathcal{A}(v_i, v_j) a_i + \langle v_j, f \rangle = 0.$$

Defining an $n \times n$ matrix A with elements $A_{ij} = \mathcal{A}(v_i, v_j)$, $1 \leq i, j \leq n$, and a vector $\mathbf{b}$ with entries $b_j = -\langle v_j, f \rangle$, $j = 1, \ldots, n$, the coefficients a_i, $i = 1, \ldots, n$ can be found by solving the system of equations given by

$$A\mathbf{a} = \mathbf{b}.$$

A is a symmetric matrix, since $\mathcal{A}$ satisfies the symmetry property, and A is also positive definite, since $\mathcal{A}$ also satisfies the non-negativity property. Depending on f, numerical techniques might be necessary to evaluate $\langle v_j, f \rangle$ to calculate $\mathbf{b}$, since no analytical solution for the integral may exist.

Choosing the basis functions carefully, we can introduce sparseness into A and exploit this when solving the system of equations. In particular, selecting basis functions which have small, finite *support* is desirable. The support of a function is the closure of the set of points where the function has non-zero values. If v_i and v_j are functions whose supports do not overlap, then

$$A_{i,j} = \mathcal{A}(v_i, v_j) = 0,$$

since $\mathcal{A}$ is defined as the integration over a given region of the product of derivatives of the two functions and these products vanish, if the supports are separated.

Because of the properties of symmetry, bi-linearity and non-negativity, $\mathcal{A}$ can be viewed as a distance measure between two functions u and v by calculating $\mathcal{A}(u - v, u - v)$. Let $\hat{u}$ be the true solution to the partial differential equation, or equivalently to the variational problem which lies in the infinite dimensional space for which $I(\cdot)$ is well defined. We are interested in how far the solution $u \in S$ is from the true solution $\hat{u}$.

Theorem 8.7. *Let u be the element of S that minimizes $I(u)$ for all elements of S. Further, let $\hat{u}$ be the element of the infinite dimensional space minimizing $I(u)$ over this space. Then u also minimizes*

$$\mathcal{A}(\hat{u} - u, \hat{u} - u)$$

for all $u \in S$. That is, u is the best solution in S with regards to the distance measure induced by $\mathcal{A}$.

Proof. Consider

$$I(\hat{u} - \lambda(\hat{u} - u)) = I(\hat{u}) + 2\lambda \left[\mathcal{A}(\hat{u}, \hat{u} - u) + \langle \hat{u} - u, f \rangle \right] + \lambda^2 \mathcal{A}(\hat{u} - u, \hat{u} - u).$$

This is quadratic in λ and has a minimum for $\lambda = 0$. This implies that the linear term has to vanish, and thus

$$I(\hat{u} - \lambda(\hat{u} - u)) = I(\hat{u}) + \lambda^2 \mathcal{A}(\hat{u} - u, \hat{u} - u),$$

especially when letting $\lambda = 1$,

$$I(u) = I(\hat{u} - (\hat{u} - u)) = \mathcal{A}(\hat{u} - u, \hat{u} - u).$$

Since $I(\hat{u})$ is already minimal, minimizing $I(u)$ minimizes $\mathcal{A}(\hat{u} - u, \hat{u} - u)$ at the same time. $\qquad\square$

We continue our one-dimensional example, solving $u''(x) = f(x)$ with zero boundary conditions. Let $x_0, \ldots, x_{n+1}$ be nodes on the interval $[0, 1]$ with $x_0 = 0$ and $x_{n+1} = 1$. Let $h_i = x_i - x_{i-1}$ be the spacing. For $i = 1, \ldots, n$ we choose

$$v_i(x) = \begin{cases} \dfrac{x - x_{i-1}}{h_i}, & x \in (x_{i-1}, x_i), \\ \dfrac{x_{i+1} - x}{h_{i+1}}, & x \in [x_i, x_i + 1), \\ 0 & \text{otherwise} \end{cases}$$

as basis functions. These are similar to the linear B-splines displayed in Figure 3.7, the difference being that the nodes are not equally spaced and we restricted the basis to those splines evaluating to zero at the boundary, because of the zero boundary conditions.

Recall that in this example

$$\mathcal{A}(u, v) = \int_0^1 u'(x)v'(x)dx$$

and the scalar product is

$$\langle u, f \rangle = \int_0^1 u(x)f(x)dx.$$

For the chosen basis functions

$$v_i'(x) = \begin{cases} \dfrac{1}{h_i}, & x \in (x_{i-1}, x_i), \\ \dfrac{-1}{h_{i+1}}, & x \in [x_i, x_i + 1), \\ 0 & \text{otherwise} \end{cases}$$

and hence

$$A_{i,j} = \mathcal{A}(v_i, v_j) = \begin{cases} \dfrac{-1}{h_{i-1}}, & j = i - 1, \\ \dfrac{1}{h_i} + \dfrac{1}{h_{i+1}}, & j = i, \\ \dfrac{-1}{h_{i+1}}, & j = i + 1, \\ 0 & \text{otherwise} \end{cases}$$

Thus A is tridiagonal and symmetric.

The scalar product of the basis functions with f is, on the other hand,

$$\langle v_j, f \rangle = \int_{x_{j-1}}^{x_{j+1}} v_j(x)f(x)dx.$$

If f is continuous, we can write

$$\langle v_j, f \rangle = f(\xi_j) \int_{x_{j-1}}^{x_{j+1}} v_j(x)dx$$

for some $\xi_j \in (x_{j-1}, x_{j+1})$. We can simplify

$$
\begin{aligned}
\langle v_j, f \rangle &= f(\xi_j) \left[\int_{x_{j-1}}^{x_j} \frac{x - x_{j-1}}{h_j} dx + \int_{x_j}^{x_{j+1}} \frac{x_{j+1} - x}{h_{j+1}} \right] \\
&= f(\xi_j) \left[\frac{(x - x_{j-1})^2}{2h_j} \right]_{x_{j-1}}^{x_j} - f(\xi_j) \left[\frac{(x_{j+1} - x)^2}{2h_{j+1}} \right]_{x_j}^{x_{j+1}} \\
&= \frac{h_j + h_{j+1}}{2} f(\xi_j).
\end{aligned}
$$

The solution is

$$ u(x) = \sum_{j=1}^n a_j v_j(x) $$

where the coefficients a_j are determined by the equations

$$
\begin{aligned}
-\left(\frac{1}{h_1} + \frac{1}{h_2} \right) a_1 + \frac{1}{h_2} a_2 &= \frac{h_1 + h_2}{2} f(\xi_1), \\
\frac{1}{h_j} a_{j-1} - \left(\frac{1}{h_j} + \frac{1}{h_{j+1}} \right) a_j + \frac{1}{h_{j+1}} a_{j+1} &= \frac{h_j + h_{j+1}}{2} f(\xi_j), \\
\frac{1}{h_n} a_{n-1} - \left(\frac{1}{h_n} + \frac{1}{h_{n+1}} \right) a_n &= \frac{h_n + h_{n+1}}{2} f(\xi_n).
\end{aligned}
$$

If $h_j = h$ for $j = 1, \ldots, n+1$ this becomes

$$
\begin{aligned}
\frac{1}{h} (-2a_1 + a_2) &= hf(\xi_1), \\
\frac{1}{h} (a_{j-1} - 2a_j + a_{j+1}) &= hf(\xi_j), \\
\frac{1}{h} (a_{n-1} - 2a_n) &= hf(\xi_n).
\end{aligned}
$$

This looks very similar to the finite difference approximation to the solution of the Poisson equation in one dimension on an equally spaced grid. Let u_m approximate $u(x_m)$. We approximate the second derivative by applying the central difference operator twice and dividing by h^2. Then, using the zero boundary conditions, the differential equation can be approximated on the grid by

$$
\begin{aligned}
\frac{1}{h^2} (-2u_1 + u_2) &= f(x_1), \\
\frac{1}{h^2} (u_{m-1} - 2u_m + u_{m+1}) &= f(x_m), \\
\frac{1}{h^2} (u_{n-1} - 2u_n) &= hf(x_n).
\end{aligned}
$$

However, the two methods are two completely different approaches to the same problem. The finite difference solution calculates approximations to function values on grid points, while the finite element method produces a function as a solution which is the linear combination of basis functions. The right-hand sides in the finite difference technique are f evaluated at the grid points. The right-hand sides in the finite element method are scalar products of the basis functions with f. Finite element methods are chosen, when it is important to have a continuous representation of the solution. By choosing appropriate basis functions such as higher-order B-splines, the solution can also be forced to have continuous derivatives up to a certain degree.

8.8   Revision Exercises

Exercise 8.12. *The diffusion equation*

$$\frac{\partial u}{\partial t} = \frac{\partial^2 u}{\partial x^2}$$

is discretized by the finite difference method

$$u_m^{n+1} - \frac{1}{2}(\mu - \alpha)\left(u_{m-1}^{n+1} - 2u_m^{n+1} + u_{m+1}^{n+1}\right)$$
$$= u_m^n + \frac{1}{2}(\mu + \alpha)\left(u_{m-1}^n - 2u_m^n + u_{m+1}^n\right),$$

where u_m^n approximates $u(m\Delta x, n\Delta t)$ and $\mu = \Delta t/(\Delta x)^2$ and α are constant.

(a) Show that the order of magnitude (as a power of Δx) of the local error is $O((\Delta x)^4)$ for general α and derive the value of α for which it is $O((\Delta x)^6)$. State which expansions and substitutions you are using.

(b) Define the Fourier transform of a sequence u_m^n, $m \in \mathbb{Z}$. Investigate the stability of the given finite difference method by Fourier technique and its dependence on α. In the process define the amplification factor. (Hint: Express the amplification factor as $1 - \ldots$)

Exercise 8.13. *The diffusion equation*

$$\frac{\partial u}{\partial t} = \frac{\partial}{\partial x}\left(a(x)\frac{\partial u}{\partial x}\right), \qquad 0 \le x \le 1, \qquad t \ge 0,$$

with the initial condition $u(x,0) = \phi(x)$, $0 \le x \le 1$ and zero boundary conditions for $x = 0$ and $x = 1$, is solved by the finite difference method $(m = 1, \ldots, M)$

$$u_m^{n+1} = u_m^n + \mu\left[a_{m-\frac{1}{2}}u_{m-1}^n - (a_{m-\frac{1}{2}} + a_{m+\frac{1}{2}})u_m^n + a_{m+\frac{1}{2}}u_{m+1}^n\right],$$

where $\mu = \Delta t/(\Delta x)^2$ is constant, $\Delta x = \frac{1}{M+1}$, and u_m^n approximates $u(m\Delta x, n\Delta t)$. The notation $a_\alpha = a(\alpha\Delta x)$ is employed.

(a) Assuming sufficient smoothness of the function a, show that the local error of the method is at least $O((\Delta x)^3)$. State which expansions and substitutions you are using.

(b) Remembering the zero boundary conditions, write the method as

$$\mathbf{u}^{n+1} = A\mathbf{u}^n$$

giving a formula for the entries $A_{k,l}$ of A. From the structure of A, what can you say about the eigenvalues of A?

(c) Describe the eigenvalue analysis of stability.

(d) Assume that there exist finite positive constants a_- and a_+ such that for $0 \le x \le 1$, $a(x)$ lies in the interval $[a_-, a_+]$. Prove that the method is stable for $0 < \mu \le \frac{1}{2a_+}$. (Hint: You may use without proof the Gerschgorin theorem: All eigenvalues of the matrix A are contained in the union of the Gerschgorin discs given for each $k = 1, \ldots, M$ by

$$\{z \in \mathbb{C} : |z - A_{k,k}| \le \sum_{l=1, l\neq k}^{M} |A_{k,l}|\}.)$$

Exercise 8.14. *(a) The computational stencil given by*

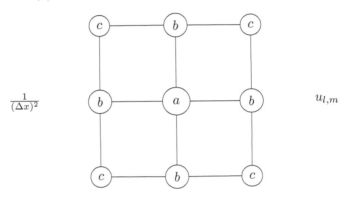

$\frac{1}{(\Delta x)^2}$ $u_{l,m}$

is used to approximate

$$\frac{\partial^2 u}{\partial x^2} + \frac{\partial^2 u}{\partial y^2}.$$

(i) Translate the computational stencil into a formula.

(ii) Express a and b in terms of c such that the computational stencil provides an approximation with error $O((\Delta x)^2)$.

(iii) Find a value of c such that the approximation has an error of $O((\Delta x)^4)$ in the case where the function u satisfies

$$\frac{\partial^2 u}{\partial x^2} + \frac{\partial^2 u}{\partial y^2} = 0.$$

(b) We now consider the partial differential equation

$$\frac{\partial^2 u}{\partial x^2} + \frac{\partial^2 u}{\partial y^2} = f, \qquad -1 \le x, y \le 1$$

where f is analytic and obeys periodic boundary conditions

$$f(-1, y) = f(1, y), \qquad -1 \le y \le 1, \qquad f(x, -1) = f(x, 1), \qquad -1 \le x \le 1.$$

Additionally, we impose the following periodic boundary conditions on u

$$u(-1, y) = u(1, y), \quad -1 \leq x \leq 1, \quad u_x(-1, y) = u_x(1, y), \quad -1 \leq y \leq 1,$$
$$u(x, -1) = u(x, 1), \quad -1 \leq x \leq 1, \quad u_y(x, -1) = u_y(x, 1), \quad -1 \leq y \leq 1$$

and a normalization condition:

$$\int_{-1}^{1} \int_{-1}^{1} u(x, y) dx dy = 0.$$

(i) Write down the Fourier expansion for both f and u.

(ii) Define convergence at spectral speed and state the two properties it depends on.

(iii) Which Fourier coefficient can be calculated from the normalization condition and what is its value?

(iv) Assuming that the Fourier coefficients of f are known, calculate the Fourier coefficients of u.

Exercise 8.15. *Consider the advection equation*

$$\frac{\partial u}{\partial t} = \frac{\partial u}{\partial x}$$

for $x \in [0, 1]$ and $t \in [0, T]$.

(a) What does it mean if a partial differential equation is well-posed?

(b) Define stability for time-marching algorithms for PDEs.

(c) Derive the eigenvalue analysis of stability.

(d) Define the forward difference operator Δ_+, the central difference operator δ, and the averaging operator μ_0, and calculate the operator defined by $\delta\mu_0$.

(e) In the solution of partial differential equations, matrices often occur which are constant on the diagonals. Let A be an $M \times M$ matrix of the form

$$A = \begin{pmatrix} a & b & & \\ -b & a & & \\ & \ddots & \ddots & b \\ & & -b & a \end{pmatrix},$$

that is, $A_{i,i} = a$, $A_{i,i+1} = b$, $A_{i+1,i} = -b$ and $A_{i,j} = 0$ otherwise. The eigenvectors of A are $\mathbf{v}_1, \ldots, \mathbf{v}_M$ where the jth component of $\mathbf{v}_k$ is given by $(\mathbf{v}_k)_j = \imath^j \sin \frac{\pi jk}{M+1}$, where $\imath = \sqrt{-1}$. Calculate the eigenvalues of A by evaluating $A\mathbf{v}_k$ (Hint: $\sin(x \pm y) = \sin x \cos y \pm \cos x \sin y$).

(f) *The advection equation is approximated by the following Crank–Nicolson scheme*

$$u_m^{n+1} - u_m^n = \frac{1}{4}\mu(u_{m+1}^{n+1} - u_{m-1}^{n+1}) + \frac{1}{4}\mu(u_{m+1}^n - u_{m-1}^n),$$

where $\mu = \Delta t/\Delta x$ and $\Delta x = 1/(M+1)$. Assuming zero boundary conditions, that is, $u(0,t) = u(1,t) = 0$, show that the scheme can be written in the form

$$B\mathbf{u}^{n+1} = C\mathbf{u}^n,$$

where $\mathbf{u}^n = \begin{pmatrix} u_1^n & \cdots & u_M^n \end{pmatrix}^T$. Specify the matrices B and C.

(g) *Calculate the eigenvalues of $A = B^{-1}C$ and their moduli.*

(h) *Deduce the range of μ for which the method is stable.*

Exercise 8.16. *The advection equation*

$$\frac{\partial u}{\partial t} = \frac{\partial u}{\partial x}$$

for $x \in \mathbb{R}$ and $t \geq 0$ is solved by the leapfrog scheme

$$u_m^{n+1} = \mu(u_{m+1}^n - u_{m-1}^n) + u_m^{n-1},$$

where $\mu = \Delta t/\Delta x$ is the Courant number.

(a) *Looking at the scheme, give the finite difference approximations to $\frac{\partial u}{\partial t}$ and $\frac{\partial u}{\partial x}$ which are employed, and state the order of these approximations.*

(b) *Determine the local error for the leapfrog scheme.*

(c) *The approximations u_m^n, $m \in \mathbb{Z}$, are an infinite sequences of numbers. Define the Fourier transform $\hat{u}^n(\theta)$ of this sequence.*

(d) *Define a norm for the sequence u_m^n, $m \in \mathbb{Z}$, and for the Fourier transform $\hat{u}^n(\theta)$. State Parseval's identity and prove it.*

(e) *How can Parseval's identity be used in the stability analysis of a numerical scheme?*

(f) *Apply the Fourier transform to the leapfrog scheme and derive a three-term recurrence relation for $\hat{u}^n(\theta)$.*

(g) *Letting $\theta = \pi/2$ and $\mu = 1$, express $\hat{u}^2(\pi/2), \hat{u}^3(\pi/2)$ and $\hat{u}^4(\pi/2)$ in terms of $\hat{u}^0(\pi/2)$ and $\hat{u}^1(\pi/2)$. Hence deduce that the scheme is not stable for $\mu = 1$.*

Exercise 8.17. *Assume a numerical scheme is of the form*

$$\sum_{k=-r}^{s} \alpha_k u_{m+k}^{n+1} = \sum_{k=-r}^{s} \beta_k u_{m+k}^{n}, \qquad m \in \mathbb{Z}, n \in \mathbb{Z}^+,$$

where the coefficients α_k and β_k are independent of m and n.

(a) *The approximations u_m^n, $m \in \mathbb{Z}$, are an infinite sequences of numbers. Define the Fourier transform $\hat{u}^n(\theta)$ of this sequence.*

(b) *Derive the Fourier analysis of stability. In the process give a definition of the amplification factor.*

(c) *Prove that the method is stable if the amplification factor is less than or equal to 1 in modulus.*

(d) *Find the range of parameters μ such that the method*

$$(1 - 2\mu)u_{m-1}^{n+1} + 4\mu u_m^{n+1} + (1 - 2\mu)u_{m+1}^{n+1} = u_{m-1}^n + u_{m+1}^n$$

is stable, where $\mu = \Delta t/\Delta x^2 > 0$ is the Courant number. (Hint: Substitute $x = \cos\theta$ and check whether the amplification factor can become unbounded and consider the gradient of the amplification factor.)

(e) *Suppose the above method is used to solve the heat equation*

$$\frac{\partial u}{\partial t} = \frac{\partial^2 u}{\partial x^2}.$$

Express the local error as a power of Δx.

Exercise 8.18. (a) *Define the central difference operator δ and show how it is used to approximate the second derivative of a function. What is the approximation error?*

(b) *Explain the method of lines, applying it to the diffusion equation*

$$\frac{\partial u}{\partial t} = \frac{\partial^2 u}{\partial x^2}$$

for $x \in [0, 1]$, $t \geq 0$ using the results from (a).

(c) *Given space discretization step Δx and time discretization step Δt, the diffusion equation is approximated by the scheme*

$$u_m^{n+1} - u_m^n - \mu(u_{m+1}^{n+1} - 2u_m^{n+1} + u_{m-1}^{n+1}) = 0,$$

where u_m^n approximates $u(m\Delta x, n\Delta t)$ and $\mu = \Delta t/(\Delta x)^2$ is constant. Show that the local truncation error of the method is $O(\Delta x^4)$, stating the first error term explicitly. Thus deduce that a higher order can be achieved for a certain choice of μ.

(d) *The scheme given in (c) is modified by adding the term*

$$\alpha \left(u_{m+2}^n - 4u_{m+1}^n + 6u_m^n - 4u_{m-1}^n + u_{m-2}^n \right).$$

on the left side of the equation. How does this change the error term calculated in (c)? For which choice of α depending on μ can a higher order be achieved?

(e) *Perform a Fourier stability analysis on the scheme given in (c) with arbitrary value of μ stating for which values of μ the method is stable. (Hint: $\cos\theta - 1 = -2\sin^2\theta/2$.)*

Exercise 8.19. *We consider the diffusion equation with variable diffusion coefficient*

$$\frac{\partial u}{\partial t} = \frac{\partial}{\partial x}\left(a\frac{\partial u}{\partial x}\right),$$

where $a(x)$, $x \in [-1,1]$ is a given differentiable function. The initial condition for $t = 0$ is given, that is, $u(x,0) = u_0(x)$, and we have zero boundary conditions for $x = -1$ and $x = 1$, that is, $u(-1,t) = 0$ and $u(1,t) = 0$, $t \geq 0$.

(a) *Given space discretization step Δx and time discretization step Δt, the following finite difference method is used,*

$$u_m^{n+1} = u_m^n + \mu \left[a_{m-1/2}u_{m-1}^n - (a_{m-1/2} + a_{m+1/2})u_m^n + a_{m+1/2}u_{m+1}^n \right],$$

where $a_{m\pm1/2} = a(-1 + m\Delta x \pm \Delta x/2)$ and u_m^n approximates $u(-1 + m\Delta x, n\Delta t)$, and $\mu = \Delta t/(\Delta x)^2$ is constant. Show that the local error is at least $O(\Delta x^4)$.

(b) *Derive the matrix A such that the numerical method given in (a) is written as*

$$\mathbf{u}^{n+1} = A\mathbf{u}^n.$$

(c) *Since the boundary conditions are zero, the solution may be expressed in terms of periodic functions. Therefore the differential equation is solved by spectral methods letting*

$$u(x,t) = \sum_{n=-\infty}^{\infty} \hat{u}_n(t)e^{i\pi n x} \quad \text{and} \quad a(x) = \sum_{n=-\infty}^{\infty} \hat{a}_n e^{i\pi n x}.$$

Calculate the first derivative of u with regards to x.

(d) *Using convolution, calculate the product*

$$a(x)\frac{\partial u(x,t)}{\partial x}.$$

(e) *By differentiating the result in (d) again with regards to x and truncating, deduce the system of ODEs for the coefficients $\hat{u}_n(t)$. Specify the matrix B such that*

$$\frac{d}{dt}\hat{\mathbf{u}}(t) = B\hat{\mathbf{u}}(t).$$

(f) *Let $a(x)$ be constant, that is, $a(x) = \hat{a}_0$. What are the matrices A and B with this choice of $a(x)$?*

(g) *Let*

$$a(x) = \cos \pi x = \frac{1}{2}(e^{i\pi x} + e^{-i\pi x}).$$

What are the matrices A and B with this choice of $a(x)$? (Hint: $\cos(x - \pi) = -\cos x$ and $\cos(x - y) + \cos(x + y) = 2\cos x \cos y$.)

Exercise 8.20. *We consider the square $[0, 1] \times [0, 1]$, which is divided by $M + 1$ in both directions, with a grid spacing of Δx in both directions. The computational stencil given by*

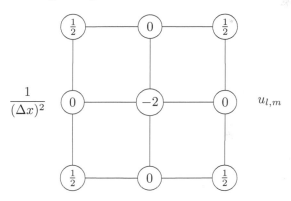

is used to approximate the right hand side of

$$\frac{\partial^2 u}{\partial x^2} + \frac{\partial^2 u}{\partial y^2} = f.$$

The function $u(x, y)$ satisfies zero boundary conditions.

(a) *Translate the computational stencil into a formula.*

(b) *Show that the computational stencil provides an approximation with error $O(\Delta x^2)$ and state the Δx^2 term explicitly.*

(c) *What is meant by natural ordering? Write down the system of equations arising from the computational stencil in this ordering. Use block matrices to simplify notation stating the matrix dimensions explicitly.*

(d) *Using the vectors* $\mathbf{q}_1, \ldots, \mathbf{q}_M$ *where the* j^{th} *component of* $\mathbf{q}_k$ *is given by* $(\mathbf{q}_k)_j = \sin \frac{\pi jk}{M+1}$, *calculate the eigenvalues of the block matrices in (c).* *(Hint:* $\sin(x \pm y) = \sin x \cos y \pm \cos x \sin y$.)

(e) *Describe how the result in (d) can be used to transform the system of equations in (c) into several uncoupled systems of equations.*

(f) *Define the inverse discrete Fourier transform of a sequence* $(x_l)_{l \in \mathbb{Z}}$ *of complex numbers with period* n, *i.e.,* $x_{l+n} = x_l$. *Show how this is related to calculating* $\mathbf{q}_k^T \mathbf{v}$ *for some general vector* $\mathbf{v}$.

Bibliography

[1] George A. Baker and Peter Graves-Morris. *Padé Approximants*. Cambridge University Press, 1996.

[2] Christopher M. Bishop. *Pattern Recognition and Machine Learning*. Springer (SIE), 2013.

[3] John C. Butcher. *The numerical analysis of ordinary differential equations: Runge–Kutta and general linear methods*. Wiley, 1987.

[4] Charles W. Curtis. *Linear Algebra: an introductory approach*. Springer, 1999.

[5] Germund Dahlquist and Björck Åke. *Numerical Methods*. Dover Publications, 2003.

[6] Biswa N. Datta. *Numerical Linear Algebra and Applications*. SIAM, 2010.

[7] James W. Demmel. *Applied Numerical Linear Algebra*. SIAM, 1997.

[8] Alexandre Ern and Jean-Luc Guermond. *Theory and Practice of Finite Elements*. Springer, 2004.

[9] Roger Fletcher. *Practical Methods of Optimization*. Wiley, 2000.

[10] Walter Gautschi. *Numerical Analysis*. Birkhuser, 2012.

[11] Peter Henrici. *Discrete Variable Methods in Ordinary Differential Equations*. Wiley, 1962.

[12] Israel Koren. *Computer Arithmetic Algorithms*. A K Peters/CRC Press, 2001.

[13] Thomas W. Körner. *Fourier Analysis*. Cambridge University Press, 1989.

[14] Randall J. LeVeque. *Finite Volume Methods for Hyperbolic Problems*. Cambridge University Press, 2002.

[15] Harald Niederreiter. *Random Number Generation and Quasi-Monte Carlo Methods*. SIAM, 1987.

[16] Jorge Nocedal and Stephen Wright. *Numerical Optimization.* Springer, 2006.

[17] Anthony Ralston and Philip Rabinowitz. *A First Course in Numerical Analysis.* Dover Publications, 2001.

[18] J. Stoer and R. Bulirsch. *Introduction to Numerical Analysis.* Springer, 2002.

[19] John C. Strikwerda. *Finite Difference Schemes and Partial Differential Equations.* SIAM, 2004.

[20] Richard S. Varga. *Matrix Iterative Analysis.* Springer, 2000.

[21] Daniel Zwillinger. *Handbook of Differential Equations.* Academic Press, 1997.

Index

Printed and bound by CPI Group (UK) Ltd, Croydon, CR0 4YY

24/10/2024

01778283-0008